W9-BBP-047

Advances in Numerical Analysis

Advances in Numerical Analysis

VOLUME I

Nonlinear Partial Differential Equations and Dynamical Systems

---

Edited by
WILL LIGHT
Senior Lecturer
Mathematics Department
Lancaster University

CLARENDON PRESS · OXFORD 1991

Oxford University Press, Walton Street, Oxford OX2 6DP
Oxford New York Toronto
Delhi Bombay Calcutta Madras Karachi
Petaling Jaya Singapore Hong Kong Tokyo
Nairobi Dar es Salaam Cape Town
Melbourne Auckland
and associated companies in
Berlin Ibadan

Oxford is a trade mark of Oxford University Press

Published in the United States
by Oxford University Press, New York

British Library Cataloguing in Publication Data
A catalogue record for this book
is available from the British Library
ISBN 0–19–853438–8

Library of Congress Cataloging in Publication Data
(Data available)
ISBN 0–19–853438–8

Typeset in TEX by Aldcliffe Enterprises,
Lancaster, England

Printed in Great Britain
by Biddles Ltd
Guildford and King's Lynn

Preface

This book is the first of a two volume collection of proceedings of the fourth Summer School in Numerical Analysis, which was held at Lancaster University from 15th July to 3rd August, 1990. The meeting was sponsored by the Science and Engineering Research Council of Great Britain, and was attended by approximately 120 participants from 14 countries. Many of the participants delivered seminars at the meeting, but this volume contains only the contributions of six of the nine invited main speakers.

Each week of the School was devoted to a specific topic within numerical analysis, the topics for 1990 being

- nonlinear partial differential equations
- dynamical systems
- multivariate approximation.

Each main speaker was asked to give a series of five one hour lectures, with the exposition pitched at such a level that researchers and graduate students could both gain something useful from the courses – a demanding brief, but one which was met very well by all speakers. The six chapters which form this volume are an account of the happennings in weeks one and two. The proceedings of week three are contained in volume II, which bears the title *Advances in Numerical Analysis II: Wavelets, Subdivision Algorithms and Radial Functions*, and is also published by Oxford University Press.

The selection of topics for the School reflects current trends in research in numerical analysis. Since the mid-1970s, mathematical research into non-linear phenomena has grown, and that growth is just as vigorous in numerical analysis as in other branches of mathematics. The first three chapters of this book present well-formed discussions of non-linear phenomena in evolution equations, free boundary problems and spectral methods. In the 1980s, one of the major new thrusts in mathematical research has been the study of dynamical systems. This work has had a major impact on the study of the numerical solution of differential equations, with new ideas, particularly about the notion of stability, emerging. The second three chapters give snapshots of the activities that are taking place in this area, with descriptions of numerical methods for dynamical systems, nonlinear stability and differential equations on manifolds.

In compiling these proceedings, I have striven for as uniform a presentation as possible, without encroaching on the rights of the individual author to present things as he sees fit. A decimal system of notation is adopted for sectioning, with the label 2.1.3 representing subsection 3 of section 1 of chapter 2. The equation numbering is done within sections, so that equation 2.3 is the third numbered equation in section 2 of the current chapter. Theorems, Lemmas, etc., are numbered sequentially within sections. The postal address of the author is listed at the end of the chapter.

Finally, a few acknowledgements are in order. The organisation of the School was carried out jointly with my good friend and colleague John Gilbert. I thank him for his help and support. Sue Hubbard assisted in the day to day running of the School and did some preliminary 'TEXing' of manuscripts. The Science and Engineering Research Council once again provided generous support. Their contribution covered all the organisational and running costs of the meeting as well as the expenses of the speakers, and the accommodation and subsistence expenses of up to twenty participants each week. The organisation of each weekly scientific programme was undertaken by people who were entitled 'local experts'. These were Professor Charles Elliott, Professor Alastair Spence and Dr. John Gregory. There was some puzzlement on the part of participants as to the choice of words in this title, but I am unable to remember whether the questions centred around the use of the word 'expert' or the word 'local'! It is a pleasure to acknowledge their contributions to the overall success of the meeting.

Will Light

Contents

3 An Introduction to Spectral Methods for Partial Differential Equations

4 Two Topics in Nonlinear Stability

5 Numerical Methods for Dynamical Systems

1
Finite Element Methods for Evolution Equations

Lars B. Wahlbin

1.1 Preface

This chapter will treat the "standard" finite element space discretization as applied to two classes of evolutionary problems: parabolic problems and integro-differential equations. The further discretization of the time variable will also be treated. The development will be mainly descriptive and only occasionally will we dig into proofs.

1.2 Parabolic Theory

1.2.1 Setting the stage

We shall consider the so-called h-version of the finite element method (in contrast to the p-version or combined $h-p$-version). With Ω a basic bounded domain in $\mathbb{R}^d$ having smooth boundary $\partial\Omega$ we are approximating functions $u : \Omega \to \mathbb{R}$, in the following way. With h to be thought of as the "typical" diameter of an element in a triangulation of Ω (or, in more complicated situations, $N^{-1/d}$, where N is the number of degrees of freedom in the approximating system), we have a family S_h of approximating function spaces, typically constructed as piecewise polynomial functions of a fixed degree on each element. Various so-called "isoparametric" modifications are typical close to the boundary $\partial\Omega$. Most results will be asymptotic as h tends to zero.

In this set of lectures only (at most) second order partial differential operators in the space variable will be considered, and it is then natural to assume that the spaces S_h consists of functions which are continuous across inter–element boundaries so that, in particular, they belong to $H^1(\Omega)$. The problem also comes equipped with boundary conditions, e.g., $u = 0$ on $\partial\Omega$ or $(\frac{\partial u}{\partial n}) = 0$ on $\partial\Omega$. We shall only consider the first case, homogeneous

Dirichlet boundary conditions. (The second case is slightly easier technically.) We assume that our finite element spaces S_h satisfy those boundary conditions.

Approximation properties of S_h are typically stated in the following way. With a fixed integer $r \geq 2$ ($r-1$ can be thought of as the polynomial degree) there exist a constant C, independent of h, such that the following holds. Given $u \in W_p^k(\Omega)$, $u = 0$ on $\partial\Omega$, $k \leq r$, $1 \leq p \leq \infty$

$$\inf_{\chi \in S_h} \|u - \chi\|_{W_p^s(\Omega)} \leq C h^{k-s} \|u\|_{W_p^k(\Omega)}, \quad s = 0 \text{ or } 1,\ k \geq s. \tag{2.1}$$

The top order r is referred to as "optimal" order. Such an estimate depends on a certain orderly progression in the underlying family of triangulations. (Equation (2.1) should in practice, for $s = 1$, $k > 2$, actually only involve a meshdomain Ω_h on the left. For simplicity, we shall not consider this "technical detail" further.)

We shall actually frequently demand even more orderliness, in the sense of the following "inverse hypothesis": There exists a constant C, independent of h, such that for any $\chi \in S_h$,

$$\|\chi\|_{W_p^1(\Omega)} \leq C h^{-1} \|\chi\|_{L_p(\Omega)}. \tag{2.2}$$

Such an estimate typically demands that the elements are roughly all of the same size h, and uniformly fat. (Questions about refinements of meshes near singularities will not be addressed here.)

There are various ways, "interpolants" or "quasi-interpolants", of constructing particular $\chi = Op(u)$ such that (2.1) holds with the infimum removed. We assume that you know about such "interpolants". Recall that they are given locally; the value of it at a point is influenced only by the values of u over a few adjoining elements. For our purposes two non-local operators will be of more importance. The first is the L_2-projection into S_h, to be denoted $P_h u$, given by

$$(P_h u - u\,, \chi) = 0, \quad \text{for } \chi \in S_h. \tag{2.3}$$

Here $(f, g) = \int_\Omega fg$ is the usual L_2 inner product and we consider only real valued functions. It can then be proven (see [14] and [16]) that there exists a constant C, independent of h, such that for $1 \leq p \leq \infty$,

$$\|P_h u - u\|_{L_p(\Omega)} \leq C \inf_{\chi \in S_h} \|u - \chi\|_{L_p(\Omega)}. \tag{2.4}$$

Of course, this is trivial for $p = 2$. (Convention: Constants C are allowed to change at each place they appear but are always independent of h and u.)

The second important operator is the Ritz-projection (or elliptic projection) into S_h, denoted $R_h u$, given (in a model case) by

$$(\nabla(R_h u - u), \nabla\chi) = 0 \quad \text{for } \chi \in S_h. \tag{2.5}$$

Note that if we were trying to solve for u in the problem

$$\begin{cases} -\Delta u &= f \quad \text{in } \Omega \\ u &= 0 \quad \text{on } \partial\Omega, \end{cases} \tag{2.6}$$

by solving the discrete system for u_h,

$$(\nabla u_h, \nabla\chi) = (f, \chi) \quad \text{for all } \chi \in S_h, \tag{2.7}$$

then $u_h = R_h u$. Let us write down estimates for two common cases, L_∞ and L_2. Here there exists a constant C, independent of h and u, such that,

$$\begin{cases} \|R_h u - u\|_{L_\infty(\Omega)} &\leq C(\ell n 1/h)^{\bar{r}} \inf_{\chi\in S_h} \|u - \chi\|_{L_\infty(\Omega)} \\ \|R_h u - u\|_{L_2(\Omega)} &\leq Ch \inf_{\chi\in S_h} \|u - \chi\|_{H^1(\Omega)}. \end{cases} \tag{2.8}$$

Here $\bar{r} = 1$ if $r = 2$; for $r \geq 3$ the logarithmic factor is not there ($\bar{r} = 0$). We refer to Schatz and Wahlbin[55] for the maximum norm estimate. For the L_2 result, we note that it is in general not true that

$$\|R_h u - u\|_{L_2(\Omega)} \leq C \inf_{\chi\in S_h} \|u - \chi\|_{L_2(\Omega)},$$

see Babuska and Osborn[1, pg. 58], for a simple counterexample. We note that we may consider such elliptic projections based on more general elliptic operators,

$$A = \sum_{i,j=1}^{d} -\frac{\partial}{\partial x_i}\left(a_{ij}(x)\frac{\partial}{\partial x_j}\right) + \sum_{k=1}^{d} b_k(x)\frac{\partial}{\partial x_k} + c(x), \tag{2.9}$$

with the same result as long as the usual ellipticity condition holds and there is uniqueness in the corresponding problem $Au = f$, $u = 0$ on $\partial\Omega$ (and h is small enough), cf. [53].

For simplicity we shall in the future demand that $\partial\Omega$ is smooth and, unless specificially pointed out, that coefficients of operators are smooth. Also, for technical ease, we shall often do the exposition only in L_2-based spaces ($p = 2$ above) and write then

$$\|f\| = \|f\|_{L_2(\Omega)}, \qquad \|f\|_k = \|f\|_{H^k(\Omega)}. \tag{2.10}$$

1.2.2 The semi-discrete problem

Consider the problem for $u = u(t,x)$, $t \geq 0$, $\chi \in \Omega$,

$$\begin{cases} \dfrac{\partial u}{\partial t} + Au &= f(x,t), \\ u &= 0 \quad \text{on } \partial\Omega\ , \\ u(0,x) &= v(x). \end{cases} \tag{2.11}$$

Here A is of the form (2.9), and, for ease of exposition, we assume that its coefficients are independent of time. We now first consider discretization in space, i.e., we seek $u_h(t,x)$ in S_h for each t, by the recipe

$$\begin{cases} (u_{h,t},\chi) + A(u_h,\chi) &= (f,\chi) \quad \text{for } \chi \in S_h, \\ u_h(0) &= v_h. \end{cases} \tag{2.12}$$

Here $A(\cdot,\cdot)$ is the bilinear form corresponding to the operator A, and v_h is some suitable approximation in S_h to v. The basic analysis of the error between u and u_h goes back to Wheeler[67]. Her main idea is to look at the difference between $u_h(t)$ and $R_h u(t)$ (a fictitious quantity, note), since much is known about $u - R_h u$. So, write

$$u_h - u = (u_h - R_h u) + (R_h u - u) \equiv \theta + \rho. \tag{2.13}$$

We then have for θ,

$$\begin{aligned} (\theta_t,\chi) + A(\theta,\chi) &= (u_{h,t},\chi) + A(u_h,\chi) - ((R_h u)_t,\chi) - A(R_h u,\chi) \\ &= (f,\chi) - (R_h u_t,\chi) - A(u,\chi) \\ &= (u_t - R_h u_t,\chi), \qquad \text{for } \chi \in S_h. \end{aligned} \tag{2.14}$$

Here we used that A has time-independent coefficients and the definitions of u_h and R_h. We may take $\chi = \theta$, and denoting $u_t - R_h u_t = \rho_t$,

$$\frac{1}{2}\frac{d}{dt}\|\theta\|^2 + A(\theta,\theta) = (\rho_t,\theta). \tag{2.15}$$

Let us now further assume that $A(\theta,\theta) \geq 0$. Using the Cauchy-Schwarz inequality,

$$\frac{1}{2}\frac{d}{dt}\|\theta\|^2 \leq \|\rho_t\|\,\|\theta\|, \tag{2.16}$$

so that

$$\frac{d}{dt}\|\theta\| \leq \|\rho_t\|. \tag{2.17}$$

Since we know how to treat ρ and ρ_t, if we assume the solution is sufficiently smooth and that $\|v - v_h\| \leq Ch^r$, we will obviously attain an optimal order estimate

$$\|u(t) - u_h(t)\| \leq K(u,t)h^r, \tag{2.18}$$

where K is independent of h.

I want to avoid going too much into detail about basic parabolic analysis. A treatise on the subject is Thomée[59].

1.2.3 About pointwise estimates

Deriving estimates for $\|u_h(t) - u(t)\|_\infty = \|u_h(t) - u(t)\|_{L_\infty(\Omega)}$ quickly becomes rather technical, for general d, cf. Nitsche and Wheeler[45] for the case of $d = 3$. However, the case of $d \leq 2$ is comparatively easy if one is not too fussy about the exact regularity needed of $u(t,x)$. Assume that $A(f,f) \geq C\|f\|_1^2$, for some $C > 0$. Go back to (2.14) and take there $\chi = \theta_t$. As you can easily see using (2.8), you will obtain an estimate of the form

$$\|\theta(t)\|_1 \leq C\|\theta(0)\|_1 + C(t,u)h^r \tag{2.19}$$

under appropriate smoothness demands, and hence if say $v_h = R_h v$ is taken as initial data,

$$\|\theta(t)\|_1 \leq C(t,u)h^r. \tag{2.20}$$

A basic two-dimensional fact (one dimension left to you) is that

$$\|\chi\|_\infty \leq C(\ell n 1/h)^{1/2}\|\chi\|_1, \text{ for } \chi \in S_h. \tag{2.21}$$

See [54, Lemma 1.1]. See also Sobolev's inequality in two dimensions. Thus,

$$\|\theta(t)\|_\infty \leq C(t,u)(\ell n 1/h)^{1/2}h^r \tag{2.22}$$

and combined with (2.8) for $p = \infty$ we have an optimal order estimate in L_∞ (apart from a mild logarithmic factor).

1.2.4 Nonsmooth initial data

Consider, for simplicity,

$$\begin{cases} u_t - \Delta u &= 0 \quad t \geq 0,\ x \in \Omega, \\ u &= 0 \quad \text{on } \partial\Omega, \\ u(0) &= v. \end{cases} \tag{2.23}$$

Going back to the estimate (2.18), for example, and tracing the dependence on the smoothness of u, one has, for example,

$$\|u(t) - u_h(t)\| \leq \|v_h - v\| + Ch^r\left\{\|v\|_r + \int_0^t \|u_t\|_r\, ds\right\}. \tag{2.24}$$

Even if by "luck" $\|v_h - v\|$ was small without v being smooth, the term $\|v\|_r$ and $\int_0^t \|u_t\|_r$ would be very large and destroy the estimate.

However, for the problem (2.23), provided $\partial\Omega$ is smooth, we know that the solution $u(t)$ is very smooth for $t > 0$. To put that quantitatively, let $\lambda_j, \varphi_j, j = 1, \ldots,$ be the eigenvalues and eigenfunctions respectively of $-\Delta$. Then in L_2-based spaces, since

$$u(t,x) = \sum_{j=1}^{\infty} \beta_j e^{-\lambda_j t} \varphi_j(x), \qquad \beta_j = (v, \varphi_j), \tag{2.25}$$

we have that, for example,

$$|u(t)|_\ell \leq C t^{-(\ell-k)/2} |v|_k, \qquad \ell > k, \tag{2.26}$$

where

$$|f|_\ell = \left(\sum_{j=1}^{\infty} k_j^2 \lambda_j^\ell \right)^{1/2}, \qquad k_j = (f, \varphi_j). \tag{2.27}$$

It is known that the norm $|\cdot|_\ell$ also incorporates compatibility conditions on $\partial\Omega$. For ℓ a non-negative integer, the corresponding Hilbert space $\dot{H}^\ell$ consists of those functions w in $H^\ell(\Omega)$ which satisfy $\Delta^j w = 0$ on $\partial\Omega$ for $j < \ell/2$ and if this is the case, $|w|_\ell$ is equivalent to the usual norm $\|w\|_\ell$, see e.g., Bramble and Thomée[6, Lemma 2.2]. Hence, to avoid blow-up in the estimate (2.24), not only is smoothness required of v, but also boundary compatibility.

We now embark on a heuristic investigation concerning how to choose v_h when approximating the problem (2.23) with nonsmooth or incompatible v. Imagine that v consists of one low Fourier mode and one high, say

$$v = \beta_1 \varphi_1 + \beta_j \varphi_j, \qquad j \text{ large}. \tag{2.28}$$

With λ_j^k, φ_j^h the corresponding discrete L_2-orthonormal eigenvalues and eigenfunctions,

$$v_h = \sum_{j=1}^{N} \beta_j^h \varphi_j^h, \qquad \beta_j^h = (v_h, \varphi_j^h). \tag{2.29}$$

Thus

$$\begin{aligned} u_h(t) - u(t) &= \sum_{j \text{ low}} (\beta_j^h - \beta_j) e^{-\lambda_j^h t} \varphi_j^h + \sum_{j \text{ low}} \beta_j \left(e^{-\lambda_j^h t} \varphi_j^h - e^{-\lambda_j t} \varphi_j \right) \\ &\quad + \sum_{j \text{ high}} \beta_j^h e^{-\lambda_j^h t} \varphi_j^h - \sum_{j \text{ high}} \beta_j e^{-\lambda_j t} \varphi_j \\ &\equiv I_1 + I_2 + I_3 + I_4 \end{aligned} \tag{2.30}$$

Here I_2 is small because, as is known, cf. Bramble and Osborn[4], $\lambda_j^h \sim \lambda_j$ and $\varphi_j^h \sim \varphi_j$ for low j. Also, I_3 and I_4 are small because λ_j^h and λ_j are large. So, we have to consider further I_1. Here we need that $\beta_j^h - \beta_j$ is small for low j. Write

$$\beta_j^h - \beta_j = (v_h - v, \varphi_j^h) - (v, \varphi_j - \varphi_j^h), \quad \text{low } j, \tag{2.31}$$

where now the second member is small. Thus, loosely speaking, we need

$$(v_h - v, \varphi_j^h) \quad \text{small for low } j. \tag{2.32}$$

If we selected v_h as some "interpolant" of v, (2.32) would hold for v smooth. For rough v, however, systematic sign errors may cause this to fail.

An obvious idea is now to take v_h to be the L_2-projection $P_h v$, causing $(v_h - v, \phi_j^h)$ to vanish. This choice of discrete initial data will always be made below. For the moment it is assumed that we can evaluate this $P_h v$ without errors!

The above heuristic ideas can be made exact to show that the error $u_h(t) - u(t)$ is of optimal order h^r for $t > 0$ even if v is merely in L_2, provided $v_h = P_h v$. Early thoughts about this can be found in Blair[3] and a major result is Helfrich[28], cf. also Thomée[58] and Fujita and Mizutani[21]. (There is also an earlier tradition in the finite difference setting on non-smooth data results, going back to Juncosa and Young[35].) A proof of the following result can be found in [59, Chapter 3]. For $v_h = P_h v$,

$$\|u_h(t) - u(t)\| \le C h^r t^{-r/2} \|v\|, \tag{2.33}$$

where C does not depend on h, t or v. The proof there is essentially that of Bramble, Schatz, Thomée and Wahlbin[5]; not particularly hard but a bit too lengthy to be included here.

Extension to nonself-adjoint and time-dependent operators can be found in Huang and Thomée[30], Luskin and Rannacher[44], Sammon[49] and Lasiecka[40].

Extension of the estimate (2.33) to pointwise error estimates are given in Bramble, Schatz, Thomée and Wahlbin[5, Section 4]. More precise pointwise estimates (tighter accounting for negative powers of t) can be found in two space dimensions in Schatz, Thomée and Wahlbin[54], and in one space dimension in Thomée and Wahlbin[61].

1.2.5 Evaluation of $P_h v$: data preparation

In the previous section it was assumed that $P_h v$ was exactly evaluated. In general, to approximately evaluate it requires numerical quadrature in the right hand side of

$$(P_h v, \chi) = (v, \chi), \quad \text{for } \chi \in S_h. \tag{2.34}$$

And, small errors in numerical quadrature typically demand smoothness of v! It appears that we have a vicious circle. A study of the problem (including time-discretization) is given in Wahlbin[65] in a one-dimensional case.

The results can be briefly summarized as follows. If v has piecewise r derivatives in L_1 on each subinterval of the mesh (some relaxing of this condition towards endpoints) and a certain integration rule is used in the right hand side of (2.34), then optimal order error is preserved for positive time.

In practice it is not uncommon that v has isolated singularities of a known form,

$$v = v_{\text{smooth}} + v_{\text{sing}}. \tag{2.35}$$

In evaluating

$$(v_{\text{smooth}}, \chi) + (v_{\text{sing}}, \chi) \tag{2.36}$$

one would then use a standard quadrature rule on the first term and do something special to evaluate the last piece. Due to linearity, separate analyzes recombine.

1.2.6 Semilinear problems, smooth solutions

Consider the problem

$$\begin{cases} u_t - \Delta u &= f(u) \quad \text{in } \Omega \times [0,T], \\ u &= 0 \quad\quad\ \text{on } \partial\Omega, \\ u(0) &= v, \end{cases} \tag{2.37}$$

where it is assumed that $u(t,x)$ is smooth. Assume further for now that f is actually globally Lipschitz-continuous,

$$|f(v) - f(w)| \leq C|v - w|. \tag{2.38}$$

Let u_h be the semi-discrete solution, defined by

$$\begin{cases} (u_{h,t}, \chi) + A(u_h, \chi) &= (f(u_h), \chi), \quad \text{for } \chi \in S_h, \\ u_h(0) &= v_h. \end{cases} \tag{2.39}$$

The error analysis of this case presents very little trouble over the linear case, and was contained already in Wheeler[67]. One introduces $R_h u$ and writes an equation for $\theta = R_h u - u_h$ which will now look like, with $\rho_t = u_t - R_h u_t$,

$$\begin{aligned}(\theta_t, \chi) + A(\theta, \chi) &= (\rho_t, \chi) + (f(u_h) - f(R_h u), \chi) \\ &\quad + (f(R_h u) - f(u), \chi)\,. \end{aligned} \tag{2.40}$$

Setting $\chi = \theta$ one obtains by use of the Lipschitz continuity of f,

$$\frac{1}{2}\frac{d}{dt}\|\theta\|^2 + C\|\theta\|_1^2 \leq \|\rho_t\|\,\|\theta\| + C\|\theta\|^2 + C\|\rho\|\,\|\theta\|. \tag{2.41}$$

One easily obtains by use of (2.8),

$$\frac{d}{dt}\|\theta\|^2 \leq Ch^{2r} + C\|\theta\|^2 \tag{2.42}$$

so that by Gronwall's lemma, provided $\|\theta(0)\| \leq Ch^r$,

$$\|\theta(t)\| \leq C(t,u)h^r, \tag{2.43}$$

an optimal order estimate.

For the case of f not globally Lipschitz, if one knows that u stays in a bounded region, say $\|u\|_\infty \leq K$, for $0 \leq t \leq T$, which follows from smoothness, and one knew a priori that $\|u_h\|_\infty$ also was bounded, then f could be altered outside a compact set to satisfy (2.38). In certain cases such as

$$r > d/2, \tag{2.44}$$

the argument is now easily concluded. If $\|u_h\|_\infty \geq 2K$ say, then that happens at a first time t_0. But during that time, we would have known that $\|\theta(t)\| \leq Ch^r$, $0 \leq t \leq t_0$. It follows essentially by the arguments leading to (2.2) that

$$\|\chi\|_\infty \leq Ch^{-d/2}\|\chi\| \qquad \text{for } \chi \in S_h. \tag{2.45}$$

Applying this to θ,

$$\|\theta(t)\|_\infty \leq Ch^{r-d/2}, \quad 0 \leq t \leq t_0, \tag{2.46}$$

and if $r > d/2$ one concludes easily that

$$\begin{aligned} \|u_h(t)\|_\infty &\leq \|\theta(t)\|_\infty + \|R_h u(t)\|_\infty \\ &\leq \|\theta(t)\|_\infty + \|\rho(t)\|_\infty + \|u(t)\|_\infty \\ &\leq Ch^{r-d/2} + Ch^r(\ell n 1/h)^{\overline{r}} + K \\ &< 2K, \quad 0 \leq t \leq t_0, \end{aligned} \tag{2.47}$$

for h small enough. This contradiction establishes that u_h stays within the appropriate bound and one concludes that

$$\|(u - u_h)(t)\| \leq C(t,u)h^r. \tag{2.48}$$

Elaboration on the above theme occurs in Thomée and Wahlbin[60] and it is certainly now a standard argument in nonlinear evolution problems with smooth solutions. It is ugly but useful!

The restriction (2.44) can, for example, be improved to $r > (d-2)/2$ by working in H^1.

An analysis of the equation

$$u_t - \mathbf{div}\,(a(u)\nabla u) = f(u) \tag{2.49}$$

can be found in Thomée[59, Chapter 10] (based on Wheeler[67]).

1.2.7 Semilinear problems, nonsmooth data

We consider the semilinear problem (2.37) in the case when $v \in L_2(\Omega)$ merely. This time, following the lead from the linear case, we take $v_h = P_h v$.

For our analysis below, it is convenient to introduce some notation for the linear homogeneous case. Let thus $E(t)v$ be the solution of

$$\begin{cases} u_t - \Delta u &= 0, \\ u(0) &= v, \end{cases} \tag{2.50}$$

and $E_h(t)v_h$ that of

$$\begin{cases} (u_{h,t}, \chi) + A(u_h, \chi) &= 0 \qquad \chi \in S_h, \\ u_h(0) &= v_h. \end{cases} \tag{2.51}$$

Letting $F_h(t)v = E_h(t)P_h v - E(t)v$ the results of Section 1.2.4 can then be written as

$$\|F_h(t)v\| \le Ch^k t^{-k/2}\|v\|\,, \qquad 0 \le k \le r. \tag{2.52}$$

We want to show that for the semilinear problem, if $v_h = P_h v$,

$$\|u_h(t) - u(t)\| \le Ch^2\left(t^{-1} + |\ell n(h^2/t)|\right), \qquad 0 \le t \le t^*, \tag{2.53}$$

where $C = C(B, t^*, \|v\|\,)$ with B the Lipschitz constant for f. (We also assume f bounded.) First note that for $t \le h^2$ the estimate deteriorates; actually energy arguments easily show (left to you) that both $\|u\|$ and $\|u_h\|$ are bounded and that is enough. Using E and E_h, Duhamel's principle gives us that

$$u(t) = E(t)v + \int_0^t E(t-s)f(u(s))\,ds, \tag{2.54}$$

and

$$u_h(t) = E_h(t)P_h v + \int_0^t E_h(t-s)P_h f(u_h(s))\,ds. \tag{2.55}$$

Subtracting the two, with $e(t) = (u_h - u)(t)$,

$$e(t) = F_h(t)v + \int_0^t E_h(t-s)P_h\left[f(u_h(s)) - f(u(s))\right] ds + \int_0^t F_h(t-s)f(u(s))\, ds. \tag{2.56}$$

Hence from the case $k = 2$ of (2.52),

$$\begin{aligned}
\|e(t)\| &\le Ch^2t^{-1} + C\left(\int_0^{h^2} + \int_{h^2}^t\right)\|e(s)\|\, ds \\
&\quad + \left(\int_0^{t-h^2} + \int_{t-h^2}^t\right)\|F_h(t-s)f(u(s))\|\, ds \\
&\le Ch^2t^{-1} + Ch^2 + C\int_{h^2}^t \|e(s)\|\, ds \\
&\quad + Ch^2 \int_0^{t-h^2} \frac{ds}{t-s} + Ch^2 \\
&\le Ch^2t^{-1} + Ch^2\log(t/h^2) + C\int_{h^2}^t \|e(s)\|\, ds.
\end{aligned} \tag{2.57}$$

Letting $\varphi(t) = \int_{h^2}^t \|e(s)\|\, ds$ we conclude that

$$\begin{cases} \varphi'(t) - C\varphi(t) \le Ch^2t^{-1} + Ch^2\log(t/h^2) & \text{for } h^2 \le t \le t^*, \\ \varphi(h^2) = 0, \end{cases} \tag{2.58}$$

and it follows that

$$\begin{aligned}
\varphi(t) &\le C\int_{h^2}^t e^{C(t-s)}\left(h^2s^{-1} + h^2\log(s/h^2)\right) ds \\
&\le Ch^2\log(t/h^2).
\end{aligned} \tag{2.59}$$

Inserting this into (2.57) completes the proof of (2.53).

We note that in contrast to the linear case (2.52) the estimate (2.53) is *not* of optimal order if $r > 2$. Perhaps the most interesting thing about this is that it is correct; no estimate of this type for the semilinear case can predict order higher than 2, in spite of the fact that u is $\mathcal{C}_\infty$ for positive time (if f and $\partial\Omega$ are). To phrase this precisely there exist f such that, assuming that an estimate of the form

$$\|u_h(t_0) - u(t_0)\| \le C(B, M, t_0)h^\sigma \tag{2.60}$$

holds for any u such that $|u(x,t)| \le B$, then $\sigma \le 2$.

The complete proof of this is given in Johnson, Larsson, Thomée and Wahlbin[32, section 6], in which paper also the above analysis can be found. The gist of the example can be given in the context of Fourier series (for which r is unlimited!). Let $f(y) = 4y^2$ for $|y| \le 1$, and smooth and bounded outside ($|y| > 1$ will not enter into the problem). Consider the following problem with periodic boundary conditions on $[-\pi, \pi]$, and $u = (u_1, u_2)$,

$$\left\{ \begin{array}{rcl} u_{1,t} & = & u_{1,xx} + f(u_2), \\ u_{2,t} & = & u_{2,xx}, \\ u_1(0) & = & 0, \qquad u_2(0) = v_2 \end{array} \right. \tag{2.61}$$

Taking $v_2(x) = \cos(nx)$ we have

$$\begin{array}{rcl} u_2(t,x) & = & \exp(-n^2 t) \\ u_1(t,x) & = & \dfrac{1 - \exp(-2n^2 t)}{n^2} \left[1 + \exp(-2n^2 t)\cos(2nx)\right], \end{array} \tag{2.62}$$

so that by nonlinear aliasing an initial high Fourier mode has resulted in a low, indeed constant, mode.

To approximate this problem, let $h = 1/n$, n a positive integer, and

$$S_h = \operatorname{span}\big[1, \cos(x), \sin(x), \ldots, \cos((n-1)x), \sin((n-1)x)\big].$$

Since then $P_h v \equiv 0$, the Galerkin solution u_h vanishes identically and hence, for any positive time t_0,

$$\|u_h(t_0) - u(t_0)\| = \|u(t_0)\| \simeq \frac{\sqrt{2\pi}}{n^2} = \sqrt{2\pi} h^2 \tag{2.63}$$

indicating that $\sigma > 2$ is impossible in a result such as (2.60).

In Crouzeix, Thomée and Wahlbin[13], sharp results are given for the case when initial data are somewhat smoother and compatible. In that paper the L_2-error estimates are already dimension dependent, only $d = 1, 2, 3$ are treated. (By now you would expect the corresponding L_∞ estimates to be.) There are also other "unnatural" (perhaps) restrictions in this work.

1.2.8 Long time-range estimates

Due to use of Gronwall arguments in the previous section, the error estimates become useless as $t \to \infty$. In the latter part of the 1980's it seemed to occur to many people that nonsmooth data estimates for finite time of the type given in the previous section, when combined with some type of stability of the original problem under perturbations, can give estimates

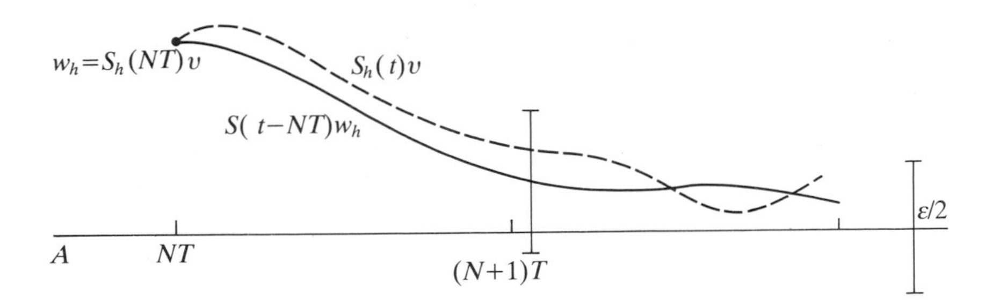

Fig. 1.1.

which are valid on the whole time interval $(0, \infty)$. Let me refer to Hale, Lin and Raugel[26], Heywood and Rannacher[29], Kloeden and Lorenz[37] and also to Larsson[38, pg. 359] in particular.

We shall take a brief look at the central idea, following Hale, Lin and Raugel. Let $S(t)$ be a nonlinear semigroup and $S_h(t)$ approximating semigroups, $0 < h < H$, with the following non-smooth data type approximating property: Given $\delta > 0$, $0 < t_* < t^* < \infty$, $M > 0$, there exists $h_0 = h_0(\delta, t_*, t^*, M) > 0$ such that

$$\begin{aligned} \|S_h(t)w - S(t)w\| \leq \delta, \quad & \text{for } t_* \leq t \leq t^*, \\ \|w\| \leq M\ , \quad & h \leq h_0. \end{aligned} \tag{2.64}$$

Let A be an attractor for a neighborhood U. Let A_ε denote $A+w$, $\|w\| \leq \varepsilon$. Then given any $\varepsilon > 0$ there exists T and $h_0 > 0$ such that

$$S_h(t)\,U \subseteq A_\varepsilon, \quad \text{for } h \leq h_0,\, t \geq T. \tag{2.65}$$

For a proof of this, assume that $U \subseteq \{\|x\| \leq M\}$. First select T such that $S(t)U \subseteq A_{\varepsilon/2}$ for $t \geq T$ (the attracting property). Then select h_0 so that $h_0 = h_0(\varepsilon/2, T, 2T, M)$. The proof is now contained in Figure 1.1 with $v \in U$.

1.2.9 On discretization in time

This is a fairly enormous subject and we refer to Thomée[59] for an in-depth study. Our intention here is simply briefly to describe and to bring out a few problems.

Let us start by considering the backward Euler Galerkin method. This comes about by defining $U^n \in S_h$ as a supposed approximation to $u(t_n)$ via

$$\left(\frac{U^n - U^{n-1}}{k_n}, \chi\right) + A(U^n, \chi) = (f(t_n), \chi), \quad \text{for } \chi \in S_h. \tag{2.66}$$

Here $k_n = t_n - t_{n-1}$. A "standard" analysis of this method proceeds by comparing U^n to $R_h u(t_n)$ (and thus the semi-discrete result is not used!). For $\theta = U^n - R_h u(t_n)$ and with $\partial_t \theta^n = (\theta^n - \theta^{n-1})/k_h$ one then has

$$(\partial_t \theta^n, \chi) + A(\theta^n, \chi) = -(\omega^n, \chi), \tag{2.67}$$

with

$$\omega^n = R_h \partial_t u(t_n) - u_t(t_n). \tag{2.68}$$

One has control over ω^n and then takes $\chi = \theta^n$ in (2.67). One ends up with a typical estimate of the form

$$\begin{aligned} \|U^n - u(t_n)\| \leq \|v_h - v\| + Ch^r \left\{ \|v\|_r + \int_0^{t_n} \|u_t\|_r \, ds \right\} \\ + k \int_0^{t_n} \|u_{tt}\| \, ds, \end{aligned} \tag{2.69}$$

see [59, Theorem 1.3]. Here, $k = \max k_m$, $m \leq n$. Recently, see Eriksson and Johnson[19], and Johnson, Nie and Thomée[33], such estimates have been criticized. For example, they seem useless as $t_n \to \infty$ and they give very little guidance for how to choose k_n adaptively. Instead, for a slightly perturbed variant of (2.66), $(f(t_n), \chi)$ is replaced by $k_n^{-1} \int_{t_{n-1}}^{t_n} (f(s), \chi) \, ds$, an error estimate of the form

$$\begin{aligned} \|U^n - u(t_n)\| \leq C \left(1 + \ell n \left(\tfrac{t_n}{k_n}\right)\right)^{1/2} \\ \cdot \left(\max_{m \leq n} k_m \|u_t\|_{\infty, I_m} + \max_{t \leq t_n} h^r \|u(t)\|_r \right) \end{aligned} \tag{2.70}$$

is derived using a parabolic duality argument. This is now suitable for a posteriori error control, since C is independent of t, k and h. If one desires the temporal part of the error to be controlled by a tolerance δ, one sets (forgetting about the logarithmic factor and assuming C estimated "somehow")

$$Ck_m \|u_t\|_{\infty, I_m} \simeq \delta, \tag{2.71}$$

or, replacing

$$k_m \|u_t\|_{\infty, I_m} \simeq \|U^m - U^{m-1}\|, \tag{2.72}$$

one takes adaptively

$$\|U^m - U^{m-1}\| \simeq \frac{\delta}{C}. \tag{2.73}$$

Theoretical results and numerical experiments are presented in Johnson, Nie and Thomée[33]. More work by these authors and Eriksson is to appear.

Analyses similar to (2.69), i.e., by comparing U^n with $R_h u(t_n)$, are standard also for the Crank-Nicolson method (and of course forward Euler).

In Baker, Bramble and Thomée[2], a class of one-step methods based on rational approximation $r(\lambda)$ to $\exp(\lambda)$ was introduced and analyzed in the case that A is time-independent. The analysis now proceeds by comparing U^n to the semidiscrete approximation $u_h(t_n)$, i.e., comparing $\exp(-A_h t_n)$ to $r(-kA_h)^n$. Smooth and nonsmooth data estimates are given.

(A side comment: The representation on top of pg. 822 of that paper is highly useful and should be in the standard arsenal.)

In Sammon[50] the analysis for rational approximations to the exponential (and backward differentiation formulae) were extended to the case when A has time-dependent coefficients. A very interesting "anomaly" occurs in the (somewhat crude) numerical experiments, where methods which are order k^4 for nonsmooth initial data for A time-independent actually only perform to order k^2 for time-dependent coefficients. This looks much like an "order-reduction" phenomenon, known in inhomogeneous problems (with A time-independent), see Sanz-Serna and Verwer[52] and references there to earlier work, and also Keeling[36] for homogeneous problems with time-dependent operators. However, in the works just mentioned the solutions considered are uniformly smooth down to $t = 0$. The connection between Sammon's example and order-reduction is not clear, at least not to me.

A perhaps less well-known method for time discretization is the "discontinuous Galerkin method", cf. Jamet[31] and Thomée[59, Chapter 9]. Let with $I_n = [t_{n-1}, t_n]$,

$$P_q(I_n) = \left\{ w : I_n \to S_h, \quad w(t) = \sum_{i=0}^{q} \chi_i(x) t^i, \quad \chi_i \in S_h \right\}, \tag{2.74}$$

and

$$\begin{aligned} w_+^n &= \lim_{s\to 0+} w(t_n + s), \qquad w_-^n = \lim_{s\to 0-} w(t_n - s), \\ [w^n] &= w_+^n - w_-^n, \qquad \text{the jump of } w \text{ at } t_n. \end{aligned}$$

We now seek U such that $U|_{I_n} \in P_q(I_n)$ and with U_-^{n-1} given,

$$\begin{aligned} &\int_{I_n} (U_t, w) + A(U, w)\, ds + ([U^{n-1}], w_+^{n-1}) \\ &\qquad = \int_{I_n} (f, w)\, ds, \quad \text{for } w \in P_q(I_n) \\ &U_-^0 = P_h v \qquad \text{(say)}. \end{aligned} \tag{2.75}$$

To see that this has a solution it suffices to check uniqueness when $U_-^{n-1} = f = 0$. Thus, taking $w = U$,

$$\int_{I_n} (U_t, U) + A(U, U)\, ds + \|U_+^{n-1}\|^2 = 0. \tag{2.76}$$

Since

$$\int_{I_n} (U_t, U) = \tfrac{1}{2} \int_{I_n} \frac{d}{dt} \|U\|^2 = \tfrac{1}{2}\|U_-^n\|^2 - \tfrac{1}{2}\|U_+^{n-1}\|^2 \tag{2.77}$$

we thus have

$$\tfrac{1}{2}\|U_-^n\|^2 + \tfrac{1}{2}\|U_+^{n-1}\|^2 + \int_{I_n} A(U, U)\, ds = 0. \tag{2.78}$$

Assuming A positive definite, uniqueness follows.

One may also get the feeling from (2.78) that the term $([U^{n-1}], w_+^{n-1})$ helps in controlling the jump of U over time levels t_n.

If one takes $q = 0$ in (2.75) one actually obtains the (modified) backward Euler method occurring in (2.70). For $q = 1$ one has a connection with the subdiagonal $(2, 1)$ Padé approximant to $\exp(\lambda)$.

The analysis such as in Thomée and Zhang[59] gives $O(k^{q+1})$ estimates and, at the time-levels t_n, $O(k^{2q+1})$ estimates, which are combined with the semi-discrete approximation in the same manner as in Baker, Bramble and Thomée[2].

Finally, the results for the semi-linear semi-discrete nonsmooth data case of Johnson, Larsson, Thomée and Wahlbin[32], (cf. 1.2.7 above) have been extended to certain time-discretizations in Crouzeix and Thomée[12]. Just as the order in the semidiscrete case is limited to $O(h^2)$ for positive time when $v \in L_2$ only, the time discretization is limited to first order, $O(k)$, no matter what its accuracy is on a linear problem.

1.3 Partial Integro-Differential Equations

1.3.1 Prelude

We shall consider partial integro-differential equations of the following "parabolic" type,

$$\begin{cases} u_t + A(t)u + \displaystyle\int_0^t B(t, \tau) u(\tau)\, d\tau &= f(t) \qquad t > 0,\ x \in \Omega\,, \\ u &= 0 \qquad \text{on } \partial\Omega, \\ u(0) &= v. \end{cases} \tag{3.1}$$

Here $A(t)$ is a second order, elliptic, symmetric, positive definite, linear (mostly) partial differential operator in the space variables and $B(t,\tau)$ is an at most second order partial differential operator, not necessarily elliptic. We shall give some brief attention to the "hyperbolic" counterpart,

$$u_{tt} + A(t)u + \int_0^t B(t,\tau)u(\tau)\,d\tau = f(t). \tag{3.2}$$

Of course, these two types do not exhaust the subject. The integral operator may, for example, be in the spatial variables such as in

$$u_{tt} = \left(1 + \int_0^\pi (u_x)^2\,dx\right) u_{xx}, \qquad t > 0, \quad 0 < x < \pi. \tag{3.3}$$

As usual, before finite elements there were finite differences (and sometimes they coincide). Some examples of finite difference methods applied to partial integro-differential equations can be found in Budak and Pavlov[7], D'Jakonov[15], Douglas and Jones[17], Galeone, Mastroserio and Montrone[22], Habetler and Schiffman[25], Lopez-Marcos[43], Pavlov[47], Tavernini[57] and Thompson[63].

A finite element method is given in Oden and Armstrong[46] for a visco-elastic problem involving partial integro-differential equations, without error analysis.

In Christie and Sanz-Serna[11] a Galerkin-finite element method is applied to the equation (3.3), and in Camino[8] (cf. Sanz-Serna[51]) to a visco-elastic problem,

$$u_t + uu_x = \int_0^t (t-\tau)^{-1/2} u_{xx}(\tau)\,d\tau. \tag{3.4}$$

The numerical analysis of finite element methods for (3.1) and (3.2) was started by Yanik and Fairweather[68], cf. Greenwell[24]. Their analysis applied to the case when B is of at most first order. Since then methods have been developed which treat also B of second order with "ease". Thus we will not reference the above two works further but acknowledge their historical priority.

1.3.2 Semidiscrete approximations, smooth solutions

$A(t;\cdot,\cdot)$ and $B(t,\tau;\cdot,\cdot)$ be the bilinear forms on H^1 coming from A and B. Then seek $u_h = u_h(t) \in S_h$ from the ordinary Volterra equation

$$\begin{cases} (u_{h,t},\chi) + A(t;u_h,\chi) + \displaystyle\int_0^t B(t,\tau;u_h(\tau),\chi)\,d\tau \\ \qquad\qquad\qquad = (f,\chi), \quad \text{for } \chi \in S_h, \\ u_h(0) = v_h. \end{cases} \tag{3.5}$$

In Thomée and Zhang[62] the following optimal order result was shown:

$$\begin{aligned}\|u_h(t) - u(t)\| \\ \leq C(T)\left\{\|v_h - v\| + h^r\left(\|v\|_r + \int_0^t \|u_t\|_r \, d\tau\right)\right\}, \\ \text{for } t \in [0, T].\end{aligned} \tag{3.6}$$

Let us go a little bit into the analysis! Following ideas from the parabolic case, introduce $R_h u$. For $\theta = u_h - R_h u$ we easily derive the error equation, $\rho = R_h u - u$, $e = u_h - u = \theta + \rho$,

$$(\theta_t, \chi) + A(\theta, \chi) = -(\rho_t, \chi) + \int_0^t B\left(t, \tau; e(\tau), \chi\right) d\tau. \tag{3.7}$$

Then write $\theta = \theta^1 + \theta^2$, where

$$\left\{\begin{array}{rcl}(\theta_t^1, \chi) + A(\theta^1, \chi) & = & -(\rho_t, \chi), \\ \theta^1(0) & = & \theta(0),\end{array}\right. \tag{3.8}$$

and

$$\left\{\begin{array}{rcl}(\theta_t^2, \chi) + A(\theta^2, \chi) & = & \displaystyle\int_0^t B(t, \tau; e(\tau), \chi) \, d\tau, \\ \theta^2(0) & = & 0.\end{array}\right. \tag{3.9}$$

Now (3.8) is exactly (2.14) and hence the analysis there gives

$$\|\theta^1(t)\| \leq C\|v_h - v\| + Ch^r\left\{\|v\|_r + \int_0^t \|u_t\|_r \, d\tau\right\}, \tag{3.10}$$

and we are left with estimating θ^2. Let $A_h : S_h \to S_h$ be defined by

$$(A_h\varphi, \chi) = -A(\varphi, \chi), \qquad \text{for } \chi \in S_h, \tag{3.11}$$

and let

$$T_h = A_h^{-1}. \tag{3.12}$$

Set $\chi = T_h\theta_t^2$ in (3.9). Then

$$\begin{aligned}(T_h\theta_t^2, \theta_t^2) + \frac{1}{2}\frac{d}{dt}\|\theta^2\|^2 &= \int_0^t B\left(t, \tau; e(\tau), T_h\theta_t^2(t)\right) d\tau \\ &= \frac{d}{dt}\int_0^t B\left(t, \tau; e(\tau), T_h\theta^2(t)\right) d\tau \\ &\quad - B\left(t, t; e(t), T_h\theta^2(t)\right) \\ &\quad - \int_0^t B_t\left(t, \tau; e(\tau), T_h\theta^2(t)\right) d\tau.\end{aligned} \tag{3.13}$$

Now $(T_h f, f) \geq 0$ and we throw the first term away. Notice how the use of T_h gave us the derivative of $\|\theta^2\|^2$ from the second term – a useful trick. Integrating with respect to t,

$$\begin{aligned} \|\theta^2(t)\|^2 &\leq C\int_0^t \{ |B(t,\tau;e(\tau),T_h\theta^2(t))| \\ &\qquad + |B(\tau,\tau;e(\tau),T_h\theta^2(\tau))| \}\, d\tau \\ &\qquad + C\int_0^t\int_0^\tau |B_t(\tau,s;e(s),T_h\theta^2(\tau))|\, ds\, d\tau \\ &\equiv Q(t). \end{aligned} \tag{3.14}$$

One proceeds to show that $Q(t)$ thus defined satisfies

$$\begin{aligned} Q(t) \leq C\Big\{ &\|v_h - v\| + h^r \left(\|v\|_r + \int_0^t \|u_t\|_r\, d\tau \right) \\ &+ \int_0^t \|e\|\, d\tau \Big\} \sup_{\tau\leq t} \|\theta^2(\tau)\|. \end{aligned} \tag{3.15}$$

Assuming this for the moment we find from (3.14) that

$$\begin{aligned} \|\theta^2(t)\| \leq C\Big\{ &\|v_h - v\| + h^r \left(\|v\|_r + \int_0^t \|u_t\|_r\, d\tau \right) \\ &+ \int_0^t \|e\|\, d\tau \Big\}. \end{aligned} \tag{3.16}$$

Combining this with the estimate

$$\|\rho(t)\| \leq Ch^r \|u(t)\|_r \leq Ch^r \left\{ \|v\|_r + \int_0^t \|u_t\|_r\, d\tau \right\}, \tag{3.17}$$

and with (3.10) we have

$$\begin{aligned} \|e(t)\| &\leq \|\rho(t)\| + \|\theta^2(t)\| + \|\theta^2(t)\| \\ &\leq C\|v_h - v\| + Ch^r \left\{ \|v\|_r + \int_0^t \|u_t\|_r\, d\tau \right\} \\ &\qquad + C\int_0^t \|e\|\, d\tau. \end{aligned} \tag{3.18}$$

Gronwall's lemma settles it!

It remains to prove (3.15). The following is a simple little result:

$$|B(t,\tau;f,T_h g)| \leq C\,(h\|f\|_1 + \|f\|)\,\|g\|. \tag{3.19}$$

(Think about the continuous case; we forego the proof.) From this we have

$$Q(t) = C \int_0^t (h\|e\|_1 + \|e\|) \, d\tau \, . \sup_{\tau \le t} \|\theta^2(\tau)\|. \tag{3.20}$$

Since by the inverse property (2.2) and properties of R_h,

$$\begin{aligned} \|e\|_1 &\le \|\theta\|_1 + \|\rho\|_1 \\ &\le Ch^{-1}\|\theta\|_0 + Ch^{r-1}\left\{\|v\|_r + \int_0^t \|u_t\|_r\right\} \\ &\le Ch^{-1}\|e\| + Ch^{r-1}\left\{\|v\|_r + \int_0^t \|u_t\|_r\right\}, \end{aligned} \tag{3.21}$$

we obtain (3.15).

Actually, in [62] they avoid the use of the inverse property. Instead they show directly by plugging in $\chi = \theta(t)$ in the error identity (3.7) and integrating (which gives $\int_0^t A(\theta, \theta)$, comparable to $\int_0^t \|\theta\|_1^2$, on the left) and fooling around a bit that

$$\int_0^t \|e\|_1 \le C\|v_h - v\| + Ch^{r-1}\left\{\|v\|_r + \int_0^t \|u_t\|_r\right\}, \tag{3.22}$$

which also, via (3.20) would give (3.15).

Now, in Cannon and Lin[9] an alternative approach to that of comparing u_h with $R_h u$ was proposed. As you have seen above the integral term in the error equation (3.7) makes it fairly obscure how to go about things. Cannon and Lin's approach can (with hindsight) be thought of as introducing an intermediate step to break up the analysis into more manageable parts. Introduce $(V_h u)(t) : [0, T] \to S_h$ via

$$A(t; V_h u - u, \chi) - \int_0^t B\,(t, \tau; (V_h - u)(\tau), \chi) \, d\tau = 0, \quad \text{for } \chi \in S_h. \tag{3.23}$$

Note the distinction to $R_h u$. If we now redefine θ as $\theta = u_h - V_h u$ we have the new error equation, with $\rho = V_h u - u$,

$$(\theta_t, \chi) + A(t; \theta, \chi) + \int_0^t B\,(t, \tau; \theta(\tau), \chi) \, d\tau = -(\rho_t, \chi). \tag{3.24}$$

Compare this with the old one, written in the form

$$\begin{aligned} (\theta_t\,, \chi) + A(\theta, \chi) + \int_0^t B\,(t, \tau; \theta(\tau), \chi) \, d\tau \\ = \quad -(\rho_t, \chi) + \int_0^t B\,(t, \tau; \rho(\tau), \chi) \, d\tau. \end{aligned} \tag{3.25}$$

You can feel in your bones how that extra thing on the right messes up L_2-estimates if they are to be of correct optimal order!

Of course, R_h is not dead yet! It will come into play as a technical tool when analysing the error in $V_h u - u$.

Let us check out our intuition that the new θ is easier: stuff $\chi = \theta$ into (3.24). It results that

$$\frac{1}{2}\frac{d}{dt}\|\theta\|^2 + C\|\theta\|_1^2 \leq \|\rho_t\|\,\|\theta\| + C\int_0^t \|\theta(\tau)\|_1\,d\tau\,\|\theta(t)\|_1. \tag{3.26}$$

Integrate, and use kickback:

$$\begin{aligned}\|\theta(t)\|^2 + \frac{C}{2}\int_0^t \|\theta\|_1^2 \leq \|\theta(0)\|^2 + \int_0^t \|\rho_t\|\,\|\theta\| \\ +C\int_0^t\int_0^s \|\theta\|_1^2\,d\tau\,ds.\end{aligned} \tag{3.27}$$

Gronwall's lemma now implies

$$\begin{aligned}\|\theta(t)\|^2 + \int_0^t \|\theta\|_1^2 &\leq C\left\{\|\theta(0)\|^2 + \int_0^t \|\rho_t\|\,\|\theta\|\,d\tau\right\} \\ &\leq C\|\theta(0)\|^2 + \tfrac{1}{2}\sup_{\tau\leq t}\|\theta(\tau)\|^2 \\ &\qquad + C\left(\int_0^t \|\rho_t\|\,d\tau\right)^2.\end{aligned} \tag{3.28}$$

Since the right hand side is monotonely increasing in t, we conclude that

$$\sup_{\tau\leq t}\|\theta(\tau)\| \leq C\left\{\|\theta(0)\| + \int_0^t \|\rho_t\|\,d\tau\right\}. \tag{3.29}$$

Actually, $\theta(0) = v_h - R_h v$ and we are back to estimating ρ_t.

Let me show how an H^1 analysis for ρ itself goes. The defining equation (3.23) may, by the definition of R_h, be written

$$A(t; V_h u - R_h u, \chi) = \int_0^t B\left(t,\tau; V_h u - R_h u + R_h u - u, \chi\right)\,d\tau. \tag{3.30}$$

Set $\chi = V_h u - R_h u \equiv \psi$:

$$\|\psi\|_1^2 \leq C\int_0^t \left(\|\psi(\tau)\|_1 + \|u - R_h u\|_1\right)\,d\tau\|\psi\|_1. \tag{3.31}$$

Cancelling one of the terms $\|\psi\|_1$ from each side, and using Gronwall,

$$\begin{aligned} \|\psi\|_1(t) &\leq C\int_0^t \|u - R_h u\|_1 \, d\tau \\ &\leq Ch^{r-1}\int_0^t \|u\|_r \, d\tau. \end{aligned} \tag{3.32}$$

A note: actually, the above analysis takes place only over the meshdomain Ω_h. This is important for $r > 2$; the result as stated would be false for isoparametrics. Why? (Cf. the parenthetical remark following (2.1).)

Now, to go to L_2, use a rather standard type duality argument. To knock off ρ_t, differentiate the defining equation; the extra junk terms are easily handled inductively.

Cannon and Lin called the above animal V_h a "nonclassical H^1 projection". The analysis above is actually more as given in Lin, Thomée and Wahlbin[42] where the animal was rechristened a "Ritz-Volterra projection". The analysis in [42], also gives pointwise L_∞-error estimates for $d = 1, 2, 3, 4$.

The reader may wonder: why not use the full-blown parabolic theory and Duhamel's principle. That is, writing

$$u(t) = E(t)v + \int_0^t E(t-s)\int_0^s B(s,\tau)u(\tau)\, d\tau\, ds \tag{3.33}$$

with E the parabolic solution operator, and similarly for $u_h(t)$ etc., ought one not then be able to bring into play all the nice results from Section 1.2? The approach works up to a point; e.g., pointwise error estimates seem very hard to get in any generality with this approach. But, maybe it hasn't been used to its full potential yet!

1.3.3 More about the Ritz-Volterra projection

Note first that if we were treating the "hyperbolic" integro-differential equation (3.2) we could still introduce the very same $V_h u$; half the work would be done and we would end up with a much simpler equation to analyze for $u_h - V_h u$ than that for $u_h - R_h u$, which would be a standard way to proceed. See Lin, Thomée and Wahlbin[42], for details of the hyperbolic case.

Consider next a Sobolev type pseudo-parabolic equation,

$$\left\{ \begin{aligned} A(t)u_t + B(t) &= f, \\ u &= 0 \quad \text{on } \partial\Omega, \\ u(0) &= v, \end{aligned} \right. \tag{3.34}$$

and its semi-discrete analogue

$$\begin{cases} A(t;u_{h,t},\chi)+B(t;u_h,\chi) = (f,\chi), & \chi\in S_h, \\ u_h(0) = v_h. \end{cases} \tag{3.35}$$

Then with $e=u_h-u$,

$$A(t;e_t,\chi)+B(t;e,\chi)=0, \tag{3.36}$$

and writing $e(t)=e(0)+\int_0^t e_t\,d\tau$,

$$A\left(t;e_t(\tau),\chi\right)+\int_0^t B\left(t;e_t(\tau),\chi\right)\,d\tau=-B\left(t;e(0),\chi\right). \tag{3.37}$$

The quantity e_t thus behaves as the error in a slightly modified Ritz-Volterra projection, where the modification is in terms of initial data only. This modification is handled in [42] and, e.g., after integration, (almost) optimal order pointwise error estimates result for smooth solutions.

For the record, Sobolev type equations present in themselves fairly few problems in a convergence analysis buttressed with present day technology of proofs. An early analysis is in Ford[20].

Finally, Lin, Thomée and Wahlbin apply the results for the Ritz-Volterra projection to furnish an easy analysis of the error for the equation

$$u_{tt}+A(t)u_t+B(t)u=f, \tag{3.38}$$

cf. Section 1.3.7 below.

1.3.4 Nonsmooth data

Thomée and Zhang[62] address the question of what happens when $v\in L_2$ merely. They first show a smoothing property in the continuous problem (3.1) of "parabolic" type, but it is limited. Essentially, if β is the order of the differential operator B, then

$$\|u(t)\|_\alpha\le Ct^{-\alpha/2}\|v\| \qquad \text{for } 0\le\alpha\le 4-\beta. \tag{3.39}$$

They give an example to show that the restriction $\alpha\le 4-\beta$ is necessary.

Taking $v_h=P_hv$ just as in the parabolic case they then proceed to show (for $f=0$) that

$$\|u_h(t)-u(t)\|\le Ch^\gamma t^{-\gamma/2}\|v\|, \qquad \gamma=\min(4-\beta,r). \tag{3.40}$$

An error analysis of finite element methods for another equation with a limited smoothing property, the Euler-Poisson-Darboux equation,

$$\begin{cases} u_{tt} + \dfrac{2p+1}{t} u_t + Au = 0 & t > 0, \\ u(0) = v, & \\ u_t(0) = 0, & \end{cases} \tag{3.41}$$

is furnished in Genis[23]. Here the limit of the smoothing property depends on p ; for $0 \leq s \leq p + 1/2$ one has

$$\|u(t)\|_s \leq Ct^{-s}\|v\|. \tag{3.42}$$

Le Roux and Thomée[41] extend the results for semilinear parabolic equations of Johnson, Larsson, Thomée and Wahlbin[32] to the integro-differential case, showing (essentially) an $O(h^2)$ estimate for positive time for $v \in L_2$ and showing that no better rate than second order is possible, cf. Section 1.2.7.

1.3.5 Time-discretizations

To discretize (3.5) further in time, consider your favorite time-discretization method for the parabolic problem, treating $\int_0^t B(t, \tau; u_h(\tau), \chi)\, d\tau$ as a source term. Assume that this method has a basic meshsize, say k, for simplicity uniform. Then, of course, some quadrature formula needs to be applied to actually approximately evaluate this time-integral. Now, that quadrature formula should obviously only use values of $U^n \simeq u(nk)$, but need not use all of them. Indeed, if the quadrature formula operated with the same mesh-scale k as the basic time-stepping method, all previous values would need to be stored. The idea of Sloan and Thomée[56] is then to use (pieces of) a quadrature rule which is comparatively of higher order than the basic time-stepping method. For example, if the basic time-stepping is backward Euler, an $O(k)$ method in itself, then use the trapezoidal rule in the quadrature, with step ℓ; this is an $O(\ell^2)$ method. Clearly one should then choose $\ell \simeq \sqrt{k}$, and clearly if we are sitting at a general time nk not also an integer multiple of ℓ, something special needs to be done on a last interval near nk, of length at most ℓ, say the rectangle rule with step k. In this way one ends up with a combined method with storage requirements $O(T/\sqrt{k})$ (as opposed to $O(T/k)$ if all previous values were stored) and of order of accuracy $O(k)$. The last of course remains to be shown!! This is done for various combinations of time-stepping methods and quadrature formulae in Sloan and Thomée. In Le Roux and Thomée[41] the above idea is combined with that of Crouzeix and Thomée[12] to furnish an (almost) $O(h^2 + k)$ estimate for positive time in the semilinear integro-differential case with initial data merely in L_2. Further results are found in Zhang[69].

1.3.6 A singular kernel

Chen, Thomée and Wahlbin[10] consider the problem (3.1) with

$$\int_0^t B(t;\tau)u(\tau)\,d\tau = \int_0^t K(t-\tau)Bu(\tau)\,d\tau, \tag{3.43}$$

where B is time-independent second order and

$$|K(s)| \le Cs^{-\alpha}, \qquad 0 \le \alpha < 1. \tag{3.44}$$

Under appropriate regularity assumptions they first show that

$$\begin{aligned} u &\in \mathcal{C}([0,T];H^2\cap H_0^1) \\ u_t &\in \mathcal{C}([0,T];L_2)\cap L_1([0,T];H^2\cap H_0^1) \\ \text{and}\qquad u_{tt} &\in L_1([0,T];L_2). \end{aligned}$$

In general, u_{tt} will blow up like $t^{-\alpha}$ as $t \to 0^+$.

Using the Ritz-Volterra projection they first show for the semi-discrete error,

$$\|u_h(t)-u(t)\| \le Ch^2\left\{\|v\|_2 + \int_0^t \|u_t\|_2\right\}. \tag{3.45}$$

For time-discretization, backward Euler with a fixed step k is used, and for quadrature, a product integration rule on meshsize k. (Thus no attempt is made to consider the storage saving techniques of Section 1.3.5; neither is a mesh graded for the singularity at the origin considered.) For the integral

$$J_n(\Phi) = \int_0^{t_n} K(t_n-\tau)\Phi(\tau)\,d\tau, \qquad \Phi(\tau) = B\left(u_h(\tau),\chi\right), \tag{3.46}$$

Φ is considered constant on each time-interval, and thus

$$J_n(\Phi) \simeq \sum_{j=0}^{n-1} K_{n-j}\Phi(t_j), \qquad K_\ell = \int_{t_{\ell-1}}^{t_\ell} K(s)\,ds. \tag{3.47}$$

It is shown that

$$\|U^n - u(t_n)\| \le C(h^2+k)\left\{\|v\|_2 + \int_0^{t_n} \left(\|u_{tt}\| + \|u_t\|_2\right)d\tau\right\}, \tag{3.48}$$

the main point being that this estimate matches the known regularity of the solution.

1.3.7 The strongly damped wave equation

Another case study is given in Larsson, Thomée and Wahlbin[38]. The equation under analysis is

$$\begin{cases} u_{tt} + \alpha A u_t + Au = 0, \\ u = 0 \quad \text{on } \partial\Omega, \\ u(0) = \Phi, \quad u_t(0) = \psi. \end{cases} \tag{3.49}$$

Looking back at (3.38) we recall that results for this were already included in Lin, Thomée and Wahlbin[42]. The point is now to make a sharper analysis with respect to regularity and compatibility demands on data Φ and ψ. Such demands will be measured in $H^\ell(\Omega)$ and in $\dot{H}^\ell(\Omega)$, cf., (2.27). Thus H^ℓ measures smoothness whereas $\dot{H}^\ell$ measures also compatibility. As a result of $B = A$ one may Fourier analyze to ones heart's content. It turns out that (the first order system equivalent for (u, u_t)) generates an analytic semi-group in time in $\dot{H}^r \times \dot{H}^{r-2}$. For spatial regularity, $D_t^j u(t)$ is in H^r for $t > 0$ if and only if $\Phi \in H^r$, $\psi \in H^{r-2}$. In this case $\|u(t)\|_r$ will be unbounded as t tends to zero, unless $\Phi \in H^r \cap \dot{H}^{r-2}$, $\psi \in \dot{H}^{r-2}$.

With the time-discretization given by a rational approximation $r(\lambda)$ of order p to the exponential (applied to the 2×2 system form for (u, u_t)), a typical smooth data estimate is

$$\|U^n - u(t_n)\| \le Ch^r\{|\Phi|_r + |\psi|_{r-2}\} + Ck^p\{|\Phi|_{2p} + |\psi|_{2p-2}\}. \tag{3.50}$$

For methods with $|r(\infty)| < 1$, a typical non-smooth data result is, with initial data L_2-projected,

$$\begin{aligned} &\|U^n - u(t_n)\| \\ &\quad \le C(t_n^{-1})(h^r + k^p)\{\|\Phi\|_r + |\Phi|_{\mathrm{Max}(r-2,1)} + |\psi|_{r-2}\}, \qquad t_n > 0. \end{aligned} \tag{3.51}$$

Note that here the regularity demands are concerned with r; the time-approximation order p does not enter. This reflects both the analytic character of the semi-group with respect to the time variable and the fact that there is only spatial smoothing with respect to the compatibility conditions, and that a limited one.

1.4 Concluding Remarks about Rates of Convergence

In the whole of this chapter, the reader has only seen integral rates of convergence, typically $h^r + k^p$ in linear situations and smooth nonlinear, and $h^2 + k$ in the nonsmooth semilinear case ($v \in L_2$). (Well, maybe the rates were $\ell n(1/h)h^r$ or $h^{r-\epsilon}$ in some cases, but that is probably due to nonperfect proof techniques.)

An unwary person may be tempted to (at least) formulate the following two conjectures:

1. In linear problems with very smooth solutions, the rate is always optimal (or close).
2. $\ell n(1/h)$ should not occur in the natural habitat of numerical analysis.

Both these conjectures are false.

For the second conjecture, in (2.8) for $p = \infty$ and $r = 2$, $d = 2$, the logarithmic term is necessary, see Haverkamp[27]. (Another well-known example is that of the Lebesgue constant in the Fourier series.)

As for the first conjecture, it may be true in the realm of smooth elliptic or parabolic problems (cf. Wahlbin[66] for the fact that it is not true "locally" for elliptic problems on corner domains). However, in the realm of first order hyperbolic problems it is not true. Dupont[18] published an example pertaining to the space-periodic problem

$$\begin{cases} u_t + u_x &= 0, \\ u(0) &= v. \end{cases} \tag{4.1}$$

Letting v be any nonconstant analytic function of period 1, he showed that for the semicontinuous Galerkin method based on Hermite cubics with uniform subdivision ($r = 4$), whatever initial data approximation one takes,

$$\max_{0 \leq t \leq T} \|u_h(t) - u(t)\| \geq Ch^3, \tag{4.2}$$

where C is a positive constant independent of h. Thus we have a suboptimal order of approximation.

Dupont caps this off by showing that if we use instead for S_h a subspace of the Hermite cubics, i.e., with fewer degrees of freedom, namely the smooth $\mathcal{C}^2$ cubics, then we do have optimal order,

$$\|u_h(t) - u(t)\| \leq Ch^4. \tag{4.3}$$

Lest someone now is tempted to conjecture that in smooth problems, at least we have (close to) integral order, the hyperbolic problem

$$\begin{cases} u_t &= 0 \qquad \text{on } [0,1] \times [0,1] = \Omega, \\ u(0,x) &= v(x), \end{cases} \tag{4.4}$$

furnishes a counterexample. Peterson[48] shows that (with a very regular mesh) the discontinuous Galerkin method with piecewise constants (optimal order $r = 1$) actually gives

$$\|u_h - u\|_{L_\infty(\Omega)} \geq Ch^{1/2}, \tag{4.5}$$

for $v = x$ with C positive independent of h. He then reports precision numerical experiments (up to 200 million elements) showing that for piecewise linears ($r = 2$) and $v = x^2$, for a certain carefully constructed family of quasi-uniform meshes, actually

$$\|u_h - u\|_{L_2(\Omega)} \geq Ch^{3/2}, \tag{4.6}$$

thus giving numerical evidence that the convergence estimates of Johnson and Pitkäranta[34] are sharp.

(In [64] I proved convergence rates of 3.5 and $10/3$ for smooth solutions. My friends used to kid me for that. Since then there have been enough non-integer rates of convergence published that people in the field don't raise an eyebrow any more.)

Acknowledgement

I thank Professor Graeme Bailey for his help with this manuscript.

References

[1] I. Babuska, and J. Osborn, *Analysis of finite element methods for second order boundary value problems using mesh dependent norms*, Numer. Math. **34**, (1980), 41–62.

[2] G. A. Baker, J. H. Bramble and V. Thomée, *Single step Galerkin approximations for parabolic problems*, Math. Comp. **31**, (1977), 818–847.

[3] J. Blair, *Approximate solution of elliptic and parabolic boundary value problems*, Thesis, UC Berkeley, 1970.

[4] J. H. Bramble and J. E. Osborn, *Rate of convergence estimates for non-selfadjoint eigenvalue approximation*, Math. Comp. **27**, (1973), 525–549.

[5] Bramble, J. H., A. H. Schatz, V. Thomée, and L. B. Wahlbin, *Some convergence estimates for semi-discrete Galerkin type approximations for parabolic equations*, SIAM J. Numer. Anal. **14**, (1977), 218–241.

[6] J. H. Bramble and V. Thomée, *Discrete time Galerkin methods for a parabolic boundary value problem*, Ann. Mat. Pura Appl. **101**, (1974), 115–152.

[7] B. M. Budak and A. R. Pavlov, *A difference method of solving boundary-value problems for a quasi-linear integrodifferential equation of parabolic type*, Soviet Math. Dokl. **14**, (1973), 565–569.

[8] P. Camino, *A numerical method for a partial differential equation with memory*, Proc. 9th CEDYA, Secretariado de Publicaciones, Universidad de Valladolid, Valladolid, 107–112.

[9] J. R. Cannon and Y. Lin, *Nonclassical H^1 projection and Galerkin methods for nonlinear parabolic integro-differential equations*, Calcolo **25**, (1988), 187–201.

[10] C. Chen, V. Thomée and L. B. Wahlbin, *Finite element approximation of a parabolic integro-differential equation with a weakly singular kernel*, to appear (preprint Chalmers).

[11] I. Christie and J. M. Sanz-Serna, *A Galerkin method for a nonlinear integro-differential wave system*, Comput. Methods Appl. Mech. Engineering **44**, (1984), 229–237.

[12] M. Crouzeix and V. Thomée, *On the discretization in time of semilinear parabolic equations with nonsmooth initial data*, Math. Comp. **49**, (1987), 359–377.

[13] M. Crouzeix, V. Thomée and L. B. Wahlbin, *Error estimates for spatially discrete approximations of semilinear parabolic equations with initial data of low regularity*, Math. Comp. **53**, (1989), 25–41.

[14] J. Descloux, *On finite element matrices*, SIAM J. Numer. Anal. **9**, (1972), 260–265.

[15] E. G. D'Jakonov, *On the stability of difference schemes for some nonstationary problems*, in *Topics in Numerical Analysis*, J. J. H. Miller, Ed., Academic Press 1970, 63–87.

[16] J. Douglas Jr., T. Dupont and L. B. Wahlbin, *The stability in L^q of the L^2 projection into finite element function spaces*, Numer. Math. **23**, (1975), 193–197.

[17] J. Douglas Jr. and B. F. Jones, Jr., *Numerical methods for integro-differential equations of parabolic and hyperbolic types*, Numer. Math. **4**, (1962), 96–102.

[18] T. Dupont, *Galerkin methods for first order hyperbolics: an example*, SIAM J. Numer. Anal. **10**, (1973), 890-899.

[19] J. Eriksson and C. Johnson, *Error estimates and automatic time step control for non-linear parabolic problems I*, SIAM J. Numer. Anal. **24**, (1987), 12–22.

[20] W. H. Ford, *Galerkin approximations to non-linear pseudo-parabolic partial differential equations*, Aequationes Math. **14**, (1976), 271–291.

[21] H. Fujita and H. Mizutani, *On the finite element method for parabolic equations*, J. Math. Soc. Japan **28**, (1976), 749–771.

[22] L. Galeone, C. Mastroserio and M. Montrone, *Asymptotic stability of the numerical solution for an integro-differential reaction-diffusion system*, Numer. Methods Partial Differential Equations **5**, (1989), 79–86.

[23] A. Genis, *On finite element methods for the Euler-Poisson-Darboux equation*, SIAM J. Numer. Anal. **21**, (1984), 1080–1106.

[24] C. E. Greenwell, *Finite element methods for partial integro-differential equations*, Ph. D. Thesis, University of Kentucky, Lexington, 1982.

[25] G. T. Habetler and R. L. Schiffman, *A finite difference method for analyzing the compression of poro-elastic media*, Computing **6**, (1970), 342–348.

[26] J. Hale, X. Lin and G. Raugel, *Upper semi-continuity of attractors in dynamical systems*, Math. Comp. **50**, (1988), 89–123.

[27] R. Haverkamp, *Eine Aussage zur L_∞-stabilität und zur genauen Konvergenzordnung der H_0^1-projektion*, Numer. Math. **44**, (1984), 393–405.

[28] H.-P. Helfrich, *Fehlerabschätzungen für das Galerkin Verfahren zur Lösung von Evolutionsgleichungen*, Manuscripta Math. **13**, (1974), 219–235.

[29] J. G. Heywood and R. Rannacher, *Finite element approximation of the nonstationary Navier-Stokes problem, Part 3*, SIAM J. Numer. Anal. **25**, (1988), 489–512.

[30] M. Huang and V. Thomée, *Some convergence estimates for semi-discrete type schemes for time-dependent nonselfadjoint parabolic equations*, Math. Comp.**37**, (1981), 327–346.

[31] P. Jamet, *Galerkin-type approximations which are discontinuous in time for parabolic equations in a variable domain*, SIAM J. Numer. Anal. **15**, (1978), 912–928.

[32] C. Johnson, S. Larsson, V. Thomée and L. B. Wahlbin, *Error estimates for spatially discrete approximations of semilinear parabolic equations with nonsmooth initial data*, Math. Comp. **49**, (1987), 331–357.

[33] C. Johnson, Y.-Y. Nie, and V. Thomée, *An a posteriori error estimate and adaptive timestep control for a backward Euler discretization of a parabolic equation*, SIAM J. Numer. Anal. **27**, (1990), 277–291.

[34] C. Johnson and J. Pitkäranta, *An analysis of the discontinuous Galerkin method for a scalar hyperbolic equation*, Math. Comp. **46**, (1986), 1–26.

[35] M. L. Juncosa and D. M. Young, *On the Crank-Nicolson procedure for solving parabolic partial differential equations*, Proc. Cambridge Philos. Soc. **53**, (1957), 448–461.

[36] S. L. Keeling, *Galerkin/Runge-Kutta discretizations for parabolic equations with time-dependent coefficients*, Math. Comp. **52**, (1989), 561–586.

[37] P. E. Kloeden and J. Lorenz, *Lyaponov stability and attractors under discretization*, Proc. Equadiff. 1987, Dafermos, C. D., Ladas, G. and Papanicolaou, G., Eds., Marcel Dekker, 1987, 361–368.

[38] S. Larsson, *The long-time behavior of finite-element approximations of solutions to semilinear parabolic problems*, SIAM J. Numer. Anal. **26**, (1989), 348–365.

[39] S. Larsson, V. Thomée and L. B. Wahlbin, *Finite element methods for a strongly damped wave equation*, IMA J. Numer. Anal., to appear (preprint Chalmers).

[40] I. Lasiecka, *Convergence estimates for semidiscrete approximations of nonselfadjoint parabolic equations*, SIAM J. Numer. Anal. **21**, (1984), 894–909.

[41] M.-N. Le Roux and V. Thomée, *Numerical solution of semilinear integro-differential equations of parabolic type with nonsmooth data*, SIAM J. Numer. Anal. **26**, (1989), 1291–1309.

[42] Y.-P. Lin, V. Thomée, and L. B. Wahlbin, *Ritz-Volterra projections to finite element spaces and applications to integro-differential and related equations*, SIAM J. Numer. Anal., to appear (preprint MSI, Cornell).

[43] J. C. Lopéz-Marcos, *A difference scheme for a nonlinear partial integro-differential equation*, SIAM J. Numer. Anal. **27**, (1990), 20–31.

[44] M. Luskin and R. Rannacher, *On the smoothing property of the Galerkin method for parabolic equations*, SIAM J. Numer. Anal. **19**, (1982), 93–113.

[45] J. A. Nitsche and M. F. Wheeler, *L_∞ boundedness of the finite element Galerkin operator for parabolic problems*, Numer. Funct. Anal. Optim. **4**, (1981), 325–353.

[46] J. T. Oden and W. H. Armstrong, *Analysis of nonlinear, dynamic coupled thermoviscoelasticity problems by the finite element method*, Computers and Structures **1**, (1971), 603–621.

[47] A. P. Pavlov, *Stability and convergence of the method of nets for an integro-differential equation of parabolic type*, (in Russian), Trudy Irkutsk. Gos. Univ. **26**, (1968), 312–334.

[48] T. Peterson, *A note on the convergence of the discontinuous Galerkin method for a scalar hyperbolic equation*, SIAM J. Numer. Anal., to appear (preprint MSI, Cornell).

[49] P. H. Sammon, *Convergence estimates for semidiscrete parabolic equation approximations*, SIAM J. Numer. Anal. **19**, (1982), 68–92.

[50] P. Sammon, *Fully discrete approximation methods for parabolic problems with nonsmooth initial data*, SIAM J. Numer. Anal. **20**, (1983), 437–470.

[51] J. M. Sanz-Serna, *A numerical method for a partial integro-differential equation*, SIAM J. Numer. Anal. **25**, (1988), 319–327.

[52] J. M. Sanz-Serna and J. G. Verwer, *Stability and convergence of the PDE/Stiff ODE interface*, Appl. Numer. Math. **5**, (1989), 117–132.

[53] A. H. Schatz, *An observation concerning Ritz-Galerkin methods with indefinite bilinear forms*, Math. Comp. **28**, (1974), 959–962.

[54] A. H. Schatz, V. Thomée and L. B. Wahlbin, *Maximum norm stability and error estimates in parabolic finite element equations*, Comm. Pure Appl. Math. **33**, (1980), 265–304.

[55] A. H. Schatz and L. B. Wahlbin, *On the quasi-optimality in L_∞ of the $\overset{\circ}{H}{}^1$-projection into finite element spaces*, Math. Comp. **38**, (1982),1–21.

[56] I. Sloan and V. Thomée, *Time discretization of an integro-differential equation of parabolic type*, SIAM J. Numer. Anal. **23**, (1986), 1052–1061.

[57] L. Tavernini, *Finite difference approximations for a class of semilinear Volterra evolution problems*, SIAM J. Numer. Anal. **14**, (1977), 931–949.

[58] V. Thomée, *Some convergence results for Galerkin methods for parabolic boundary value problems*, Mathematical Aspects of Finite Elements in Partial Differential Equations, C. de Boor, Ed., Academic Press, 1974, 55–88.

[59] V. Thomée, *Galerkin Finite Element Methods for Parabolic Problems*, Springer Lecture Notes in Mathematics **1054**, Heidelberg, 1984.

[60] V. Thomée and L. B. Wahlbin, *On Galerkin methods in semilinear parabolic problems*, SIAM J. Numer. Anal. **12**, (1975), 378–389.

[61] V. Thomée and L. B. Wahlbin, *Maximum-norm stability and error estimates in Galerkin methods for parabolic equations in one space variable*, Numer. Math. **41**, (1983), 345–371.

[62] V. Thomée and N.-Y. Zhang, *Error estimates for semidiscrete finite element methods for parabolic integro-differential equations*, Math. Comp. **53**, (1989), 121–139.

[63] R. J. Thompson, *Difference approximations for some functional differential equations, Numerische, insbesondere approximationstheoretische Behandlung von Funktionalgleichungen*, Springer Lecture Notes in Mathematics **333**, Heidelberg, 1973, 263–273.

[64] L. B. Wahlbin, *A modified Galerkin procedure with Hermite cubics for hyperbolic problems*, Math. Comp. **29**, (1975), 978–984.

[65] L. B. Wahlbin, *A remark on parabolic smoothing and the finite element method*, SIAM J. Numer. Anal. **17**, (1980), 33–38.

[66] L. B. Wahlbin, *On the sharpness of certain local estimates for* $\overset{\circ}{H}^1$*-projections into finite element spaces: Influence of a reentrant corner*, Math. Comp. **42**, (1984), 1–8.

[67] M. F. Wheeler, *A priori* L_2 *error estimates for Galerkin approximations to parabolic partial differential equations*, SIAM J. Numer. Anal. **10**, (1973), 723–759.

[68] E. G. Yanik and G. Fairweather, *Finite element methods for parabolic and hyperbolic partial integro-differential equations*, Nonlinear Anal. (Theory, Methods and Applications) **12**, (1988), 785–809.

[69] N.-Y. Zhang, *On the discretization in time and space of parabolic integro-differential equations*, Thesis, Chalmers University of Technology, Göteborg, 1990.

Professor L. B. Wahlbin
Department of Mathematics
Cornell University
Ithaca
New York, NY 14853
USA

2

Finite Element Methods for Parabolic Free Boundary Problems

R. H. Nochetto

2.1 Introduction

In this chapter we discuss the numerical approximation of parabolic free boundary problems by finite element methods. We have chosen the classical solid-liquid phase transition as a model example but the underlying ideas apply to other free boundary problems as well. Such a transition is described by the so-called two-phase Stefan problem.

In Section 2.2 we present a physical motivation of the two-phase Stefan problem and obtain its classical formulation. Its drawbacks are then discussed along with the proper remedy, namely the weak (or enthalpy) formulation, which in turn leads to the fixed domain methods to be examined.

In Section 2.3 we study a couple of (continuous) approximation procedures produced by either regularization or phase relaxation. They serve to motivate the two main techniques for error analysis of degenerate parabolic problems, namely the inversion of the Laplacian and the integral method. They are discussed in a simple setting without discretization.

The discrete schemes are presented in Section 2.4, where we first introduce discrete-time schemes because of their simplicity and the fact that they still capture the essential features. We then formulate the fully discrete methods, which are easy to implement. They are the usual nonlinear method, for which the nonlinear constitutive relation is enforced, and two linear methods obtained by a suitable violation of such a constraint. Since free boundary problems typically exhibit a lack of regularity, these linearization techniques are not based on Taylor's formula.

Several a priori estimates are derived in Section 2.5 for the fully discrete schemes. Apart from their intrinsic interest, the resulting stability estimates are extremely useful in the subsequent error analysis. This is in striking contrast to the analysis of mildly nonlinear parabolic equations, where the standard approach assumes regularity on the continuous solution.

The error analysis is carried out in Section 2.6. We illustrate the use of the Laplace inverse method for the nonlinear scheme and the integral method for the nonlinear Chernoff formula, which is one of the linear algorithms. It is to be emphasized that this analysis requires regularity properties compatible with the free boundary problem at hand, as expressed in the estimates of Section 2.5. We also comment on the approximation of free boundaries.

Even though the interface does not play any explicit role in the methods above, it is responsible for the global numerical pollution observed on the discrete solution. This difficulty can be overcome by mesh adaptation, as discussed in Section 2.7. We give a heuristic motivation for the local refinement strategy, which in turn leads to highly graded meshes. We finally examine stability and accuracy and present several numerical experiments.

2.2 Classical and Weak Formulations

Let Ω be a regular bounded domain of $\mathbb{R}^d$ ($d \geq 1$) occupied by water and ice and set $Q := \Omega \times (0,T)$, where $0 < T < \infty$ is fixed. We denote the *enthalpy* per unit volume (or energy density) by u, the relative temperature by θ and the water concentration by χ. Hence, $\chi = 1$ in the liquid phase $Q^+ := \{\theta > 0\}$ whereas $\chi = 0$ in the solid phase $Q^- := \{\theta < 0\}$; χ is thus a *phase variable.* We also denote the heat flux by $\boldsymbol{q}$, the intensity of distributed heat sources or sinks by f, the specific heat per unit volume by c and the heat conductivity by k; all these quantities may depend on θ. Heat transfer within each phase is governed by the energy balance equation

$$u_t + \mathrm{div}\boldsymbol{q} = f(\theta), \tag{2.1}$$

which coupled with the Fourier's law

$$\boldsymbol{q} = -k(\theta)\nabla\theta, \tag{2.2}$$

and the constitutive relation

$$u(\theta) := \int_0^\theta c(s)ds + \chi(\theta), \tag{2.3}$$

gives rise to the mildly nonlinear heat equation

$$c(\theta)\theta_t - \mathrm{div}(k(\theta)\nabla\theta) = f(\theta), \tag{2.4}$$

which holds separately in each phase. In the classical situation, the *interface* $I(t)$ or surface of separation between solid and liquid phases is assumed to be smooth and given by the condition $\theta = 0$, which in turn neglects supercooling and surface tension effects:

$$I(t) := \{x \in \Omega : \theta(x,t) = 0\}. \tag{2.5}$$

Let ν_x indicate the unit vector normal to $I(t)$ at x directed towards the solid phase and let $V(x)$ denote the interface velocity normal to $I(t)$ at x. If $L = 1$ is the normalized latent heat per unit volume, the energy balance about x leads to the so-called *Stefan Condition*:

$$[\![\boldsymbol{q}(x,t)]\!] \cdot \nu_x := [\boldsymbol{q}^+(x,t) - \boldsymbol{q}^-(x,t)] \cdot \nu_x = V(x). \tag{2.6}$$

The last three equations constitute the classical formulation of the Stefan problem. The drawback of this setting is that $I(t)$ may actually exhibit singularities such as cusps and mushy regions, the latter being zones in which $0 < \chi < 1$. The extra condition (2.6) has been used anyhow to track the motion of $I(t)$ and then to solve decoupled heat equations (2.4) in each phase, giving rise to the so-called Front Tracking Methods. We will not discuss these methods here, however.

We will instead seek a global formulation that incorporates (2.4) to (2.6) into a single nonlinear equation, thus making $I(t)$ disappear as an explicit unknown; the ensuing method is called a *Fixed Domain Method*. Toward this end, we note that $\chi \in H(\theta)$ where H is the Heaviside graph, introduce the Kirchhoff transformation $v = K(\theta) := \int_0^\theta k(s)ds$ and finally observe that (2.4) can be written equivalently as $u_t - \Delta v = f(K^{-1}(v))$. For the sake of simplicity, we will still call v the temperature and denote it by θ, and also write f in place of $f \circ K^{-1}$. Let us now introduce the maximal monotone graph

$$\gamma(s) := \int_0^s c(z)dz + H(s), \tag{2.7}$$

and observe that $0 < \Gamma_1 \le \gamma'(s) \le \Gamma_2 < \infty$ for all $s \neq 0$. Set $\beta := \gamma^{-1}$ and note that $0 \le \beta'(s) \le L_\beta := \Gamma_1^{-1}$ for almost all $s \in \mathbb{R}$ and $\beta'(s) \ge \Gamma_2^{-1}$ for almost all $|s| > 1$; in particular $\beta(z) = -z^- + (z-1)^+$ in the case that all material properties are normalized to one. The weak or enthalpy formulation reads

$$u_t - \Delta\theta = f(\theta), \qquad \theta = \beta(u), \tag{2.8}$$

where the partial differential equation is to be interpreted in the sense of distributions in the entire domain Q. Let the initial enthalpy $u_0 \in L^2(\Omega)$, let f be globally Lipschitz continuous (i.e. $|f(a) - f(b)| \le L_f|a-b|$ for all $a, b \in \mathbb{R}$) and, just for simplicity, let $\theta = 0$ on $\partial\Omega$. The weak form of (2.8) reads as follows:

Problem (P): *Seek* $\{u, \theta\}$ *such that*

$$u \in L^\infty(0,T; L^2(\Omega)) \cap H^1(0,T; H^{-1}(\Omega)), \theta \in L^2(0,T; H_0^1(\Omega)), \tag{2.9}$$

$$\theta(x,t) = \beta(u(x,t)) \quad \textit{for almost all} \quad (x,t) \in Q, \tag{2.10}$$

$$u(\cdot, 0) = u_0 \tag{2.11}$$

and for almost all $t \in (0,T)$ *and for all* $\phi \in H_0^1(\Omega)$ *the following equation holds,*

$$\langle u_t, \phi \rangle + \langle \nabla\theta, \nabla\phi \rangle = \langle f(\theta), \phi \rangle. \tag{2.12}$$

Here, $\langle \cdot, \cdot \rangle$ denotes both the inner product in $L^2(\Omega)$ and the duality pairing between $H_0^1(\Omega)$ and $H^{-1}(\Omega)$. It is worth noting that (2.8) can be formally rewritten as

$$u_t - \operatorname{div}(\beta'(u)\nabla u) = f(\beta(u)), \tag{2.13}$$

which in turn reveals the *degenerate* character of the problem at hand in that $\beta'(u)$ vanishes for $0 < u < 1$. By simply changing the definition of β while still keeping its monotonicity, other problems can be recast within this framework. Relevant examples are the porous medium equation, the Hele-Shaw problem and nonstationary filtration.

Exercises

2.1 Prove that a classical solution is also a weak solution of (P).

2.2 Assume there exists a weak solution θ of (P) which is smooth in each phase and there exists a smooth function Φ such that $I := \cup_{0<t<T} I(t) = \{(x,t) \in Q : \Phi(x,t) = 0\}$ with $\nabla_x \Phi \neq 0$ on $I(t)$ and $\Phi < 0$ (> 0) in the solid (liquid) phase. Prove that θ is a classical solution. Hint: Integrate (2.12) from 0 to T and then use Green's formula in each phase. Use and justify the expression: $V(x) = \Phi_t(x,t)/|\nabla_x \Phi(x,t)|$ for $x \in I(t)$.

Existence and uniqueness are well known for (P); see [11,12,13] and the references therein. Since (P) is not strictly parabolic, the solutions $\{u, \theta\}$ exhibit a global lack of regularity. In fact, both u and $\nabla\theta$ are expected to have jump discontinuities across $I(t)$, as corresponds to (2.4) and (2.6). This fact makes the design and analysis of numerical methods for degenerate parabolic problems much harder than for mildly nonlinear ones. It thus becomes imperative to know the proper functional setting in which to measure global regularity. This can be done, for instance, by approximating β with a smooth and strictly increasing function and then proving a priori estimates for the corresponding classical solutions that are uniform in the regularization parameter. Since the underlying techniques will be useful later on, we reproduce them here along with the following exercises.

Exercises

2.3 Prove the a priori estimates

$$\|u\|_{L^\infty(0,T;L^2(\Omega))} + \|\theta\|_{L^2(0,T;H_0^1(\Omega))} + \|u\|_{H^1(0,T;H^{-1}(\Omega))} \leq C. \quad (2.14)$$

Hint: Take $\theta(t) \in H_0^1(\Omega)$ as a test function in the regularized analogue of (2.12) and integrate the resulting expression from 0 to $t \leq T$. Use also the fact

$$u_t\theta = u_t\beta(u) = \frac{d}{dt}\int_0^u \beta(s)ds, \quad (2.15)$$

and set $\Phi_\beta(s) := \int_0^s \beta(z)dz$. Prove the estimates

$$\frac{1}{2L_\beta}\beta(s)^2 \leq \Phi_\beta(s) \leq \frac{L_\beta}{2}s^2, \quad \text{for all } s \in \mathbb{R}. \quad (2.16)$$

2.4 Let $\theta_0 := \beta(u_0) \in H_0^1(\Omega)$. Prove the a priori estimates

$$\|\theta_t\|_{L^2(Q)} + \|\theta\|_{L^\infty(0,T;H_0^1(\Omega))} + \|u_t\|_{L^\infty(0,T;H^{-1}(\Omega))} \leq C. \quad (2.17)$$

Hint: Take $\theta_t \in H_0^1(\Omega)$ as a test function in the regularized analogue of (2.12) and integrate the ensuing equality from 0 to t. Use the inequality $u_t\theta_t = \gamma'(\theta)\theta_t^2 \geq \Gamma_1\theta_t^2$.

2.5 Let $\varepsilon > 0$ be the regularization parameter and assume $\beta'(s) \geq \varepsilon$ for almost all $s \in \mathbb{R}$. Prove the a priori estimate

$$\|\nabla_x u^\varepsilon\|_{L^2(Q)} \leq C\varepsilon^{-1/2}. \quad (2.18)$$

Hint: Take u^ε as a test function in the regularized analogue of (2.12) and justify that u^ε attains a homogeneous Dirichlet condition on $\partial\Omega$, i.e. $u^\varepsilon \in H_0^1(\Omega)$. Note that (2.7) is also a by-product of this argument.

2.6 In addition to the assumptions in Exercise 5, let $\theta_0 := \beta(u_0) \in H_0^1(\Omega)$. Let $u_0^\varepsilon := \gamma_\varepsilon(\theta_0)$ be the initial enthalpy. Prove the a priori estimates

$$\|u_t^\varepsilon\|_{L^2(Q)} + \|\theta^\varepsilon\|_{L^2(0,T;H^2(\Omega))} \leq C\varepsilon^{-1/2}. \quad (2.19)$$

Suppose now $u_0^\varepsilon = u_0$. Is (2.19) valid? Hint: Proceed as in Exercise 2.4 but this time using the expression $u_t^\varepsilon\theta_t^\varepsilon = \beta'(u^\varepsilon)(u_t^\varepsilon)^2 \geq \varepsilon(u_t^\varepsilon)^2$.

2.7 Let $u_0 \in L^\infty(\Omega)$. Prove that $u \in L^\infty(\Omega)$. Hint: Use the Maximum Principle.

2.8 Let $\Delta\theta_0 \in L^1(\Omega)$. Let $M(\Omega)$ be the space of finite regular Baire measures. Prove

$$\Delta\theta \in L^\infty(0,T;M(\Omega)). \tag{2.20}$$

Hint: Differentiate the regularized analogue of (2.8) once more in time. Then multiply by a proper regularization of $\text{sgn}(\theta_t)=\text{sgn}(u_t)$. The above assumption on θ_0 may be weakened somewhat by taking $\Delta\theta_0 = l + \mu$ where $l \in L^1(\Omega)$ and $\mu \in M(\Omega)$, provided that the initial interface $I(0)$ is smooth and $\text{supp}(\mu) \subset I(0)$. This allows jump discontinuities of $\nabla\theta_0$ across $I(0)$. The proof proceeds as before after a suitable regularization.

In view of (2.18) and (2.19) we readily deduce that

$$\text{meas}(\{(x,t) \in Q : |u_t^\varepsilon|, |\nabla_x u^\varepsilon| \geq C\varepsilon^{-1}\}) \leq C\varepsilon.$$

If we think of the regularization procedure as a convolution operation and also add the fact that u is discontinuous across the interface I, we realize that I is spread out into a thin transition layer of width ε. In addition, we see that both (2.18) and (2.19) are sharp. Moreover, the singularity of u on I yields $u_t \notin L^2(Q)$ and thus $\theta \notin L^2(0,T;H^2(\Omega))$. However, we might expect u to be in some intermediate space between $L^2(Q)$ and $H^1(0,T;L^2(\Omega))$ that allows discontinuities; this statement will be made precise in Section 2.5.

In light of (2.14) and (2.17), we conclude that the natural energy spaces to examine accuracy are $L^\infty(0,T;H^{-1}(\Omega))$ for u and $L^2(Q)$ for θ. Morever, since

$$u(t) - \Delta\int_0^t \theta(s)ds = u_0 + \int_0^t f(\theta(s))ds, \tag{2.21}$$

we infer that, regarding the flux $\boldsymbol{q}$ $(= -\nabla\theta)$, the natural space for $\int_0^t \boldsymbol{q}(s)ds$ is $L^\infty(0,T;L^2(\Omega))$.

We close this section with the concept of nondegeneracy, which is important in the approximation of interfaces as well as in the analysis of the regularization process. We say that (P) satisfies a *nondegeneracy property* provided

$$\text{meas}\,(\{(x,t) \in Q : 0 \leq \theta(x,t) \leq \varepsilon\}) \leq C\varepsilon. \tag{2.22}$$

This condition is known to hold under certain qualitative properties of the data [18]. It prescribes a linear growth for θ away from the interface. Note that mushy regions are not allowed under this condition.

2.3 Regularization and Phase Relaxation

The purpose of this section is to study a couple of approximation procedures which result from changing the constitutive relation $\theta = \beta(u)$ (or equivalently $u \in \gamma(\theta)$). Despite their own interest, they will motivate the two main techniques for error analysis in a clean and simple setting. Such techniques will be often used in the subsequent sections.

2.3.1 Regularization

Let $\varepsilon > 0$ indicate the regularization parameter. Set

$$\beta_\varepsilon(s) := \begin{cases} \beta(s) & \text{if } s < 0 \text{ or } s > s_\varepsilon \\ \varepsilon s & \text{if } 0 \le s \le s_\varepsilon, \end{cases} \tag{3.1}$$

where $s_\varepsilon = 1 + C\varepsilon$ is such that $\beta(s_\varepsilon) = \varepsilon s_\varepsilon$. Set $\gamma_\varepsilon := \beta_\varepsilon^{-1}$ and note that both γ_ε and γ coincide everywhere but on the interval $I_\varepsilon := [0, C\varepsilon]$ and that $\gamma_\varepsilon'(s) \ge \Gamma_1$ for almost all $s \in \mathbb{R}$. Let $\{u^\varepsilon, \theta^\varepsilon\}$ denote the solutions to the problem (P_ε), obtained from (P) by replacing β with β_ε. Set $e_u := u - u^\varepsilon$, $e_\theta := \theta - \theta^\varepsilon$, $\boldsymbol{e}_q := \boldsymbol{q} - \boldsymbol{q}^\varepsilon$ and

$$A_\varepsilon := \{(x,t) \in Q : 0 \le \theta(x,t) \le C\varepsilon\} = \{(x,t) \in Q : 0 \le u(x,t) \le s_\varepsilon\}.$$

Let $G : H^{-1}(\Omega) \to H_0^1(\Omega)$ denote the Green's operator (inverse of the Laplacian), that is

$$\langle \nabla G\psi, \nabla\varphi \rangle = \langle \psi, \varphi \rangle, \qquad \text{for all } \varphi \in H_0^1(\Omega). \tag{3.2}$$

The norm in $H^{-1}(\Omega)$ can thus be represented in terms of G as follows:

$$\|\psi\|_{H^{-1}(\Omega)} = \|\nabla G\psi\|_{L^2(\Omega)} = \langle \psi, G\psi \rangle^{1/2}. \tag{3.3}$$

The error equation, assuming $f = 0$ for simplicity, reads

$$\langle \frac{d}{dt} e_u, \phi \rangle + \langle \nabla e_\theta, \nabla\phi \rangle = 0, \qquad \text{for all } \phi \in H_0^1(\Omega), \tag{3.4}$$

or equivalently,

$$\langle \frac{d}{dt} e_u, G\psi \rangle + \langle e_\theta, \psi \rangle = 0, \qquad \text{for all } \psi \in H^{-1}(\Omega). \tag{3.5}$$

Lemma 3.1. *The following error estimates are valid*

$$\|e_\theta\|_{L^2(\Omega)} + \|e_u\|_{L^\infty(0,T;H^{-1}(\Omega))} + \left\| \int_0^t \boldsymbol{e}_q \right\|_{L^\infty(0,T;L^2(\Omega))} \le C\,(\varepsilon\, meas(A_\varepsilon))^{1/2}. \tag{3.6}$$

The general rate of convergence is $O(\varepsilon^{1/2})$, which includes the formation of mushy regions [12,15]. If the nondegeneracy property (2.21) is valid, then the rate becomes $O(\varepsilon)$, which is optimal according to the regularity being dealt with (see [15]).

Proof. Take $\psi = e_u$ as a test function in (3.5) and integrate from 0 to $t \le T$. Exploiting the symmetry of G, namely $\langle G\phi, \psi\rangle = \langle \phi, G\psi\rangle$, as well as the property

$$\frac{d}{dt}\|e_u\|^2_{H^{-1}(\Omega)} = 2\langle \frac{d}{dt}e_u, Ge_u\rangle,$$

we arrive at

$$\|e_u(t)\|^2_{H^{-1}(\Omega)} + \int_0^t \langle e_\theta(s), e_u(s)\rangle ds = \|e_u(0)\|^2_{H^{-1}(\Omega)} = 0.$$

Let us now examine the middle term. Consider the product $(\gamma(a) - \gamma_\varepsilon(b))(a-b)$ for any $a, b \in \mathbb{R}$. If $a \notin I_\varepsilon$, then $\gamma(a) = \gamma_\varepsilon(a)$ and

$$(\gamma(a) - \gamma_\varepsilon(b))(a-b) \ge \Gamma_1(a-b)^2,$$

because $\gamma'_\varepsilon(s) \ge \Gamma_1$ for almost all $s \in \mathbb{R}$. The same conclusion can be reached when $b \notin I_\varepsilon$, this time arguing with γ instead of γ_ε. It only remains to consider the case where $a, b \in I_\varepsilon$. We readily have

$$|(\gamma(a) - \gamma_\varepsilon(b))(a-b)| \le C\varepsilon,$$

which leads to the desired estimate for u and θ. Let $\phi \in H_0^1(\Omega)$ be fixed in (3.4) and integrate from 0 to $t \le T$ to end up with

$$\langle e_u(t), \phi\rangle + \langle \int_0^t \nabla e_\theta, \nabla\phi\rangle = 0, \quad \text{for all } \phi \in H_0^1(\Omega).$$

Invoking the estimate for u and taking $\phi = \int_0^t e_\theta(s)ds \in H_0^1(\Omega)$, we can write

$$\left\|\int_0^t \nabla e_\theta\right\|^2_{L^2(\Omega)} \le \|e_u(t)\|_{H^{-1}(\Omega)} \left\|\int_0^t e_\theta\right\|_{H_0^1(\Omega)}.$$

The Poincaré inequality, in conjunction with the above estimate for u, yields the asserted bound for $\boldsymbol{e}_q$. ■

Exercises

3.1 Prove the Lemma 3.1 with $f \neq 0$.

3.2 Let $u_0^\varepsilon = \gamma_\varepsilon(\theta_0)$ and $\text{meas}\{x \in \Omega : 0 \le \theta_0(x) \le \varepsilon\} = O(\varepsilon)$. Prove the following error estimate for the initial enthalpies

$$\|u_0 - u_0^\varepsilon\|_{H^{-1}(\Omega)} \le C\varepsilon|\log\varepsilon|^{1/2}.$$

Hint: Use the initial nondegeneracy property to derive

$$\|u_0 - u_0^\varepsilon\|_{L^p(\Omega)} \le C\varepsilon^{1/p}, \quad \text{for } 1 \le p < \infty.$$

Then apply the 2D Poincaré-Sobolev inequality:

$$\|\phi\|_{L^p(\Omega)} \le Cp^{1/2}\|\phi\|_{H_0^1(\Omega)}, \quad \text{for all } \phi \in H_0^1(\Omega).$$

3.3 Under the assumptions of Exercise 3.2, prove that Lemma 3.1 is still valid if the initial enthalpy for (P_ε) is replaced by $u_0^\varepsilon = \gamma_\varepsilon(\theta_0)$.

3.4 Prove that if the nondegeneracy condition (2.22) holds, then

$$\|u - u^\varepsilon\|_{L^2(Q)} \leq C\varepsilon^{1/2}.$$

Hint: Write $\langle e_\theta, e_u \rangle = \langle \beta(u) - \beta(u^\varepsilon), u - u^\varepsilon \rangle + \langle \beta(u) - \beta(u), u - u^\varepsilon \rangle$ and use the property $\beta' \geq \varepsilon$.

2.3.2 Phase Relaxation

Let $0 < \mu \leq \Gamma_1$ be a relaxation parameter. Then γ can be split as follows:

$$\gamma = \mu I + H, \tag{3.7}$$

where I is the identity and H is still a maximal monotone graph which, in the simplest situation, coincides with the Heaviside graph. Let $\chi := u - \mu\theta$ stand for the phase variable. The classical constitutive relation reads

$$\chi \in H(\theta) \qquad \text{or} \qquad \theta \in \Lambda(\chi) := H^{-1}(\chi). \tag{3.8}$$

The following dynamic phase condition was proposed by Visintin [35] as a substitute for the stationary relation (3.8):

$$\varepsilon\chi_t^\varepsilon + \Lambda(\chi^\varepsilon) \ni \theta^\varepsilon, \tag{3.9}$$

where $\varepsilon > 0$ is a small relaxation parameter. Therefore, (3.9) incorporates a time delay in the constitutive relation. The original partial differential equation (2.8) is now replaced by the system (P_ε)

$$\begin{cases} \mu\theta_t^\varepsilon + \chi_t^\varepsilon - \Delta\theta^\varepsilon = 0 \\ \varepsilon\chi_t^\varepsilon + \Lambda(\chi^\varepsilon) \ni \theta^\varepsilon, \end{cases} \tag{3.10}$$

which is to be interpreted in weak sense. We still suppose, for simplicity, $f = 0$ and that the initial conditions are the same as for (P), namely θ_0 and $\chi_0 = u_0 - \mu\theta_0$. Apart from its own physical relevance, (3.10) can be viewed as an approximation to the Stefan problem. In this light, we may wonder about the rate of convergence of such an approximation as $\varepsilon \downarrow 0$. To this end, we first need a priori estimates for (P_ε), which in turn will be a common tenet in the sequel. Since $\chi^\varepsilon \to \chi$, we cannot expect $\chi^\varepsilon \in H^1(0,T;L^2(\Omega))$. Here is the precise way χ^ε degenerates as $\varepsilon \downarrow 0$. Note that (3.11) is sharp.

Exercises

3.5 Prove the a priori estimate

$$\|\chi^\varepsilon\|_{L^\infty(0,T;L^2(\Omega))} + \varepsilon^{1/2}\|\chi_t^\varepsilon\|_{L^2(Q)} \leq C. \qquad (3.11)$$

Hint: Multiply the first partial differential equation in (3.10) by θ^ε and the second one by χ_t^ε. Let $\Phi : \mathbb{R} \to \mathbb{R} \cup \{+\infty\}$ be a lower semicontinuous convex function such that its subdifferential coincides with Λ and $\Phi(0) = 0$; Φ is thus a generalized antiderivative of Λ. Since $|H(s)| \leq C_1|s| + C_2$ with $C_1, C_2 \geq 0$, then $s^2/2 \leq C_1\Phi(s) + C_2|s|$. Prove that

$$\int_\Omega \int_0^t \Lambda(\chi^\varepsilon)\chi^\varepsilon = \int_\Omega \Phi(\chi^\varepsilon(t)) - \int_\Omega \Phi(\chi^\varepsilon(0)) \geq C\|\chi^\varepsilon(t)\|_{L^2(\Omega)}^2 - C.$$

We are now in a position to derive the desired error estimates. The error equations, written in weak form, read as follows:

$$\mu\langle \frac{d}{dt}e_\theta, \phi\rangle + \langle \frac{d}{dt}e_\chi, \phi\rangle + \langle \nabla e_\theta, \nabla\phi\rangle = 0, \qquad \text{for all } \phi \in H_0^1(\Omega), \quad (3.12)$$

$$\langle e_\theta, \psi\rangle = \langle \Lambda(\chi) - \Lambda(\chi^\varepsilon), \psi\rangle - \langle \varepsilon\chi_t^\varepsilon, \psi\rangle, \qquad \text{for all } \psi \in L^2(\Omega). \quad (3.13)$$

Integrate (3.12) from 0 to $t \leq T$ with ϕ fixed to get

$$\langle \mu e_\theta(t), \phi\rangle + \langle e_\chi(t), \phi\rangle + \langle \int_0^t \nabla e_\theta(s), \nabla\phi\rangle = 0 \quad \text{for all } \phi \in H_0^1(\Omega). \quad (3.14)$$

Next choose $\phi = e_\theta(t) \in H_0^1(\Omega)$ as a test function and integrate (3.14) again from 0 to $t \leq T$ to obtain

$$0 = \mu\int_0^t \|e_\theta(s)\|_{L^2(\Omega)}^2 ds + \int_0^t \langle e_\chi(s), e_\theta(s)\rangle ds$$
$$+ \int_0^t \langle \nabla e_\theta(s), \int_0^s \nabla e_\theta(z)dz\rangle ds =: I + II + III. \qquad (3.15)$$

We point out first that

$$2III = \int_\Omega \int_0^t \frac{d}{ds}\left(\int_0^s \nabla e_\theta(z)dz\right)^2 ds = \left\|\int_0^t \nabla e_\theta(s)ds\right\|_{L^2(\Omega)}^2.$$

In order to handle the middle term II we resort to the equation (3.13). Take $\psi = e_\chi \in L^2(\Omega)$ as a test function and integrate from 0 to $t \leq T$. Since Λ is monotone, namely $(\Lambda(\chi) - \Lambda(\chi^\varepsilon))(\chi - \chi^\varepsilon) \geq 0$, (3.11) yields

$$II \geq -\varepsilon\int_0^t \|\chi_t(s)\|_{L^2(\Omega)}\|e_\chi(s)\|_{L^2(\Omega)}ds \geq -C\varepsilon^{1/2}.$$

An error bound for $u(t) = \mu\theta(t) + \chi(t)$ in $H^{-1}(\Omega)$ can be finally derived by virtue of (3.14) and the error estimate for $\int_0^t \nabla e_\theta(s)ds$. We summarize the above results in the following Lemma.

Lemma 3.2. *We have*

$$\|e_\theta\|_{L^2(Q)} + \|e_u\|_{L^\infty(0,T;H^{-1}(\Omega))} + \|\int_0^t \boldsymbol{e}_q\|_{L^\infty(0,T;L^2(\Omega))} \leq C\varepsilon^{1/4}.$$

Exercises

3.6 Prove Lemma 3.2 with $f \neq 0$ and possibly different initial conditions for both problems (P) and (P$_\varepsilon$).

We can now improve upon Lemma 3.2 under additional assumptions on the initial data. Let

$$u_0 \in L^\infty(\Omega), \qquad \theta_0 \in H_0^1(\Omega), \qquad \Delta\theta_0 \in L^1(\Omega), \tag{3.16}$$

which, in view of Exercises 2.4, 2.7 and 2.8, implies

$$u \in L^\infty(Q), \qquad \theta_t \in L^2(Q), \qquad \Delta\theta \in L^\infty(0,T;M(\Omega)). \tag{3.17}$$

The above constraint on $\Delta\theta_0$ can be weakened slightly, thus allowing jump discontinuities of $\nabla\theta_0$ on $I(0)$; see Exercise 2.8.

Lemma 3.3. *If (3.16) is valid, then we have*

$$\|e_\theta\|_{L^2(Q)} + \|e_u\|_{L^\infty(0,T;H^{-1}(\Omega))} + \|\int_0^t \boldsymbol{e}_q\|_{L^\infty(0,T;L^2(\Omega))} \leq C\varepsilon^{1/2}. \tag{3.18}$$

Proof. We modify the treatment of term II above. In fact, integration by parts, in conjunction with the monotonicity of Λ, yields

$$\begin{aligned} II &\geq \varepsilon \int_0^t \langle \chi_t^\varepsilon, \chi^\varepsilon \rangle - \varepsilon \int_0^t \langle \chi_t^\varepsilon, \chi \rangle \\ &= \frac{\varepsilon}{2} \left(\|\chi^\varepsilon(t)\|_{L^2(\Omega)}^2 - \|\chi^\varepsilon(0)\|_{L^2(\Omega)}^2 \right) \\ &\qquad + \varepsilon \int_0^t \langle \chi^\varepsilon(s), \chi_t(s) \rangle ds - \varepsilon \langle \chi^\varepsilon(s), \chi(s) \rangle \big|_0^t . \end{aligned}$$

In view of (2.14), (3.11) and (3.17), we have

$$\begin{aligned} II &\geq -C\varepsilon + \varepsilon \int_0^t \langle \chi^\varepsilon(s), u_t(s)\rangle ds - \varepsilon\mu \int_0^t \langle \chi^\varepsilon(s), \theta_t(s)\rangle ds \\ &\geq -C\varepsilon - \varepsilon \int_0^t \|\chi^\varepsilon(s)\|_{C^0(\overline{\Omega})} \|u_t\|_{M(\Omega)} ds \\ &\qquad -\varepsilon\mu \int_0^t \|\chi^\varepsilon(s)\|_{L^2(\Omega)} \|\theta_t(s)\|_{L^2(\Omega)} ds \\ &\geq -C\varepsilon, \end{aligned}$$

which clearly implies the assertion. ∎

Exercises

3.7 Use the above integral method to prove Lemma 3.1.

2.4 Discretization

In order to motivate the fully discrete schemes below we first discuss discrete-time approximations. They are simpler but at the same time possess the major characteristic features of interest. We also introduce the finite element spaces to be dealt with.

2.4.1 Semidiscrete Schemes

Let $\tau := T/N$ denote the time step and set $\partial z^n := \tau^{-1}(z^n - z^{n-1})$ for any sequence $\{z^n\}_{n=0}^N$. If we simply use backward differences to discretize (2.8) in time, we obtain the following set of *nonlinear* elliptic partial differential equations:

$$\begin{cases} U^0 := u_0, \\ \Theta^n := \beta(U^n), \\ U^n - \tau\Delta\Theta^n = U^{n-1} + \tau f(\Theta^n), \quad 1 \leq n \leq N. \end{cases} \tag{4.1}$$

Upon finite element discretization in space, this scheme becomes a standard nonlinear algorithm. Its main drawback is the strong nonlinearity already present in $\Theta^n = \beta(U^n)$, which makes powerful iterative methods such as Newton's Method fail, at least in a general setting.

Linearization of nonlinear parabolic problems is an extremely useful numerical tool in that one can use efficient linear solvers to compute the solution of the ensuing *linear* algebraic systems. It success is based essentially on the regularity of the underlying solution. Such a regularity is quite unrealistic for free boundary problems, even with a preliminary regularization, because of the singularities located on the interface. We are thus forced to abandon the standard approach based on Taylor's formula. Replacing the underlying strongly nonlinear constitutive relation by a suitable one which preserves the natural stability properties turns out to be a potential linearization technique. We now discuss this crucial idea in light of two relevant examples.

We first consider the following *nonlinear Chernoff formula* ([3, 4, 13, 14, 25, 26, 29, 33]):

$$\begin{cases} U^0 := u_0 \\ \Theta^n - \frac{\tau}{\mu}\Delta\Theta^n = \beta(U^{n-1}), & 1 \le n \le N \\ U^n := U^{n-1} + \mu[\Theta^n - \beta(U^{n-1})], & 1 \le n \le N. \end{cases} \tag{4.2}$$

This scheme, which arises in the theory of nonlinear semigroups of contractions [4], consists of solving a linear elliptic problem in Θ^n and performing next an algebraic correction to account for the nonlinearity; μ is a relaxation parameter which satisfies the constraint $0 < \mu \le \Gamma_1$. To understand its physical nature, and thus identify its variational properties, we relate (4.2) to the semidiscrete phase relaxation scheme of Verdi and Visintin [34]. Such a scheme is the usual time discretization of (3.10), namely,

$$\begin{cases} \mu\partial\Theta^n + \partial\chi^n - \Delta\Theta^n = 0 \\ \varepsilon\partial\chi^n + \Lambda(\chi^n) \ni \Theta^{n-1}, \end{cases} \tag{4.3}$$

except for the presence of Θ^{n-1} rather than Θ^n in the last equation, which in turn makes it explicit in χ^n. As a result, the first equation becomes linear in Θ^n. The scheme (4.3) is stable if it is subject to the stability constraint $\tau \le \mu\varepsilon$ (see [34]). Set now $\tau = \mu\varepsilon$ and rewrite the second equation in (4.3) as

$$\chi^n + \mu\Lambda(\chi^n) \ni \mu\Theta^{n-1} + \chi^{n-1} =: U^{n-1}, \tag{4.4}$$

or equivalently as

$$\chi^n = (I + \mu\Lambda)^{-1}U^{n-1} = (I - \mu\beta)U^{n-1} = U^{n-1} - \mu\beta(U^{n-1}), \tag{4.5}$$

because $(I + \mu\Lambda)^{-1} = I - \mu\beta$. Consequently, we obtain

$$U^n = \mu\Theta^n + \chi^n = U^{n-1} + \mu[\Theta^n - \beta(U^{n-1})], \tag{4.6}$$

which is the algebraic correction in (4.2). Since the partial differential equations in both (4.2) and (4.3) obviously coincide, we realize that the nonlinear Chernoff formula is nothing else than a proper combination of phase relaxation and time discretization. This provides some insight about its underlying variational structure.

Exercises

4.1 Prove $(I+\mu\Lambda)^{-1} = I - \mu\beta$ under the sole assumption that Λ is monotone.

Of a different nature is the *extrapolation method* proposed in [19]. A preliminary regularization such as (3.1) is first used to replace H by H_ε, where now $0 \le H'_\varepsilon(s) \le \varepsilon^{-1}$ for almost all $s \in \mathbb{R}$. Here $\varepsilon > 0$ stands for a regularization parameter and $\theta^\varepsilon, u^\varepsilon, \chi^\varepsilon$ denote the regularized physical variables. Secondly, we observe that whenever a mushy region occupies the entire domain Ω, which turns out to be the most critical regularity situation, the function $H_\varepsilon(\theta^\varepsilon)$ coincides with $\varepsilon^{-1}\theta^\varepsilon$. This motivates the following extrapolation procedure ($\Theta^{-1} := \Theta^0$):

$$\chi^n := H_\varepsilon(\Theta^{n-1}) + \varepsilon^{-1}(\Theta^n - \Theta^{n-1}), \qquad 0 \le n \le N. \tag{4.7}$$

Compared with standard extrapolation, (4.7) is a very crude formula in that it entails time regularity of θ^ε rather than χ^ε; note that χ^ε has very low regularity properties (uniformly in ε). Setting now $U^n := \mu\Theta^n + \chi^n$, the resulting linear partial differential equation associated with (2.8) reads as follows, for $1 \le n \le N$:

$$(\mu + \varepsilon^{-1})\partial\Theta^n - \Delta\Theta^n = \partial[\varepsilon^{-1}\Theta^{n-1} - H_\varepsilon(\Theta^{n-1})] + f(\Theta^{n-1}). \tag{4.8}$$

Stability is guaranteed provided $\theta_0 (= \beta(u_0))$ satisfies $\theta_0 \in H^1_0(\Omega)$ which, in turn, ensures $\partial\theta^\varepsilon/\partial t \in L^2(Q)$ uniformly in ε; see Exercise 2.4. Since (4.7) can also be written as

$$\chi^n = H_\varepsilon(\Theta^{n-1}) + \frac{\tau}{\varepsilon}\partial\Theta^n, \tag{4.9}$$

we realize that a bit of time regularity of $\{\Theta^n\}_{n=1}^N$, and thus of θ^ε, is expected to enhance stability.

We conclude with some notation concerning the time discretization. Set $t^n := n\tau$ and $I^n := (t^{n-1}, t^n]$ for $1 \le n \le N$. We also set $\zeta^n := \zeta(\cdot, t^n)$ for any continuous function in time defined in Q.

2.4.2 Finite Element Spaces

Let $\{\mathcal{S}_h\}_h$ be a family of regular and quasi-uniform partitions of Ω into triangular finite elements [5, pp. 132,140]. As usual $h > 0$ stands for the meshsize. We assume, for simplicity, that $\Omega = \cup_{S \in \mathcal{S}_h} S$ and refer to [24] for a precise analysis of the discrepancy between these two sets within the framework of degenerate problems. We will be dealing with finite element spaces $\boldsymbol{V}^1_h \subset H^1_0(\Omega)$ and $\boldsymbol{V}^0_h \subset L^2(\Omega)$ which satisfy

$$\boldsymbol{V}^1_h|_S = P^1(S), \quad \boldsymbol{V}^0_h|_S = P^0(S), \qquad \text{for all } \ S \in \mathcal{S}_h, \tag{4.10}$$

where $P^i(S)$ indicates the space of polynomials of total degree not greater than i restricted to S. Let $\{x_j\}_{j=1}^J$ indicate the set of nodes of $\mathcal{S}_h$. Let $\{\phi_j\}_{j=1}^J$ denote the canonical basis of $\boldsymbol{V}_h^1$ and $\{\psi_j\}_{j=1}^J$ the set of piecewise constant functions obtained by multiplying the characteristic function of $\text{supp}(\phi_j)$ by $(d+1)^{-1}$. Let Π_h be the local Lagrange interpolation operator, that is $\Pi_h|_S : C^0(\bar{S}) \to P^1(S)$ for all $S \in \mathcal{S}_h$. We associate with $\langle \cdot, \cdot \rangle$ a discrete inner product defined as follows:

$$\langle \zeta, \phi \rangle_h := \sum_{S \in \mathcal{S}_h} \int_S \Pi_h|_S(\zeta\phi), \tag{4.11}$$

for any piecewise uniformly continuous functions ζ and ϕ. Note that the integral in (4.11) can be evaluated easily by means of the vertex quadrature rule, which is exact for piecewise linear functions [5, p. 182]. It is well known that $\langle \cdot, \cdot \rangle_h$ is an inner product in $\boldsymbol{V}_h^1$ which satisfies

$$\|\phi\|_{L^2(\Omega)}^2 \le \langle \phi, \phi \rangle_h \le C\|\phi\|_{L^2(\Omega)}^2, \qquad \text{for all} \quad \phi \in \boldsymbol{V}_h^1, \tag{4.12}$$

where $C > 1$ is a constant independent of h. The following well-known error bound accounts for the effect of numerical integration:

$$\begin{aligned} |\langle \zeta, \phi \rangle - \langle \zeta, \phi \rangle_h| &\le Ch^2 \|\nabla \zeta\|_{L^2(\Omega)} \|\nabla \phi\|_{L^2(\Omega)} \\ &\le Ch\|\zeta\|_{L^2(\Omega)} \|\nabla \phi\|_{L^2(\Omega)}, \end{aligned} \tag{4.13}$$

for all $\zeta, \phi \in \boldsymbol{V}_h^1$. We then have the following symmetric stiffness and mass matrices

$$\boldsymbol{K} := (\langle \nabla \phi_i, \nabla \phi_j \rangle)_{i,j=1}^J, \ \boldsymbol{M} := (\langle \phi_i, \phi_j \rangle_h)_{i,j=1}^J, \ \boldsymbol{N} := (\langle \phi_i, \psi_j \rangle)_{i,j=1}^J, \tag{4.14}$$

and observe that $\boldsymbol{M}$ is diagonal. We will also need some projection operators associated with the above discrete spaces. The first one, labeled R_h, is a projection operator onto $\boldsymbol{V}_h^1$ defined by

$$\langle \nabla R_h \zeta, \nabla \phi \rangle = \langle \nabla \zeta, \nabla \phi \rangle, \qquad \text{for all } \zeta \in H_0^1(\Omega), \ \phi \in \boldsymbol{V}^1. \tag{4.15}$$

Note that $\|\nabla R_h \zeta\|_{L^2(\Omega)} \le \|\nabla \zeta\|_{L^2(\Omega)}$ for all $\zeta \in H_0^1(\Omega)$. The two remaining operators, P_h^0 and P_h^1, are L^2-projections onto $\boldsymbol{V}_h^0$ and $\boldsymbol{V}_h^1$, respectively,

$$\langle P_h^i \zeta, \phi \rangle = \langle \zeta, \phi \rangle, \qquad \text{for all } \zeta \in L^2(\Omega), \ \phi \in \boldsymbol{V}_h^i. \tag{4.16}$$

We assume that the following approximation properties are valid, which in turn is guaranteed provided Ω is convex (or smooth) [2]:

$$\|R_h \zeta - \zeta\|_{H^s(\Omega)} \le Ch^{1+r-s} \|\zeta\|_{H^{1+r}(\Omega)}, \text{ for all } 0 \le s, r \le 1, \ \zeta \in H_0^{1+r}(\Omega), \tag{4.17}$$

$$\|P_h^i\zeta - \zeta\|_{H^{-s}(\Omega)} \le Ch^{r+s}\|\zeta\|_{H^r(\Omega)}, \text{ for all } 0 \le s, r \le 1,\ \zeta \in H^r(\Omega). \tag{4.18}$$

Note that (4.18) is a superconvergence estimate for $s > 0$. The discrete Green's operator $G_h : H^{-1}(\Omega) \to \boldsymbol{V}_h^1$ is then defined by $G_h := R_h \circ G$, namely,

$$\langle \nabla G_h\zeta, \nabla\phi\rangle = \langle \zeta, \phi\rangle, \quad \text{for all } \zeta \in H^{-1}(\Omega),\ \phi \in \boldsymbol{V}_h^1. \tag{4.19}$$

Note that $\|\nabla G_h\zeta\|_{L^2(\Omega)} \le \|\nabla G\zeta\|_{L^2(\Omega)} = \|\zeta\|_{H^{-1}(\Omega)}$ for all $\zeta \in H^{-1}(\Omega)$. In view of (4.17) and $\|G\zeta\|_{H^{2-s}(\Omega)} \le C\|\zeta\|_{H^{-s}(\Omega)}$, G_h satisfies the error estimate

$$\|(G - G_h)\zeta\|_{L^2(\Omega)} \le Ch^{2-s}\|\zeta\|_{H^{-s}(\Omega)}, \quad \text{for all } \zeta \in H^{-s}(\Omega),\ s = 0, 1. \tag{4.20}$$

On several occasions, to compensate for the lack of regularity, we will exploit monotonicity properties via the *Discrete Maximum Principle* (DMP) [6], which holds under an extra condition on $\mathcal{S}_h$. We say that $\mathcal{S}_h$ is weakly acute if for any pair of adjacent triangles the sum of the opposite angles relative to the common side does not exceed π. Suppose $\phi \in \boldsymbol{V}_h^1$ attains its maximum at the internal node x_j. Then

$$\langle \nabla\phi, \nabla\phi_j\rangle \ge 0. \tag{4.21}$$

Exercises

4.2 Prove that if $\mathcal{S}_h$ is weakly acute, then the stiffness matrix $\boldsymbol{K}$ satisfies

$$k_{ii} > 0, \quad k_{ij} \le 0\ (j \ne i), \quad \sum_{j=1}^{\tilde{J}} k_{ij}, \qquad \text{for all } 1 \le i \le J, \tag{4.22}$$

where $\tilde{J}$ stands for the total number of nodes, thus including those on $\partial\Omega$. Then $\boldsymbol{K}$ is an M-matrix, (see [27]).

4.3 Use (4.22) to derive (4.21).

Another consequence of (4.22) is the following statement. Let $\alpha \in W^{1,\infty}(\mathbb{R})$ satisfy $\alpha(0) = 0$ and $0 \le \alpha'(s) \le L_\alpha < \infty$ for almost all $s \in \mathbb{R}$. Then

$$\|\nabla\Pi_h[\alpha(\phi)]\|_{L^2(\Omega)}^2 \le L_\alpha\langle \nabla\phi, \nabla\Pi_h[\alpha(\phi)]\rangle, \quad \text{for all } \phi \in \boldsymbol{V}_h^1. \tag{4.23}$$

Exercises

4.4 Prove (4.23). Hint: Use and justify the equalities

$$\langle \nabla\phi, \nabla\Pi_h\alpha(\phi)\rangle = \sum_{i=1}^{J}\left(\sum_{j\neq i}\alpha_i(\phi_j-\phi_i)k_{ij} + \alpha_i\phi_i(k_{ii}+\sum_{j\neq i}k_{ij})\right),$$

and

$$\sum_{i=1}^{J}\sum_{j\neq i}\alpha_i(\phi_j-\phi_i)k_{ij} = \sum_{j=1}^{J}\sum_{i\neq j}\alpha_j(\phi_i-\phi_j)k_{ij},$$

where $\alpha_i := \alpha(\phi_i)$.

2.4.3 Fully Discrete Schemes

The finite element method associated with (4.1) reads as follows: *find* $\{U^n, \Theta^n\}_{n=1}^N \subset \boldsymbol{V}_h^1$ *such that*

$$U^0 := P_h^1 u_0, \tag{4.24}$$

$$\Theta^n := \Pi_h[\beta(U^n)], \tag{4.25}$$

and for all $\phi \in \boldsymbol{V}_h^1$ *we have*

$$\langle U^n, \phi\rangle_h + \tau\langle\nabla\Theta^n, \nabla\phi\rangle = \langle U^{n-1}, \phi\rangle_h + \tau\langle f(\Theta^n), \phi\rangle_h. \tag{4.26}$$

It is worth noting that the use of numerical quadrature makes (4.26) practical in that it is easy to implement [9,24,29]. This algebraic system is strongly nonlinear because of the constraint (4.25), and can be equivalently written in matrix form as follows

$$\boldsymbol{M}U^n + \tau\boldsymbol{K}\Theta^n = \boldsymbol{M}(U^{n-1} + \tau f(\Theta^{n-1})) =: B^{n-1} = (b_i^{n-1})_{i=1}^J. \tag{4.27}$$

Here we have identified all piecewise linear functions of $\boldsymbol{V}_h^1$ with the vectors of $\mathbb{R}^J$ containing their nodal values. As $\boldsymbol{M}$ is diagonal, (4.27) can be easily and efficiently solved via the following nonlinear SOR method [32,24,22,29]

$$\Theta_{i,m+1}^n := \Theta_{i,m}^n - \omega\left[\Theta_{i,m}^n - \zeta_i\left(b_i^{n-1} - \sum_{j=1}^{i-1}k_{ij}\Theta_{j,m+1}^n - \sum_{j=i+1}^{J}k_{ij}\Theta_{j,m}^n\right)\right], \tag{4.28}$$

for all $1 \le i \le J$, where $\Theta_1^n := \Theta^{n-1}$, $\zeta_i := (m_{ii}\gamma + \tau k_{ii}I)^{-1}$ is a nondecreasing function and $1 < \omega < 2$ is a suitably chosen relaxation parameter. If $\mathcal{S}_h$ is weakly acute, the convergence is as fast as that of the corresponding linear SOR method for the M-matrix $\Gamma_1\boldsymbol{M} + \tau\boldsymbol{K}$ [32,36].

The fully discrete *nonlinear Chernoff formula* reads as follows: *for any* $1 \le n \le N$ *find* $U^n \in \boldsymbol{V}_h^0, \Theta^n \in \boldsymbol{V}_h^1$ *such that*

$$U^0 := P_h^0 u_0 \tag{4.29}$$

$$\langle \Theta^n, \phi \rangle_h + \frac{\tau}{\mu} \langle \nabla \Theta^n, \nabla \phi \rangle = \langle \beta(U^{n-1}) + \frac{\tau}{\mu} f(\beta(U^{n-1})), \phi \rangle, \text{ for all } \phi \in \boldsymbol{V}_h^1 \tag{4.30}$$

$$U^n := U^{n-1} + \mu [P_h^0 \Theta^n + \beta(U^{n-1})]. \tag{4.31}$$

The discrete equation (4.30) is equivalent to the following linear algebraic system

$$\boldsymbol{A} \Theta^n := \left(\boldsymbol{M} + \frac{\tau}{\mu} \boldsymbol{K} \right) \Theta^n = R^{n-1}, \tag{4.32}$$

where R^{n-1} depends only on U^{n-1}. Since

$$C h^d \le \frac{\boldsymbol{x}^T \cdot \boldsymbol{A x}}{|\boldsymbol{x}|^2} \le C^{-1}(1 + \tau h^{-2}) h^d, \quad \text{for all } \boldsymbol{x} \in \mathbb{R}^J, \tag{4.33}$$

and $\tau = O(h)$ will be enforced, the condition number $k(\boldsymbol{A})$ of $\boldsymbol{A}$ verifies $k(\boldsymbol{A}) = O(h^{-1})$. The most efficient solver for a large system such as (4.32) is the preconditioned Conjugate Gradient Method (CGM), as illustrated in [25, 29], where the incomplete Cholesky factorization was used as a preconditioner to accelerate the convergence [1]. Since both U^n and $P_h^0 \Theta^n$ are piecewise constant, (4.31) may be regarded as an inexpensive elementwise algebraic correction which accounts for the nonlinearity.

In order to present the fully discrete *extrapolation method* we must first know how to select the initial temperature Θ^0 so that the following two properties hold:

$$\|\Theta^0\|_{H_0^1(\Omega)} \le C, \qquad \|U^0 - u_0\|_{H^{-1}(\Omega)} \le C h^{1/2}, \tag{4.34}$$

where U^0, the initial enthalpy, is defined by $U^0 := \mu P_h^0 \Theta^0 + H_\varepsilon(P_h^0 \Theta^0)$. The obvious choices $\Theta^0 = P_h^1 \theta_0$ and $\Theta^0 = \Pi_h \theta_0$ fail to work in this context because the corresponding U^0 does not satisfy the error bound in (4.34), unless an initial nondegeneracy property is verified [15,26]. This condition prevents (P) from having an initial mushy region. Since their presence is the major motivation for our extrapolation method, we would like to allow mushy regions from the very beginning. Such a difficulty is overcome in the following Exercise.

Exercises

4.5 Let $\theta_0 \in W_0^{1,\infty}(\Omega)$ and $u_0 \in C^{0,1/2}$ in each single phase, so u_0 may exhibit jump discontinuities across the initial interface. Set $\Theta^0(x_j) := \limsup_{x\to x_j} \beta_\varepsilon(u_0(x))$ for all nodes x_j, $1 \le j \le J$. Prove that Θ^0 and U^0 satisfy (4.34) provided ε and h are subject to the mild constraint $\varepsilon^2 \le Ch$, where $C > 0$ is an arbitrary constant.

Set $\Theta^{-1} := \Theta^0$ and $\alpha_\varepsilon(s) := \varepsilon^{-1}s - H_\varepsilon(s)$ for all $s \in \mathbb{R}$; note that α_ε satisfies $0 \le \alpha_\varepsilon'(s) \le \varepsilon^{-1}$ for almost all $s \in \mathbb{R}$. The fully discrete scheme then reads: *for any* $1 \le n \le N$ *find* $\Theta^n \in \boldsymbol{V}_h^1$ *such that for all* $\phi \in \boldsymbol{V}_h^1$

$$(\mu + \varepsilon^{-1})\langle \partial P_h^0 \Theta^n, \phi\rangle + \langle \nabla\Theta^n, \nabla\phi\rangle = \langle \partial\alpha_\varepsilon(P_h^0\Theta^{n-1}), \phi\rangle + \langle f(P_h^0\Theta^{n-1}), \phi\rangle. \tag{4.35}$$

The discrete phase variable and enthalpy are defined by

$$\chi^n := H_\varepsilon(P_h^0\Theta^{n-1}) + \varepsilon^{-1}P_h^0[\Theta^n - \Theta^{n-1}]; \quad U^n := \mu P_h^0\Theta^n + \chi^n. \tag{4.36}$$

Since $P_h^0\Theta^{n-1}$ is constant on each finite element S (it is actually the value of Θ^{n-1} at the barycenter of S), the right hand side of (4.35) can be readily and exactly evaluated. Consequently, (4.35) is a practical scheme in that it can be easily implemented on a computer. Equation (4.35) can then be written equivalently in matrix form as follows:

$$\boldsymbol{A}\Theta^n := \left(\boldsymbol{N} + \frac{\varepsilon\tau}{1+\mu\varepsilon}\boldsymbol{K}\right)\Theta^n = R^{n-1}. \tag{4.37}$$

Here R^{n-1} depends only on Θ^{n-1}. We see that this problem is *linear* and that the corresponding matrix is symmetric, positive definite and also *independent* of n. This yields unique solvability of (4.35). The condition number $k(\boldsymbol{A})$ of $\boldsymbol{A}$ is bounded uniformly in h because

$$Ch^d \le \frac{\boldsymbol{x}\cdot\boldsymbol{A}\boldsymbol{x}}{\|\boldsymbol{x}\|_2^2} \le C(1+\varepsilon\tau h^{-2})h^d, \quad \text{for all } \boldsymbol{x} \in \mathbb{R}^J, \tag{4.38}$$

and $\varepsilon\tau = o(h^2)$, as we will see in Section 2.6. The CGM is again the most effective method for the solution of (4.37).

2.4.4 Incomplete Iteration

In the case that the symmetric linear systems (4.32) and (4.37) are solved with the preconditioned CGM, the iterations can be stopped safely if a prescribed error reduction is attained. This reduces considerably the computational labor and still preserves stability and accuracy. This idea resembles that in [8] for the standard extrapolation procedure.

Let $\|\cdot\|_{\boldsymbol{A}}$ denote the norm subordinate to $\boldsymbol{A}$, namely $\|\boldsymbol{x}\|_{\boldsymbol{A}} := (\boldsymbol{x}^T \cdot \boldsymbol{A}\boldsymbol{x})^{1/2}$ for all $\boldsymbol{x} \in \mathbb{R}^J$. Let $\Theta_k^n, \Xi^n \in \boldsymbol{V}^1$ indicate the kth-iteration of the CGM ($\Theta_0^n := \Theta^{n-1}$) and the exact solution, respectively. Then, it is well known (see [1]) that

$$\|\Theta_k^n - \Xi^n\|_{\boldsymbol{A}} \leq \rho_k \|\Theta^{n-1} - \Xi^n\|_{\boldsymbol{A}}, \tag{4.39}$$

where the *error reduction* ρ_k satisfies

$$\rho_k := 2\left(\frac{k(\tilde{A})^{1/2} - 1}{k(\tilde{A})^{1/2} + 1}\right)^k = 2Q^k. \tag{4.40}$$

Here $\tilde{\boldsymbol{A}}$ stands for the preconditioned matrix, that is $\tilde{\boldsymbol{A}} = \boldsymbol{E}^{-T}\boldsymbol{A}\boldsymbol{E}^{-1}$ with $\boldsymbol{C} = \boldsymbol{E}^T\boldsymbol{E}$ being the preconditioning. Let $\Theta^n \in \boldsymbol{V}_h^1$ be any iteration satisfying

$$\|\Theta^n - \Xi^n\|_{\boldsymbol{A}} \leq 2\tau^\lambda \|\Theta^{n-1} - \Xi^n\|_{\boldsymbol{A}}, \tag{4.41}$$

where $\lambda = 1/2$ or 1 in accordance with (4.32) and (4.37). It is easily seen that

$$k \geq k_0 := \lambda \log \tau / \log Q, \tag{4.42}$$

and thus the computational labor involved is nearly optimal if $k(\tilde{\boldsymbol{A}}) = O(1)$.

It turns out that if (4.41) is valid then stability and accuracy are preserved. We refer to [19] for a full analysis of the incomplete iteration process for degenerate parabolic problems.

2.5 Stability

Our aim now is to derive several a priori estimates for the above fully discrete schemes. The techniques employed are just discrete analogues to those collected in Exercises 2.3 to 2.7. In striking contrast to mildly nonlinear parabolic partial differential equations, for which the usual approach is to assume regularity on the continuous solution, the discrete stability estimates will play a major role later on in the error analysis. We will assume, for simplicity only, that $f = 0$.

Given a Lipschitz continuous function $\lambda : \mathbb{R} \to \mathbb{R}$ so that $\lambda(0) = 0$ and $0 \leq \lambda' \leq \Lambda < \infty$, Φ_λ stands for the convex function defined by

$$\Phi_\lambda(s) := \int_0^s \lambda(z)dz, \qquad \text{for all } s \in \mathbb{R}, \tag{5.1}$$

with the following obvious properties (see (2.16))

$$\frac{1}{2\Lambda}\lambda^2(s) \leq \Phi_\lambda(s) \leq \frac{\Lambda}{2}s^2, \quad \text{for all } s \in \mathbb{R}. \tag{5.2}$$

2.5.1 The Nonlinear Method

We will argue as in Exercise 2.3. Take $\phi = \Theta^n \in \boldsymbol{V}_h^1$ as a test function in (4.26) and add the resulting expression on n from 0 to $m \le N$. We have

$$\sum_{n=1}^{m} \langle U^n - U^{n-1}, \Theta^n \rangle_h + \tau \sum_{n=1}^{m} \|\nabla \Theta^n\|_{L^2(\Omega)}^2 = 0. \tag{5.3}$$

Since $\Theta_j^n = \beta(U_j^n)$ nodewise, the convexity of Φ_β together with (5.2) leads to

$$\begin{aligned}\sum_{n=1}^{m} (U_j^n - U_j^{n-1}) \Theta_j^n &\ge \sum_{n=1}^{m} \Phi_\beta(U_j^n) - \Phi_\beta(U_j^{n-1}) \\ &= \Phi_\beta(U_j^m) - \Phi_\beta(U_j^0) \ge \frac{1}{2L_\beta}(\Theta_j^m)^2 - \frac{L_\beta}{2}(U_j^0)^2.\end{aligned}$$

Then, on using (4.12) and the fact that $\|U^0\|_{L^2(\Omega)} \le \|u_0\|_{L^2(\Omega)}$, we arrive at

$$\sum_{n=1}^{m} \langle U^n - U^{n-1}, \Theta^n \rangle_h \ge C \|\Theta^m\|_{L^2(\Omega)}^2 - C.$$

Since a bound on $\|\Theta^n\|_{L^2(\Omega)}$ readily extends to a bound on $\|U^n\|_{L^2(\Omega)}$ because $|\gamma(s)| \le C_1|s| + C_2$, we have the following *weak* result.

Lemma 5.1. *We have*

$$\max_{1 \le n \le N} \|U^n\|_{L^2(\Omega)} + \sum_{n=1}^{N} \tau \|\nabla \Theta^n\|_{L^2(\Omega)}^2 \le C.$$

Exercises

5.1 Extend Lemma 5.1 to the case $f \ne 0$.

We can also derive an estimate for U^n in intermediate spaces by reasoning as in Exercise 2.5. First, smooth γ out in the same fashion as in Section 2.3.1 but omit the subscript ε; thus $\gamma(0) = 0$. Take next $\phi = U^n = \Pi_h[\gamma(\Theta^n)] \in H_0^1(\Omega)$ as a test function in (4.26). We get

$$\sum_{n=1}^{m} \langle U^n - U^{n-1}, U^n \rangle_h + \tau \sum_{n=1}^{m} \langle \nabla \Theta^n, \nabla U^n \rangle = 0.$$

By virtue of the elementary relation

$$2a(a-b) = a^2 - b^2 + (a-b)^2, \qquad \text{for all } \; a, b \in \mathbb{R}, \tag{5.4}$$

we can handle the first term above as follows:

$$2\sum_{n=1}^{m}\langle U^n - U^{n-1}, U^n\rangle_h = \|U^m\|^2_{L^2(\Omega)} - \|U^0\|^2_{L^2(\Omega)} + \sum_{n=1}^{m}\|U^n - U^{n-1}\|^2_{L^2(\Omega)}.$$

If the mesh $\mathcal{S}_h$ is weakly acute, then we can use (4.23) with $\alpha = \beta$ to deduce

$$\tau\sum_{n=1}^{m}\langle\nabla\Theta^n, \nabla U^n\rangle = \tau\sum_{n=1}^{m}\langle\nabla\Pi_h[\beta(U^n)], \nabla U^n\rangle \geq \Gamma_1\tau\sum_{n=1}^{m}\|\nabla\Theta^n\|^2_{L^2(\Omega)}.$$

Since the above bounds are uniform in ε, we have then proved the following discrete $H^{1/2}(0,T;L^2(\Omega))$ estimate.

Lemma 5.2. *Assume $\mathcal{S}_h$ is weakly acute. Then*

$$\sum_{n=1}^{m}\|U^n - U^{n-1}\|^2_{L^2(\Omega)} \leq C. \tag{5.5}$$

Exercises

5.2 Extend Lemma 5.2 to the case $f \neq 0$.

Our next task is to derive a *strong* stability estimate. Choose $\phi = \partial\Theta^n = \tau^{-1}(\Theta^n - \Theta^{n-1}) \in H^1_0(\Omega)$ as a test function in (4.26), which in turn is a discrete version of that suggested in Exercise 2.4. On adding the resulting equality from 1 to $m \leq N$, and using (4.12) and (5.4), we get

$$\begin{aligned} 0 &= \tau\sum_{n=1}^{m}\langle\partial U^n, \partial\Theta^n\rangle_h + \sum_{n=1}^{m}\langle\nabla\Theta^n, \nabla(\Theta^n - \Theta^{n-1})\rangle \\ &\geq \Gamma_1\tau\sum_{n=1}^{m}\|\partial\Theta^n\|^2_{L^2(\Omega)} + \frac{1}{2}\|\nabla\Theta^m\|^2_{L^2(\Omega)} - \frac{1}{2}\|\nabla\Theta^0\|^2_{L^2(\Omega)}, \end{aligned}$$

which leads to the desired bound if $\|\nabla\Theta^0\|_{L^2(\Omega)} \leq C$, with $C > 0$ independent of h. Since $\Theta^0 = \Pi_h[\beta(U^0)]$, such a constraint may not be satisfied even with $\theta_0 \in H^1_0(\Omega)$. The following Exercise states sufficient, and still reasonable, conditions for such a bound to hold.

Exercises

5.3 Assume $d = 2$, $\theta_0 \in W^{1,\infty}_0(\Omega)$ and $I(0)$ is a Lipschitz curve. Then set $U^0(x_j) := \limsup_{x\to x_j} u_0(x)$. Prove

$$\|\nabla\Theta^0\|_{L^\infty(\Omega)} \leq \|\nabla\theta_0\|_{L^\infty(\Omega)} \leq C, \quad \|U^0 - u_0\|_{H^{-1}(\Omega)} \leq Ch|\log h|^{1/2}.$$

Lemma 5.3. *Let $\|\nabla\Theta^0\|_{L^2(\Omega)} \leq C$ hold uniformly in h. Then*

$$\sum_{n=1}^{N} \tau \|\partial\Theta^n\|^2_{L^2(\Omega)} + \max_{1\leq n\leq N} \|\nabla\Theta^n\|_{L^2(\Omega)} \leq C. \tag{5.6}$$

Exercises

5.4 Suppose the regularization procedure of Section 2.3.1 is used. Show that the following discrete analogue of (2.19) holds:

$$\sum_{n=1}^{N} \tau \|\partial U^n\|^2_{L^2(\Omega)} \leq C\varepsilon^{-1/2}.$$

2.5.2 The Linear Methods

We will only examine the nonlinear Chernoff formula and just propose the analysis of the extrapolation method along with several exercises.

Let $\alpha : \mathbb{R} \to \mathbb{R}$ be defined by $\alpha := I - \mu\beta$; thus $0 \leq \alpha'(s) \leq 1$ for almost all $s \in \mathbb{R}$ because $0 < \mu \leq L_\beta^{-1}$. We would now like to repeat the proof of Lemma 5.1, but the fact that the constitutive relation (4.25) is no longer enforced makes the analysis much more intricate. Note that combining (4.30) and (4.31) and using (4.16) results in

$$\langle U^n - U^{n-1}, P_h^0\phi\rangle + \tau\langle\nabla\Theta^n, \nabla\phi\rangle = \mu\langle[P_h^0 - I]\Theta^n, [I - P_h^0]\phi\rangle_h, \text{ for all } \phi \in \boldsymbol{V}_h^1. \tag{5.7}$$

Lemma 5.4. *The following weak a priori estimates are valid*

$$\max_{1\leq n\leq N} \|U^n\|_{L^2(\Omega)} + \sum_{n=1}^{N} \|U^n - U^{n-1}\|^2_{L^2(\Omega)} + \sum_{n=1}^{N} \tau\|\Theta^n\|^2_{H_0^1(\Omega)} \leq C. \tag{5.8}$$

Proof. Select $\Theta^n \in \boldsymbol{V}_h^1$ as a test function in (4.30) and add the resulting expression from 1 to $m \leq N$. We get

$$\sum_{n=1}^{N} \langle U^n - U^{n-1}, \Theta^n\rangle + \tau \sum_{n=1}^{N} \|\nabla\Theta^n\|^2_{L^2(\Omega)}$$
$$= \mu \sum_{n=1}^{N} \langle [P_h^0 - I]\Theta^n, [I - P_h^0]\Theta^n\rangle_h \leq 0. \tag{5.9}$$

We proceed to estimate each term separately. We first point out that

$$\begin{aligned} P_h^0\Theta^n &= \frac{1}{\mu}(U^n - U^{n-1}) + \beta(U^{n-1}) \\ &= \frac{1}{\mu}[U^n - \alpha(U^{n-1})] \\ &= \frac{1}{2}\beta(U^n) - \frac{1}{2\mu}\alpha(U^{n-1}) + \frac{1}{2\mu}U^n \\ &\qquad + \frac{1}{2\mu}[\alpha(U^n) - \alpha(U^{n-1})]. \end{aligned} \tag{5.10}$$

On using the convexity of both Φ_β and Φ_α as well as the identity (5.4), the first term in (5.10) becomes

$$\begin{aligned} &2\sum_{n=1}^{m}\langle U^n - U^{n-1}, P_h^0\Theta^n\rangle \\ &\quad \geq \int_\Omega \sum_{n=1}^{m}\left(\left[\Phi_\beta(U^n) - \Phi_\beta(U^{n-1})\right] + \frac{1}{\mu}\left[\Phi_\alpha(U^{n-1}) - \Phi_\alpha(U^n)\right]\right) dx \\ &\qquad + \frac{1}{2\mu}\left(\|U^m\|^2_{L^2(\Omega)} - \|U^0\|^2_{L^2(\Omega)} + \sum_{n=1}^{m}\|U^n - U^{n-1}\|^2_{L^2(\Omega)}\right). \end{aligned}$$

Here we have also used the fact that α is nondecreasing in order to eliminate the contribution of the last term in (5.10). The terms involving Φ_β and Φ_α can be further bounded by means of (5.6), namely,

$$\begin{aligned} \int_\Omega \sum_{n=1}^{m}\left[\Phi_\beta(U^n) - \Phi_\beta(U^{n-1})\right] &= \int_\Omega \left[\Phi_\beta(U^m) - \Phi_\beta(U^0)\right] \\ &\geq \frac{1}{2\mu}\|\beta(U^m)\|^2_{L^2(\Omega)} - \frac{L_\beta}{2}\|U^0\|^2_{L^2(\Omega)}, \end{aligned}$$

and

$$\begin{aligned} \frac{1}{\mu}\int_\Omega \sum_{n=1}^{m}\left[\Phi_\alpha(U^{n-1}) - \Phi_\alpha(U^n)\right] &= \frac{1}{\mu}\int_\Omega \left[\Phi_\alpha(U^0) - \Phi_\alpha(U^m)\right] \\ &\geq \frac{1}{2\mu}\|\alpha(U^0)\|^2_{L^2(\Omega)} - \frac{1}{2\mu}\|\alpha(U^m)\|^2_{L^2(\Omega)}. \end{aligned}$$

Inserting these a priori bounds into the previous expression, and recalling that $\|U^0\|_{L^2(\Omega)} \leq \|u_0\|_{L^2(\Omega)}$, leads to

$$\sum_{n=1}^{m}\langle U^n - U^{n-1}, \Theta^n\rangle \geq -C + C\|\beta(U^m)\|^2_{L^2(\Omega)} + \frac{1}{4\mu}\sum_{n=1}^{m}\|U^n - U^{n-1}\|^2_{L^2(\Omega)}.$$

Finally, the estimate $\|\beta(U^m)\|_{L^2(\Omega)} \leq C$ yields $\|U^m\|_{L^2(\Omega)} \leq C$, because β grows at least linearly at infinity, i.e. $\beta'(s) \geq \Gamma_2^{-1}$ for almost all $|s| > 1$. This concludes the argument. ■

Exercises

5.5 Prove Lemma 5.3 with $f \neq 0$.

It is also possible to prove a *strong* stability estimate that resembles that in Lemma 5.3 and is useful in analyzing the incomplete iteration process of Section 2.4.4. The choice of the initial data is again a crucial issue that we describe in the following Exercises.

Exercises

5.6 Under the same assumptions as in Exercise 5.3, define $U^0 \in \boldsymbol{V}_h^0$ to be $U^0(x_S) := \limsup_{x \to x_S} u_0(x_S)$ for all $S \in \mathcal{S}_h$, where x_S stands for the barycenter of S. Set $\Theta^0 := \Pi_h \theta_0$ and $\alpha(U^{-1}) := U^0 - \mu P_h^0 \Theta^0$, thus $\|\nabla\Theta^0\|_{H_0^1(\Omega)} \leq C$. Prove

$$\|\alpha(U^0) - \alpha(U^{-1})\|_{L^2(\Omega)} = \mu\|\beta(U^0) - P_h^0\Theta^0\|_{L^2(\Omega)} \leq Ch. \quad (5.11)$$

5.7 In the case that a preliminary regularization procedure is used, we have the following recipe for U^0 [26]. Assume, in addition to Exercise 5.6, the initial nondegeneracy $\text{meas}(\{x \in \Omega : 0 \leq \theta_0(x) \leq \varepsilon\}) \leq C\varepsilon$. Then, set $\Theta^0 := \Pi_h\theta_0$ and $U^0 := \gamma_\varepsilon(P_h^0\Theta^0)$, or equivalently $\beta(U^0) := P_h^0\Theta^0$. Hence, U^0 is easy to compute elementwise. Prove

$$\|\gamma_\varepsilon(P_h^0\Theta^0) - \gamma_\varepsilon(\theta_0)\|_{H^{-1}(\Omega)} \leq C(\varepsilon + h)|\log(\varepsilon + h)|^{1/2}.$$

Lemma 5.5. *Let the assumptions of Exercise 5.6 hold. Let the mild constraint $h^2 \leq C^*\tau$ be enforced with $C^* > 0$ arbitrary. Then, the following strong stability estimate holds*

$$\sum_{n=1}^N \tau\|\partial\Theta^n\|_{L^2(\Omega)}^2 + \sum_{n=1}^N \|\nabla(\Theta^n - \Theta^{n-1})\|_{L^2(\Omega)}^2 + \max_{1\leq n\leq N}\left(\|\nabla\Theta^n\|_{L^2(\Omega)}^2 + \tau^{-1}\|U^n - U^{n-1}\|_{L^2(\Omega)}^2\right) \leq CC^*. \quad (5.12)$$

The first two terms are essential to examine the effect of the incomplete iteration process described in Section 2.4. Note that the second and fourth estimates are discrete $H^{1/2}$-bounds.

Proof. Take $\phi = \partial\Theta^n = \tau^{-1}(\Theta^n - \Theta^{n-1}) \in \boldsymbol{V}_h^1$ in (4.7) to obtain

$$0 = \langle U^n - U^{n-1}, P_h^0\partial\Theta^n\rangle + \langle\nabla\Theta^n, \nabla(\Theta^n - \Theta^{n-1})\rangle$$
$$+ \mu\langle[I - P_h^0]\Theta^n, [I - P_h^0]\partial\Theta^n\rangle_h =: I + II + III. \quad (5.13)$$

We next observe that (4.31) is equivalent to

$$U^n = \mu P_h^0\Theta^n + \alpha(U^{n-1}) \text{ or } P_h^0\Theta^n = \mu^{-1}[U^n - \alpha(U^{n-1})],\ 0 \le n \le N,$$

where $\alpha(U^{-1}) := U^0 - \mu P_h^0\Theta^0$. This results in

$$\begin{aligned}\tau^{-1}I &= \mu\|\partial P_h^0\Theta^n\|_{L^2(\Omega)}^2 + \langle\partial\alpha(U^{n-1}), P_h^0\partial\Theta^n\rangle \\ &= \mu\|\partial P_h^0\Theta^n\|_{L^2(\Omega)}^2 + \mu^{-1}\langle\partial\alpha(U^{n-1}), \partial U^n\rangle - \mu^{-1}\|\partial\alpha(U^{n-1})\|_{L^2(\Omega)}^2,\end{aligned}$$

as well as in

$$\tau^{-1}I = \mu^{-1}\|\partial U^n\|_{L^2(\Omega)}^2 - \mu^{-1}\langle\partial U^n, \partial\alpha(U^{n-1})\rangle,$$

whence

$$2\tau^{-1}I = \mu\|\partial P_h^0\Theta^n\|_{L^2(\Omega)}^2 + \mu^{-1}\|\partial U^n\|_{L^2(\Omega)}^2 - \mu^{-1}\|\partial\alpha(U^{n-1})\|_{L^2(\Omega)}^2,$$

for all $1 \le n \le N$. By virtue of (5.4), term II becomes

$$2II = \|\nabla\Theta^n\|_{L^2(\Omega)}^2 - \|\nabla\Theta^{n-1}\|_{L^2(\Omega)}^2 + \|\nabla(\Theta^n - \Theta^{n-1})\|_{L^2(\Omega)}^2.$$

Likewise, III can be rewritten as follows:

$$2\tau\mu^{-1}II = \|[I - P_h^0]\Theta^n\|_h^2 - \|[I - P_h^0]\Theta^{n-1}\|_h^2 + \|[I - P_h^0](\Theta^n - \Theta^{n-1})\|_h^2,$$

where we have used the abbreviation $\|\cdot\|_h^2 = \langle\cdot,\cdot\rangle_h$. In view of (4.12), this norm in equivalent to $\|\cdot\|_{L^2(\Omega)}$ in $\boldsymbol{V}_h^1$. Substituting back in (5.13), adding the resulting expressions over n from 1 to $m \le N$ and recalling that $0 \le \alpha' \le 1$, we easily get

$$\begin{aligned}\|\nabla\Theta^m\|^2 + \tau^{-1}\|U^m - U^{m-1}\|_{L^2(\Omega)}^2 + \sum_{n=1}^m \tau\|P_h^0\partial\Theta^n\|_{L^2(\Omega)}^2 \\ + \sum_{n=1}^m \|\nabla(\Theta^n - \Theta^{n-1})\|_{L^2(\Omega)}^2 \\ \le \|\nabla\Theta^0\|_{L^2(\Omega)}^2 + \tau^{-1}\|\alpha(U^0) - \alpha(U^{-1})\|_{L^2(\Omega)}^2 \\ + \tau^{-1}\|[I - P_h^0]\Theta^0\|_{L^2(\Omega)}^2.\end{aligned}$$

In view of the regularity assumed on u_0 and θ_0, we have $\|\nabla\Theta^0\|_{L^2(\Omega)} \leq C$ and

$$\tau^{-1}\|\alpha(U^0) - \alpha(U^{-1})\|_{L^2(\Omega)}^2 \leq Ch^2/\tau \leq CC^*,$$

and also

$$\tau^{-1}\|[I - P_h^0]\Theta^0\|_{L^2(\Omega)}^2 \leq Ch^2\tau^{-1}\|\nabla\Theta^0\|_{L^2(\Omega)}^2 \leq C$$

as results from (4.18). To derive the first estimate in (5.12), we resort again to (4.18) to write

$$\begin{aligned}\sum_{n=1}^{m}\tau^{-1}\|[I - P_h^0](\Theta^n - \Theta^{n-1})\|_{L^2(\Omega)}^2 &\leq C\frac{h^2}{\tau}\sum_{n=1}^{m}\|\nabla(\Theta^n - \Theta^{n-1})\|_{L^2(\Omega)}^2 \\ &\leq CC^*,\end{aligned}$$

which in turn yields the assertion (5.12). ■

Exercises

5.8 Show that the following stability estimates hold for the extrapolation method (see [19]):

$$\sum_{n=1}^{N}\tau\|\partial P_h^0\Theta^n\|_{L^2(\Omega)}^2 + \max_{1\leq n\leq N}\|\nabla\Theta^n\|_{L^2(\Omega)}^2 \leq C\left(1 + \|\nabla\Theta^0\|_{L^2(\Omega)}^2\right), \tag{5.14}$$

$$\max_{1\leq n\leq N}\|U^n\|_{L^2(\Omega)} \leq C\left(1 + \left(\frac{\tau}{\varepsilon}\right)^{1/2}\right)\left(1 + \|\nabla\Theta^0\|_{L^2(\Omega)}^2\right). \tag{5.15}$$

5.9 Prove the above results with $f \neq 0$.

We now conclude this section with a stability estimate in the maximum norm for the nonlinear Chernoff formula.

Lemma 5.6. *Let $\mathcal{S}_h$ be weakly acute and $u_0 \in L^\infty(\Omega)$. Then*

$$\max_{1\leq n\leq N}\|U^n\|_{L^\infty(\Omega)} \leq \|U^0\|_{L^\infty(\Omega)}. \tag{5.16}$$

Proof. Let C_0 be a positive constant such that $-C_0 \leq U^0 = P_h^0 u_0 \leq C_0$. This is actually possible in view of the boundedness in L^∞ of P_h^0 and the assumption on u_0. We argue by induction. Suppose $-C_0 \leq U^{n-1} \leq C_0$ is valid in Ω. Since this is obviouly true for $n = 1$, let $n > 1$. Let x be a point at which Θ^n attains its maximum. Since Θ^n is piecewise linear, $x = x_j$ is clearly a node of $\mathcal{S}_h$. We can assume that x_j is interior for otherwise $\Theta^n(x_j) = 0$. Let ϕ_j be the corresponding shape function. Since

$\mathcal{S}_h$ is weakly acute, the Discrete Maximum Principle (4.21) applies. This, in conjunction with the equality $\int_\Omega \phi_j = \text{meas}(\text{supp}\phi_j)/(d+1)$, yields

$$\Theta^n(x_j)\frac{\text{meas}(\text{supp}\phi_j)}{d+1} = \langle\Theta^n, \phi_j\rangle_h \le \langle\beta(U^{n-1}, \phi_j\rangle \le \beta(C_0)\frac{\text{meas}(\text{supp}\phi_j)}{d+1},$$

because β is nondecreasing. A similar argument produces a bound from below for Θ^n, this time dealing with the minimum of Θ^n. Therefore, we get the estimate

$$\beta(-C_0) \le \Theta^n \le \beta(C_0).$$

Now, since

$$U^n = U^{n-1} + \mu[P_h^0\Theta^n - \beta(U^{n-1})] = \alpha(U^{n-1}) + \mu P_h^0\Theta^n,$$

and α is nondecreasing and verifies $\alpha + \beta = I$, we see that

$$-C_0 \le U^n \le \alpha(C_0) + \beta(C_0) = C_0,$$

which in turn completes the induction argument. ■

Exercises

5.10 Extend the maximum norm stability estimate to the case $f \neq 0$.

5.11 Prove an estimate similar to (5.16) for the nonlinear scheme.

2.6 Error Analysis

In this section we will prove error estimates for the physical variables u, θ and $\boldsymbol{q}$ in the natural energy spaces and for the various schemes above. We will also comment on the approximation of interfaces. As usual, we will assume $f = 0$.

We start out by defining the error functions

$$e_u(t) : u(t) - U^n,\ e_\theta(t) := \theta(t) - \Theta^n,\ \text{for all } t \in I^n = (t^n, t^{n-1}], \tag{6.1}$$

and recalling that $e_u^n = e_u(t^n)$. We also introduce a discrete form of (2.12) which turns out to be convenient for the error analysis. It comes from integrating (2.12) on the interval I^n:

$$\langle u^n - u^{n-1}, \phi\rangle + \langle\int_{I^n} \nabla e_\theta(t)dt, \nabla\phi\rangle = 0, \tag{6.2}$$

for all $\phi \in H_0^1(\Omega)$, $1 \le n \le N$.

2.6.1 The Nonlinear Scheme

We proceed as in Lemma 3.1. Note that we deal with minimal regularity on the initial datum, namely $u_0 \in L^2(\Omega)$. The discrete equation (4.26) can be written equivalently as

$$\begin{aligned}\langle U^n - U^{n-1}, \phi\rangle + \tau\langle \nabla\Theta^n, \nabla\phi\rangle + \langle \nabla\Theta^n, \nabla(\varphi - \phi)\rangle \\ = \langle U^n - U^{n-1}, \varphi\rangle - \langle U^n - U^{n-1}, \varphi\rangle_h \\ + \langle U^n - U^{n-1}, \phi - \varphi\rangle,\end{aligned} \tag{6.3}$$

for all $\phi \in H_0^1(\Omega)$, $\varphi \in \boldsymbol{V}^n$, $1 \leq n \leq N$. Subtraction of (6.3) from (6.2) yields the *error equation*

$$\begin{aligned}\langle e_u^n - e_u^{n-1}, \phi\rangle + \langle \int_{I^n} \nabla e_\theta(t)dt, \nabla\phi\rangle + \tau\langle \nabla\Theta^n, \nabla(\phi - \varphi)\rangle \\ = \langle U^n - U^{n-1}, \varphi\rangle_h - \langle U^n - U^{n-1}, \varphi\rangle \\ + \langle U^{n-1} - U^n, \phi - \varphi\rangle,\end{aligned} \tag{6.4}$$

for all $\varphi \in H_0^1(\Omega)$, $\varphi \in \boldsymbol{V}^n$.

Theorem 6.1. *Let $\tau = Ch$ with $C > 0$ arbitrary. Then*

$$\|e_u\|_{L^\infty(0,T;H^{-1}(\Omega))} + \|e_\theta\|_{L^2(0,T;L^2(\Omega))} + \|\int_0^t e_q\|_{L^\infty(0,T;L^2(\Omega))} \leq Ch^{1/2}. \tag{6.5}$$

Proof. Take first $\phi := Ge_u^n \in H_0^1(\Omega)$ and $\varphi := G^n e_u^n \in \boldsymbol{V}^n$ in (6.4), and next sum on n from 1 to $m \leq N$. In view of (3.2) and (4.19), we get

$$\begin{aligned} I + II &:= \sum_{n=1}^m \langle e_u^n - e_u^{n-1}, Ge_u^n\rangle + \sum_{n=1}^m \langle \int_{I^n} e_\theta(t)dt, e_u^n\rangle \\ &= \sum_{n=1}^m \left(\langle U^n - U^{n-1}, G^n e_u^n\rangle_h - \langle U^n - U^{n-1}, G^n e_u^n\rangle\right) \\ &\quad + \sum_{n=1}^m \langle U^{n-1} - U^n, (G - G^n)e_u^n\rangle \\ &=: III + IV. \end{aligned} \tag{6.6}$$

We now evaluate these four terms separately. In order to simplify notation, set

$$A_m := \max_{1\leq n\leq m} \|e_u^n\|^2_{H^{-1}(\Omega)}.$$

By virtue of (4.18), we have $\|U^0 - u_0\|_{H^{-1}(\Omega)} \le Ch\|u_0\|_{L^2(\Omega)} \le Ch$. The elementary relation (5.4), in conjunction with (3.2), then yields

$$\begin{aligned} 2I &= \|e_u^m\|^2_{H^{-1}(\Omega)} - \|u_0 - U^0\|^2_{H^{-1}(\Omega)} + \sum_{n=1}^m \|e_u^n - e_u^{n-1}\|^2_{H^{-1}(\Omega)} \\ &\ge \|e_u^m\|^2_{H^{-1}(\Omega)} - Ch^2. \end{aligned}$$

In view of the constitutive relations (2.10) and (4.25), term II can be further split as follows

$$\begin{aligned} II_1 + II_2 + II_3 &:= \sum_{n=1}^m \int_{I^n} \langle e_\theta(t), u(t^n) - u(t)\rangle dt \\ &+ \sum_{n=1}^m \int_{I^n} \langle \beta(u(t)) - \beta(U^n), u(t) - U^n\rangle dt \\ &+ \sum_{n=1}^m \int_{I^n} \langle \beta(U^n) - \Pi_h\beta(U^n), e_u(t)\rangle dt. \end{aligned}$$

We first make use of (2.14) and Lemma 5.1 to estimate II_1 as follows:

$$|II_1| \le \sum_{n=1}^m \int_{I^n} \|\nabla e_\theta(t)\|_{L^2(\Omega)} \| \int_t^{t^n} u_t(s)ds\|_{H^{-1}(\Omega)} dt \le C\tau.$$

In fact, after applying the Cauchy-Schwarz inequality, the assertion is reduced to

$$\begin{aligned} \sum_{n=1}^m \int_{I^n} \| \int_t^{t^n} u_t(s)ds\|^2_{H^{-1}(\Omega)} dt &\le \sum_{n=1}^m \int_{I^n} (t^n - t) \int_t^{t^n} \|u_t(s)\|^2_{H^{-1}(\Omega)} ds dt \\ &\le \sum_{n=1}^m \tau^2 \int_{I^n} \|u_t(t)\|^2_{H^{-1}(\Omega)} dt \\ &\le \tau^2 \|u_t\|_{L^2(0,t^m;H^{-1}(\Omega))} \\ &\le C\tau^2. \end{aligned}$$

We next use an elementary interpolation estimate that we propose as an Exercise.

Exercises

6.1 Let $\alpha : \mathbb{R} \to \mathbb{R}$ be a continuous and nondecreasing function. Prove

$$\|\alpha(\varphi) - \Pi_h\alpha(\varphi)\|_{L^p(S)} \le Ch_S \|\nabla \Pi_h \alpha(\varphi)\|_{L^p(S)}, \tag{6.7}$$

for all $S \in \mathcal{S}_h$, $\varphi \in \boldsymbol{V}_h^1$, $1 \le p \le \infty$.

Applying this inequality with $\alpha = \beta$ and $p = 2$, in conjunction with (2.14) and Lemma 5.1 and the relation $a^2 \leq 2(a-b)^2 + 2b^2$ for all $a, b \in \mathbb{R}$, we obtain

$$\begin{aligned} II_2 &\geq \Gamma_1 \sum_{n=1}^m \int_{I^n} \|\beta(u(t)) - \beta(U^n)\|^2_{L^2(\Omega)} dt \\ &\geq \frac{\Gamma_1}{2} \sum_{n=1}^m \int_{I^n} \|e_\theta(t)\|^2_{L^2(\Omega)} dt - \Gamma_1 \tau \sum_{n=1}^m \|\beta(U^n) - \Pi_h[\beta(U^n)]\|^2_{L^2(\Omega)} \\ &\geq \frac{\Gamma_1}{2} \|e_\theta\|^2_{L^2(0,t^m;L^2(\Omega))} - Ch^2. \end{aligned}$$

To bound II_3 we use again (6.7), together with (2.14) and Lemma 5.1, to arrive at

$$\begin{aligned} |II_3| &\leq \sum_{n=1}^m \int_{I^n} \|\beta(U^n) - \Pi_h[\beta(U^n)]\|_{L^2(\Omega)} \|e_u(t)\|_{L^2(\Omega)} dt \\ &\leq Ch \sum_{n=1}^m \tau \|\nabla \Theta^n\|_{L^2(\Omega)} \\ &\leq Ch. \end{aligned}$$

In summary, for term II we have obtained the lower bound

$$II \geq \frac{\Gamma_1}{2} \|e_\theta\|^2_{L^2(0,t^m;L^2(\Omega))} - Ch.$$

The analysis of III is based on (4.13) and (5.4) which we apply as follows:

$$\begin{aligned} |III| &\leq Ch \sum_{n=1}^m \|U^n - U^{n-1}\|_{L^2(\Omega)} \|\nabla G_h e_u^n\|_{L^2(\Omega)} \\ &\leq Ch \sum_{n=1}^m \|U^n - U^{n-1}\|_{L^2(\Omega)} \|e_u\|_{H^{-1}(\Omega)} \\ &\leq \eta A_m + \frac{Ch^2}{\eta\tau} \sum_{n=1}^m \|U^n - U^{n-1}\|^2_{L^2(\Omega)} \\ &\leq \eta A_m + \frac{Ch^2}{\eta\tau}, \end{aligned}$$

where $\eta > 0$ is to be selected. To examine IV we make use of (4.20) with $s = 1$ and then proceed as for III to find

$$|IV| \leq Ch \sum_{n=1}^m \|U^n - U^{n-1}\|_{L^2(\Omega)} \|e_u^n\|_{H^{-1}(\Omega)} \leq \eta A_m + \frac{Ch^2}{\eta\tau}.$$

Inserting the above estimates back into (6.6), we see that a proper choice of η, coupled with the relation $\tau = Ch$, leads to

$$\max_{1\leq n\leq m} \|e_u^n\|_{H^{-1}(\Omega)}^2 + \|e_\theta\|_{L^2(0,t^m;L^2(\Omega))}^2 \leq C(\tau + h + h^2/\tau) \leq Ch.$$

We can now replace $\max_{1\leq n\leq m} \|e_u^n\|_{H^{-1}(\Omega)}$ by $\|e_u\|_{L^\infty(0,T;H^{-1}(\Omega))}$ to obtain the asserted error estimate for u, because $u \in H^1(0,T;H^{-1}(\Omega)) \subset C^{0,1/2}(0,T;H^{-1}(\Omega))$.

To derive the remaining estimate for the flux $\boldsymbol{q} = -\nabla\theta$ we sum (6.4) on n from 1 to $m \leq N$ and next choose $\phi = \int_0^{t^m} e_\theta(t)dt \in H_0^1(\Omega)$ and $\varphi = \int_0^{t^m} R_h e_\theta(t)dt \in \boldsymbol{V}_h^1$ as test functions in the resulting expression, which becomes

$$\begin{aligned}\left\|\int_0^{t^m} \nabla e_\theta(t)dt\right\|_{L^2(\Omega)}^2 &= \langle e_u^0 - e_u^m, \int_0^{t^m} e_\theta(t)dt\rangle \\ &\quad + \langle U^m - U^0, \int_0^{t^m} R_h e_\theta(t)dt\rangle_h \\ &\quad - \langle U^m - U^0, \int_0^{t^m} R_h e_\theta(t)dt\rangle \\ &\quad + \langle U^0 - U^m, \int_0^{t^m} [I - R_h] e_\theta(t)dt\rangle.\end{aligned}$$

In view of (4.17) for $s = 0$ as well as Lemma 5.1, we can write

$$\begin{aligned}&\left\|\int_0^{t^m} \nabla e_\theta(t)dt\right\|_{L^2(\Omega)}^2 \\ &\quad \leq C(\|e_u^0\|_{H^{-1}(\Omega)} + \|e_u^m\|_{H^{-1}(\Omega)} + h)\left\|\int_0^{t^m} \nabla e_\theta(t)dt\right\|_{L^2(\Omega)},\end{aligned}$$

because $\|\nabla R_h\zeta\|_{L^2(\Omega)} \leq \|\nabla\zeta\|_{L^2(\Omega)}$ for all $\zeta \in H_0^1(\Omega)$. This completes the proof. ■

Exercises

6.2 Extend Theorem 6.1 to the case $f \neq 0$.

In case the preliminary regularization (3.1) is used, several variants of (6.5) can be derived as suggested in the following Exercises.

Exercises

6.3 Prove the error estimates

$$\|e_u\|_{L^\infty(0,T;H^{-1}(\Omega))} + \varepsilon^{1/2}\|e_u\|_{L^2(Q)} + \|e_\theta\|_{L^2(0,T;L^2(\Omega))} + \left\|\int_0^t e_q\right\|_{L^\infty(0,T;L^2(\Omega))} \leq C\left(\frac{h^2}{\varepsilon} + \frac{h^4}{\tau\varepsilon} + \tau\right)^{1/2}.$$

Hint: Use that $\beta' \geq \varepsilon$ to deduce the second error bound. Then modify the analysis of III and IV to allow $\|e_u^n\|_{L^2(\Omega)}$ instead of $\|e_u^n\|_{H^{-1}(\Omega)}$.

6.4 Derive global rates of convergence by suitably relating h, ε and τ. Repeat under the assumption of nondegeneracy (2.22). In the latter case, prove

$$\|e_u\|_{L^2(Q)} \leq Ch^{1/3}.$$

A sharp L^2-error estimate for enthalpy, namely $O(h^{1/2})$, was shown in [16] but without considering numerical integration.

2.6.2 The Linear Methods

We now examine in detail the fully discrete nonlinear Chernoff formula and propose in Exercises the corresponding results for the fully discrete extrapolation method.

We will proceed as in Lemma 3.2. Take the difference between (2.12) and (5.7), and sum over n. The resulting expression reads

$$\langle e_u^n - e_u^0, \varphi\rangle + \langle\nabla\sum_{i=1}^n\int_{I^i} e_\theta(t)dt, \nabla\varphi\rangle = \mu\langle[I-P_h^0]\sum_{i=1}^n\Theta^i, \varphi\rangle_h, \text{ for all } \varphi\in \boldsymbol{V}_h^1. \tag{6.8}$$

The next step is to choose a suitable test function φ. Let us take $\varphi := \int_{I^n}(R_h\theta(t) - \Theta^n)dt = R_h(\int_{I^n} e_\theta(t)dt) \in \boldsymbol{V}_h^1$ and sum over n from 1 to $m \leq N$. After reordering, we get

$$\begin{aligned}
I + II \;&:= \\
&\sum_{n=1}^m\int_{I^n}\langle e_u(t), e_\theta(t)\rangle dt + \sum_{n=1}^m\langle\nabla\sum_{i=1}^n\int_{I^i} e_\theta(t)dt, \nabla\int_{I^n} R_h e_\theta(t)dt\rangle \\
&= \sum_{n=1}^m\int_{I^n}\langle u(t) - u^n, e_\theta(t)\rangle dt + \sum_{n=1}^m\langle u^n - U^n, \int_{I^n}[I - R_h]\theta(t)dt\rangle \\
&+ \langle u_0 - U^0, \sum_{n=1}^m\int_{I^n} R_h e_\theta(t)dt\rangle + \mu\sum_{n=1}^m\langle[I - P_h^0]\sum_{i=1}^n\Theta^i, \int_{I^n} R_h e_\theta(t)dt\rangle_h \\
&=: \;\; III + \cdots + VI.
\end{aligned} \tag{6.9}$$

We now proceed to estimate each one of these terms in turn. To begin with, note that

$$u = \mu\theta + \alpha(u) \qquad \text{and} \qquad U^n = \mu P_h^0\Theta^n + \alpha(U^{n-1}),$$

whence

$$e_u(t) = \mu e_\theta(t) + [\alpha(u(t)) - \alpha(U^{n-1})] + \mu[I - P_h^0]\Theta^n \quad \text{for all } t \in I^n. \tag{6.10}$$

Moreover, we know that

$$e_\theta(t) = [\beta(u(t)) - \beta(U^{n-1})] - \frac{1}{\mu}(U^n - U^{n-1}) - [I - P_h^0]\Theta^n \tag{6.11}$$

for all $t \in I^n$. Then, in view of (6.10) and (6.11), term I in (6.9) can be further split as follows:

$$\begin{aligned} I \;=\; & \mu\|e_\theta\|^2_{L^2(0,t^m;L^2(\Omega))} \\ & + \sum_{n=1}^m \int_{I^n} \langle \alpha(u(t)) - \alpha(U^{n-1}), \beta(u(t)) - \beta(U^{n-1})\rangle dt \\ & + \frac{\tau}{\mu}\sum_{n=1}^m \langle U^n, U^n - U^{n-1}\rangle + \sum_{n=1}^m \int_{I^n} \langle e_\theta(t), U^n - U^{n-1}\rangle dt \\ & - \frac{1}{\mu}\sum_{n=1}^m \int_{I^n} \langle u(t), U^n - U^{n-1}\rangle dt + \sum_{n=1}^m \tau\langle [I - P_h^0]\Theta^n, U^n - U^{n-1}\rangle \\ & - \sum_{n=1}^m \int_{I^n} \langle \alpha(u(t)) - \alpha(U^{n-1}), [I - P_h^0]\Theta^n\rangle dt \\ & + \mu\sum_{n=1}^m \int_{I^n} \langle e_\theta(t), [I - P_h^0]\Theta^n\rangle dt =: I_1 + I_2 + \cdots + I_8. \end{aligned} \tag{6.12}$$

The fact that both α and β are nondecreasing functions implies that $I_2 \geq 0$. Term I_3 is handled by means of (4.29) and (5.4) as follows:

$$I_3 \geq \frac{\tau}{2\mu}\sum_{n=1}^m \left(\|U^n\|^2_{L^2(\Omega)} - \|U^{n-1}\|^2_{L^2(\Omega)}\right) \geq -\frac{\tau}{2\mu}\|U^0\|^2_{L^2(\Omega)} \geq -C\tau.$$

We now make use of (5.8) to arrive at

$$|I_4| \leq \frac{1}{4}I_1 + \tau\sum_{n=1}^m \|U^n - U^{n-1}\|^2_{L^2(\Omega)} \leq \frac{1}{4}I_1 + C\tau,$$

and also of (4.18) to deduce

$$\sum_{n=1}^{m} \tau \|[I - P_h^0]\Theta^n\|_{L^2(\Omega)}^2 \leq Ch^2. \tag{6.13}$$

Hence

$$|I_8| \leq \frac{1}{4} I_1 + Ch^2.$$

At the same time, on using (2.14) and (5.8) together with (6.13), we get $|I_7| \leq Ch$ as well as

$$|I_6| \leq \tau \sum_{n=1}^{m} \|U^n - U^{n-1}\|_{L^2(\Omega)}^2 + Ch^2 \leq C\tau + Ch^2.$$

The remaining term I_5 will be analyzed later, under various regularity assumptions on u_0. Instead, we now bound term II in (6.9). To do so, we need the following elementary identity, which is an easy consequence of (5.4):

$$2\sum_{n=1}^{m} a_n \left(\sum_{i=1}^{n} a_i\right) = \left(\sum_{n=1}^{m} a_n\right)^2 + \sum_{n=1}^{m} a_n^2 \qquad \text{for all } a_n \in \mathbb{R},\ 1 \leq n \leq N.$$

Thus, combining (4.15) and (4.17) with the regularity property $\int_0^t \theta \in L^\infty(0,T;H^2(\Omega))$, which comes from (2.21), we easily obtain

$$\begin{aligned} II &= \sum_{n=1}^{m} \left\langle \nabla \sum_{i=1}^{n} \int_{I^i} R_h e_\theta(t)dt, \nabla \int_{I^n} R_h e_\theta(t)dt \right\rangle \\ &\geq \frac{1}{2} \| \sum_{n=1}^{m} \nabla \int_{I^n} R_h e_\theta(t)dt\|_{L^2(\Omega)}^2 \geq \frac{1}{2} \|\nabla \int_0^{t^m} R_h e_\theta(t)dt\|_{L^2(\Omega)}^2 \\ &\geq \frac{1}{2} \|\nabla \int_0^{t^m} e_\theta(t)dt\|_{L^2(\Omega)}^2 - Ch^2 =: II^m - Ch^2. \end{aligned}$$

Treating III requires a duality argument between $H^{-1}(\Omega)$ and $H_0^1(\Omega)$, which in turn is similar to that in Theorem 6.1. The a priori estimates (2.14) and (5.8) lead to

$$\begin{aligned} |III| &= \left| \sum_{n=1}^{m} \int_{I^n} \langle \int_t^{t^n} u_t(s)ds, e_\theta(t) \rangle dt \right| \\ &\leq \tau \|u_t\|_{L^2(0,t^m;H^{-1}(\Omega))} \|e_\theta\|_{L^2(0,t^m;H_0^1(\Omega))} \\ &\leq C\tau. \end{aligned}$$

The other term IV is estimated by invoking (4.17) as well as (2.14) and (5.8). Indeed, we have

$$|IV| \leq \|e_u\|_{L^2(0,t^m;L^2(\Omega))} \|[I - P_h^0]\theta\|_{L^2(0,t^m;L^2(\Omega))} \leq Ch.$$

Using (4.18), we easily obtain

$$|V| \leq Ch\|u_0\|_{L^2(\Omega)} \|\nabla \int_0^{t^m} R_h e_\theta(t)dt\|_{L^2(\Omega)} \leq \frac{1}{2} II^m + Ch^2.$$

In order to get a bound for VI, we first rewrite this term in a more suitable form, namely,

$$\begin{aligned} VI &= \mu \sum_{n=1}^{m} \langle [I - P_h^0] \sum_{i=1}^{n} \Theta^i, \int_{I^n} R_h e_\theta(t)dt \rangle \\ &+ \mu \sum_{n=1}^{m} \tau \left(\left\langle \sum_{i=1}^{n} \Theta^i, \int_{I^n} R_h e_\theta(t)dt \right\rangle_h - \left\langle \sum_{i=1}^{n} \Theta^i, \int_{I^n} R_h e_\theta(t)dt \right\rangle \right). \end{aligned}$$

The estimate for the first term proceeds along the same lines as for III, with the only difference that now we exploit (4.18). The other term is an error due to the quadrature rule. So, with the aid of (4.13), we can control its contribution. Thus

$$|VI| \leq Ch^2 \sum_{n=1}^{m} \| \sum_{i=1}^{n} \Theta^i \|_{H^{-1}(\Omega)} \| \int_{I^n} R_h e_\theta(t)dt \|_{H_0^1(\Omega)} \leq C\frac{h^2}{\tau},$$

where (2.14) and (5.8) have been used again.

Collecting all the previous estimates, and inserting them into (6.9), gives

$$\begin{aligned} &\frac{\mu}{4}\|e_\theta\|^2_{L^2(0,t^m;L^2(\Omega))} + \frac{1}{4}\|\nabla \int_0^{t^m} e_\theta(t)dt\|^2_{L^2(\Omega)} \\ &\qquad \leq C\left(\tau + h + \frac{h^2}{\tau}\right) + \frac{1}{\mu}\sum_{n=1}^{m} \int_{I^n} \langle u(t), U^n - U^{n-1} \rangle dt. \end{aligned}$$

The last term will be showm to be $O(\tau^{2\nu})$, where $0 < \nu \leq 1/2$ depends on the regularity assumed on u_0 and θ_0 (see the theorems below). Therefore,

$$\|e_\theta\|_{L^2(Q)} + \| \int_0^t e_q \|_{L^\infty(0,T;L^2(\Omega))} \leq C\left(\tau^\nu + h + h\tau^{-1/2}\right) =: \sigma(h, \tau). \tag{6.14}$$

In order to derive an error estimate for u in $L^\infty(0,T;H^{-1}(\Omega))$, we first rewrite (6.8) as follows:

$$\langle e_u^n, \phi\rangle = \langle e_u^n, \phi - \varphi\rangle + \langle e_u^0, \varphi\rangle \\ - \langle \nabla \int_0^{t^n} e_\theta(t)dt, \nabla\varphi\rangle + \mu\langle [I - P_h^0] \sum_{i=1}^n \Theta^i, \varphi\rangle_h,$$

for all $\phi \in H_0^1(\Omega)$, $\varphi \in \boldsymbol{V}_h^1$. We take now $\varphi := R_h\phi$ for any $\phi \in H_0^1(\Omega)$, calculate

$$|\langle e_u^n, \phi\rangle| \le C\Big(h\|e_u^n\|_{L^2(\Omega)} + \|e_u^0\|_{H^{-1}(\Omega)} \\ + \|\nabla \int_0^{t^n} e_\theta(t)dt\|_{L^2(\Omega)} + h^2\|\sum_{i=1}^n \nabla\Theta^i\|_{L^2(\Omega)}\Big) \cdot \|\nabla\phi\|_{L^2(\Omega)} \\ \le C\sigma(h,\tau)\|\phi\|_{H_0^1(\Omega)},$$

and realize that $u \in H^1(0,T;H^{-1}(\Omega)) \subset C^{0,1/2}(0,T;H^{-1}(\Omega))$ to determine the desired error bound on u. We are now in a position to state the final error estimates.

Theorem 6.2. *Assume that*

$$u_0 \in L^\infty(\Omega), \quad \Delta\theta_0 \in L^1(\Omega), \tag{6.15}$$

and in addition that

$$\max_{1\le n\le N} \|U^n\|_{L^\infty(\Omega)} \le C. \tag{6.16}$$

Let $\tau = C^*h$, with $C^* > 0$ an arbitrary constant. Then

$$\|e_u\|_{L^\infty(0,T;H^{-1}(\Omega))} + \|e_\theta\|_{L^2(0,T;L^2(\Omega))} + \|\int_0^t e_q\|_{L^\infty(0,T;L^2(\Omega))} \le Ch^{1/2}. \tag{6.17}$$

The a priori estimate (6.16) is valid, for instance, for weakly acute meshes $\mathcal{S}_h$, as shown in Lemma 5.6. The assumption on $\Delta\theta_0$ can be slightly weakened as explained in Exercise 2.8, thus allowing discontinuities of $\nabla\theta_0$ across $I(0)$.

Proof. It only remains to demonstrate the bound

$$\left|\sum_{n=1}^m \int_{I^n} \langle u(t), U^n - U^{n-1}\rangle dt\right| \le C\tau. \tag{6.18}$$

To this end, we need the following summation by parts formula,

$$\sum_{n=1}^{m} a_n(b_n - b_{n-1}) = a_m b_m - a_0 b_0 - \sum_{n=1}^{m} b_{n-1}(a_n - a_{n-1}).$$

We can then rewrite the sum in (6.18) as follows:

$$\sum_{n=1}^{m}\int_{I^n}\langle u(t), U^n - U^{n-1}\rangle dt = \sum_{n=1}^{m}\langle\int_{I^n} u(t)dt, U^n - U^{n-1}\rangle$$
$$= \langle\int_{I^m} u(t)dt, U^m\rangle - \tau\langle u_0, U^0\rangle - \sum_{n=1}^{m}\langle\int_{I^n}[u(t) - u(t-\tau)]dt, U^{n-1}\rangle,$$

where $u(t) = u_0$ for $-\tau < t \le 0$. The first and middle terms are $O(\tau)$ as a consequence of (2.14) and (5.8). For the last term, we note that

$$\int_{I^n}[u(t) - u(t-\tau)]dt = \int_{I^n}\int_{t-\tau}^{t} u_t(s)dsdt,$$

and the fact that $u_t \in L^\infty(0,T;M(\Omega))$ as stated in (2.20). So we would like to use a duality argument between $M(\Omega)$ and $C^0(\bar{\Omega})$; unfortunately this fails because $u^n \notin C^0(\bar{\Omega})$. The remedy consists of regularizing U^n, say by convolution, in such a way that we get continuous functions U_ε^n having the properties

$$\max_{1\le n\le N}\|U_\varepsilon^n\|_{L^\infty(\Omega)} \le C, \quad U_\varepsilon^n \overset{\varepsilon\downarrow 0}{\to} U^n \text{ strongly in } L^2(\Omega).$$

Thus,

$$\left|\langle\int_{I^n}[u(t) - u(t-\tau)]dt, U^{n-1}\rangle\right| \le \left|\langle\int_{I^n}[u(t) - u(t-\tau)]dt, U_\varepsilon^{n-1}\rangle\right|$$
$$+ \left|\langle\int_{I^n}[u(t) - u(t-\tau)]dt, U^{n-1} - U_\varepsilon^{n-1}\rangle\right|$$
$$\le C\tau + C\|u\|_{L^\infty(0,T;L^2(\Omega))}\|U^{n-1} - U_\varepsilon^{n-1}\|_{L^2(\Omega)}.$$

Finally, taking the limit as $\varepsilon \downarrow 0$ and using (3.6) yields the desired estimate (6.18). ■

We now extend the previous error estimate to a weaker situation.

Theorem 6.3. *Let $u_0 \in L^2(\Omega)$ hold and τ be chosen so that $\tau = C^* h^{4/3}$ for any positive constant C^*. Then we have*

$$\|e_u\|_{L^\infty(0,T;H^{-1}(\Omega))} + \|e_\theta\|_{L^2(0,T;L^2(\Omega))} + \|\int_0^t e_q\|_{L^\infty(0,T;L^2(\Omega))} \le Ch^{1/3}. \tag{6.19}$$

Proof. The present task is that of proving the estimate

$$\left|\sum_{n=1}^{m}\int_{I^n}\langle u(t),U^n-U^{n-1}\rangle dt\right|\leq C\tau^{1/2}, \tag{6.20}$$

which in turn implies that $\nu = 1/4$ in (6.14). In order to derive (6.20), we make use of the a priori estimates (2.14) and (5.8), which yield

$$\begin{aligned}\left|\sum_{n=1}^{m}\int_{I^n}\langle u(t),U^n-U^{n-1}\rangle dt\right| &\leq C\tau^{1/2}\|u\|_{L^2(Q)}\left(\sum_{n=1}^{m}\|U^n-U^{n-1}\|^2_{L^2(\Omega)}\right)^{\frac{1}{2}}\\ &\leq C\tau^{1/2},\end{aligned}$$

and complete the proof. ■

Exercises

6.5 Extend Theorems 6.2 and 6.3 to the case $f \neq 0$.

6.6 Consider the following modification of the fully discrete nonlinear Chernoff formula:

$$U^0 := P_h^0 u_0, \qquad \Theta^0 := P_h^1\theta_0,$$

$$\begin{aligned}\langle\Theta^n,\phi\rangle_h &+ \frac{\tau}{\mu}\langle\nabla\Theta^n,\nabla\phi\rangle\\ &= \langle\beta(U^{n-1})+\frac{\tau}{\mu}f(\beta(U^{n-1})),\phi\rangle + \langle[I-P_h^0]\Theta^{n-1},\phi\rangle_h,\end{aligned}$$

$$U^n := U^{n-1} + \mu[P_h^0\Theta^n - \beta(U^{n-1})].$$

Prove that this scheme is stable and satisfies $\sigma(h,\tau) = h^{1/2}$ provided $\tau = C^*h^2$ where $C^* > 0$ is an arbitrary constant.

6.7 Using the integral method above, prove the following error estimates for the fully discrete extrapolation method:

$$\begin{aligned}\|e_u\|_{L^\infty(0,T;H^{-1}(\Omega))} &+ \|e_\theta\|_{L^2(0,T;L^2(\Omega))} + \left\|\int_0^t e_q\right\|_{L^\infty(0,T;L^2(\Omega))}\\ &\leq C\left(h\varepsilon^{-1/2}+\tau\varepsilon^{-1}+\tau^{1/2}+\|e_u^0\|_{H^{-1}(\Omega)}\right).\end{aligned}$$

Set $\varepsilon = C_1 h$ and $\tau = C_2 h^{3/2}$, where C_1, C_2 are arbitrary positive constants. If Θ^0 is chosen as indicated in Exercise 4.5, then the final rate of convergence becomes $O(h^{1/2})$. Note that τ converges faster to 0 than ε does, which in view of (4.9) shows that the constitutive relation $\chi \in H(\theta)$ is attained in the limit as $h \downarrow 0$.

2.6.3 Approximation of Interfaces

Convergence of discrete solutions may provide no information about convergence of level sets as is easily seen for flat solutions. Approximation of interfaces is in fact related also to nondegeneracy (see (2.22)).

Let v be a function defined in a domain D and let v_h be an approximation to v. Suppose they satisfy an L^p-error estimate of order r and a nondegeneracy property of order s, namely,

$$\|v - v_h\|_{L^p(D)} \leq Ch^r \qquad (1 \leq p \leq \infty), \tag{6.21}$$

$$\text{meas}\{0 \leq v \leq \varepsilon^s\} \leq C\varepsilon. \tag{6.22}$$

These properties are sufficient to guarantee approximation of the zero level set of v in a weak sense. The key idea is based upon defining the discrete interface as a suitable level set [17]. More precisely, set

$$D^+ := \{v > 0\}, \qquad D_h^+ := \{v_h > h^{rsp/(1+sp)}\}. \tag{6.23}$$

Then the symmetric difference $D^+ \Delta D_h^+$ satisfies

$$\text{meas}(D^+ \Delta D_h^+) \leq Ch^{rp/(1+sp)}. \tag{6.24}$$

In view of the error estimates (6.5) and (6.17) for which $p = 2$, $r = 1/2$, and the nondegeneracy property (2.22) for which $s = 1$, we deduce the following rate of convergence in measure for interfaces [17,24]:

$$\text{meas}(Q^+ \Delta Q_h^+) \leq Ch^{1/3}. \tag{6.25}$$

Even though this result can be improved a bit by combining (3.6) and Exercise 6.3, it is still pessimistic in most cases (see Section 2.7.7). We refer to [17] for other applications of this crucial idea.

2.7 Mesh Adaptation

In dealing with fixed domain approximations to parabolic free boundary problems, the interface does not play any explicit role but, nonetheless, is responsible for the global numerical pollution of finite element solutions [24,29]. A remedy to this undesirable situation is to be found in terms of mesh adaptation [20,21,22,23]. Its basic principles are fully discussed in light of the two-phase Stefan problem in 2 dimensions. The local refinement strategy is based on equidistributing pointwise interpolation errors as well as on specifying the so-called refined region, which in turn must contain the discrete free boundaries. This is accomplished by performing various tests on the computed temperature to extract information about discrete derivatives as well as to predict free boundary locations. A typical triangulation is coarse away from the discrete interface, where the discretization parameters satisfy a parabolic relation, whereas it is locally refined in the vicinity of the discrete interface for the relation to become hyperbolic (refined region); a drastic reduction of spatial degrees of freedom is obtained with highly graded meshes. The local truncation error in time is thus properly balanced with the local interpolation error in space. Consecutive meshes are not compatible in that they are completely regenerated rather than being produced by enrichment-coarsening strategies. Mesh changes incorporate an interpolation error which eventually accumulates in time. A suitable interpolation theory for noncompatible meshes is developed to quantify such an error. Its control imposes several constraints on admissible meshes and leads to the mesh selection algorithm. The resulting scheme is stable in various Sobolev norms. A rate of convergence of essentially $O(\tau^{1/2})$ is derived in the natural energy spaces for both temperature and enthalpy.

Let $\{\mathcal{S}^n\}_{n=1}^N$ denote a set of graded partitions of Ω into triangles, that are shape regular [5, p.132] and weakly acute uniformly with respect to $1 \le n \le N$. Given a triangle $S \in \mathcal{S}^n$, h_S stands for its size and satisfies $\lambda\tau \le h_S \le \Lambda\tau^{1/2}$, where $0 < \lambda, \Lambda$ are fixed constants. Let $\mathcal{E}^n$ be the set of interelement boundaries e of $\mathcal{S}^n$, and let h_e denote the length of e. Let $\{x_j^n\}_{j=1}^{J^n}$ indicate the set of internal nodes and $\{\phi_j^n\}_{j=1}^{J^n}$ the set of basis functions. Let $\mathcal{R}^n := \cup\{S \in \mathcal{S}^n : h_S = O(\tau)\}$ indicate the *refined region* and $F^n := \{\Theta^n = 0\}$ the discrete interface. Let $\boldsymbol{V}^n$ be the space of continuous piecewise linear finite elements over $\mathcal{S}^n$. Let $\Pi^n : C^0(\bar{\Omega}) \to \boldsymbol{V}^n$ be the usual Lagrange interpolation operator.

Let $U^0 \in \boldsymbol{V}^0 := \boldsymbol{V}^1$ be a suitable approximation to u_0 and $\Theta^0 := \Pi^1\beta(U^0)$ [21]. Given a mesh $\mathcal{S}^{n-1}$ and a discrete enthalpy $U^{n-1} \in \boldsymbol{V}^{n-1}$ for any $1 \le n \le N$, the discrete scheme then reads as follows: *select $\mathcal{S}^n$ and find $U^n, \Theta^n \in \boldsymbol{V}^n$ such that*

$$\Theta^n = \Pi^n\beta(U^n), \tag{7.1}$$

$$\hat{U}^{n-1} := \Pi^n U^{n-1}, \qquad \hat{\Theta}^{n-1} := \Pi^n[\beta(U^{n-1})], \tag{7.2}$$

$$\tau^{-1}\langle U^n - \hat{U}^{n-1}, \varphi\rangle^n + \langle\nabla\Theta^n, \nabla\varphi\rangle = \langle f(\hat{\Theta}^{n-1}), \varphi\rangle^n, \text{ for all } \varphi \in \boldsymbol{V}^n. \tag{7.3}$$

To fully describe the adaptive FEM we must indicate how to select the new mesh $\mathcal{S}^n$; this is a crucial task. We will define three local parameters over $\mathcal{S}^{n-1}$ which, conveniently postprocessed [22], give rise to a meshsize function $\boldsymbol{h}^n$ with which $\mathcal{S}^n$ is next generated by the automatic mesh generator of [30].

2.7.1 Heuristic Guidelines

We now give a heuristic motivation to the local refinement strategy. We first consider the following 1-dimensional problem discretized only in time

$$U - \tau\beta(U)_{xx} = u_0, \qquad \text{in } (-1,1), \tag{7.4}$$

where $\beta(s) := (s-1)^+ - s^-$ and $u_0(x) := \gamma_2(e^{-x}-1)+1$ if $x > 0$, $u_0(x) := \gamma_1(e^{-x}-1)$ if $x < 0$. Since $u(x,\tau) = u_0(x+V\tau)$, the interface $I(t)$, initially at 0, reaches $-V\tau$ at time τ, where $V := \gamma_2 - \gamma_1 > 0$. Let δ denote the position of the discrete-time interface F. It is not difficult but tedious to see that $(\Theta := \beta(U))$

$$\begin{aligned}
\|\Theta - \theta(\tau)\|_{L^\infty((-1,1))} &= O(\tau) \geq V^2\tau/2, && (7.5)\\
\delta &= -V\tau + O(\tau^{3/2}), && (7.6)\\
\Theta_x(\delta^-) = \Theta_x(\delta^+) &= \gamma_1 + O(\tau^{1/2}) \text{ (smearing effect!)}, && (7.7)\\
\Theta_x(0) &= \gamma_2 + O(\tau^{1/2}), && (7.8)\\
\Theta_{xx}(x) &= \tau^{-1} + O(1), \quad \text{for all } \delta < x < 0. && (7.9)
\end{aligned}$$

On the other hand, suppose $(-1,1)$ is partitioned into equal intervals of size h. Then, in view of the shape of $\theta(\tau) := \beta(u(\tau))$ and (7.9), the pointwise interpolation errors in space satisfy

$$\begin{aligned}
\|\theta(\tau) - \Pi^1\theta(\tau)\|_{L^\infty((-1,1))} &\leq Vh/2 + O(h^2),\\
\|\Theta - \Pi^1\Theta\|_{L^\infty((-1,1))} &\leq h^2\tau^{-1}/2 + O(h^2).
\end{aligned} \tag{7.10}$$

What we learn from this relevant example can be expressed as follows. Since we expect to deal with Lipschitz continuous temperatures, the local meshsize h_S near F and interface velocity V_S should verify $h_S \approx V_S\tau$ to balance the interpolation errors in space (7.10) with the truncation error in time (7.5). In addition, no condition similar to (3.10) is valid for the semidiscrete problem at F even though the free boundary moves correctly. To retrieve the proper jump condition, however, we just have to move a distance δ backwards along the normal to F because, by virtue of (7.7) and (7.8),

$$\Theta_x(0) - \Theta_x(\delta^-) = V + O(\tau^{1/2}), \tag{7.11}$$

or, equivalently, $V = \int_\delta^0 \Theta_{xx}(s)ds + O(\tau^{1/2})$. Consequently, an overrefinement near the interface is extremely dangerous in that we may lose information on the interface velocity without gaining accuracy and, as a result, we might be in troubles with regard to predicting its future position. We thus realize that enforcing these two observations would require a stepwise control of the relation $h_S \approx V_S\tau$, where V_S could be determined by means of (7.11) with δ being replaced by h_S. On the other hand, there is an interval $O(\tau)$-long behind F, namely $(\delta, 0)$, on which second derivatives are $O(\tau^{-1})$.

Away from the interface I, problem (2.1) is strictly parabolic, namely,

$$c(\theta)\theta_t - \Delta\theta = f(\theta), \tag{7.12}$$

which is a mildly nonlinear heat equation; $c(s) := 1/\beta'(\beta^{-1}(s))$ for all $s \in \mathbb{R}\backslash[0,1]$. Hence the discretization parameters should verify the usual parabolic constraint $h_S = O(\tau^{1/2})$.

These two distinct behaviors, rephrased here in terms of local regularity, must be reflected in the local refinement algorithm, for instance, as illustrated in Section 2.7.2.

Let us now explore some heuristic properties of the fully discrete scheme (7.3) which, upon integration by parts, can be written equivalently

$$\sum_{e\cap x_j^n \neq \emptyset} h_e [\![\nabla\Theta^n]\!]_e \cdot \nu_e$$
$$= \frac{2}{3}\text{meas}(\text{supp}\phi_j) \left(f(\hat{\Theta}^{n-1}(x_j^n) - \tau^{-1}(U^n - \hat{U}^{n-1})(x_j^n) \right),$$

where $[\![\cdot]\!]_e$ indicates the jump operator on e. Near the discrete interfaces, where the best we can say is

$$|(U^n - \hat{U}^{n-1})(x_j^n)| \leq C,$$

we get

$$\left| \sum_{e\cap x_j^n \neq \emptyset} h_e [\![\nabla\Theta^n]\!]_e \cdot \nu_e \right| \leq C\tau^{-1}\,\text{meas}(\text{supp }(\phi_j^n)).$$

Hence, except for a very unlikely cancellation in the above summation, we can expect discrete second derivatives D_e to verify

$$D_e := h_e^{-1} |[\![\nabla\Theta^n]\!]_e| \leq C\tau^{-1} h_e^{-2}\text{ meas}(\text{supp }\phi_j^n) \leq C\tau^{-1}. \tag{7.13}$$

This is consistent with (7.9). Moreover, away from the discrete free boundaries, we can expect

$$|(U^n - \hat{U}^{n-1})(x_j^n)| \leq C|(\Theta^n - \hat{\Theta}^{n-1})(x_j^n)| \leq C\tau,$$

because of the strict parabolicity of (7.12). Thus

$$\left| \sum_{e\cap x_j^n \neq \emptyset} h_e [\![\nabla\Theta^n]\!]_e \cdot \nu_e \right| \leq C\,\text{meas}(\text{supp }\phi_j^n).$$

Therefore, arguing as before, we conclude that

$$D_e \leq C h_e^{-2} \text{ meas}(\text{supp } \phi_j^n) \leq C, \tag{7.14}$$

for all e in the parabolic region. Heuristic observations (7.13) and (7.14) regarding D_e as well as the smearing effect (7.7) were confirmed by 2-D numerical experimentation. It also revealed the validity of the following L^1-type a priori estimate

$$\sum_{e \in \mathcal{E}^n} h_e^2 D_e \leq C. \tag{7.15}$$

This property is a discrete analogue of (2.20). It is still in good agreement with numerical evidence. Indeed, actual computations show the occurrence of a strip $O(\tau)$-wide behind the discrete interface F^n where $D_e = O(\tau^{-1})$, which in turn is consistent with (7.9) and (7.13). In this case, since the local meshsize near F^n should be $h_e = O(\tau)$, (7.15) imposes a severe regularity restriction on the interface, namely,

$$\text{length}(F^n) \leq \sum_{e \in \mathcal{E}^n : e \cap F^n \neq \emptyset} h_e \leq C \sum_{e \in \mathcal{E}^n} h_e^2 D_e \leq C. \tag{7.16}$$

Such a condition is quite reasonable for practical purposes but is not known to hold though, in a general setting. We stress that without some kind of additional regularity it is probably hopeless to improve upon the Fixed Mesh Method of §2.4.3 [9,12,15,16,24,33]. In this light, (7.15) is always assumed at the mesh changes, though it constitutes a limitation of the Adaptive Method. It is however partially justified by (7.35) which, being implicitly guaranteed by the scheme, combines with (4.26) to yield

$$\begin{aligned} \sum_{j=1}^{J^n} \left| \sum_{e \cap x_j^n \neq \emptyset} h_e [\![\nabla \Theta^n]\!]_e \cdot \nu_e \right| &\leq 2 \int_\Omega \Pi^n \left(|f(\hat{\Theta}^{n-1})| + \tau^{-1} |U^n - \hat{U}^{n-1}| \right) \\ &\leq C, \end{aligned} \tag{7.17}$$

for all time steps n between consecutive mesh changes. We then see that only a cancellation in the above summation could lead to a bound weaker than (7.15). This seems to be unlikely for locally smooth interfaces as well as for cusps because of their local character. This somehow justifies the fact that (7.15) was never violated in our numerical experiments. Designing an algorithm for which (7.15) is implicitly guaranteed constitutes a challenging open problem though. Note that mushy regions may occur as long as their boundaries are also well-behaved. Regarding first derivatives instead, the following L^2-type a priori estimate is implicitly guaranteed (see (7.34))

$$\sum_{S \in \mathcal{S}^n} h_S^2 |\nabla \Theta^n|_S|^2 \leq C. \tag{7.18}$$

We finally comment upon the effect of interpolation between noncompatible meshes. Let $\zeta : \mathbb{R} \to \mathbb{R}$ be sufficiently smooth and $S \in \mathcal{S}^n$ be a generic element. Proceed then formally as if Θ^{n-1} were smooth to deduce that

$$\|\zeta(\Theta^{n-1}) - \Pi^n[\zeta(\Theta^{n-1})]\|_{L^\infty(S)} \leq Ch_S^2 \left(\|D^2\Theta^{n-1}\|_{L^\infty(S)} + \|D\Theta^{n-1}\|^2_{L^\infty(S)}\right), \quad (7.19)$$

where D and D^2 denote discrete first and second derivatives, respectively. In Section 2.7.4 we give proper justification to (7.19). Since we want this interpolation error to be $O(\tau)$, the new local meshsize should satisfy

$$h_S \leq \tau^{1/2} \min\left(\mu_1\|D^2\Theta^{n-1}\|^{-1/2}_{L^\infty(S)}, \mu_2\|D\Theta^{n-1}\|^{-1}_{L^\infty(S)}\right). \quad (7.20)$$

This in turn allows second derivatives to blow up without violating $h_S \geq \lambda\tau$ as long as $\|D^2\Theta^{n-1}\|_{L^\infty(S)} \leq (\mu_1/\lambda)^2\tau^{-1}$, which is consistent with (7.13). First derivatives may also degenerate without violating $h_S \geq \lambda\tau$ provided $\|D\Theta^{n-1}\|_{L^\infty(S)} \leq (\mu_2/\lambda)\tau^{-1/2}$. Such a degeneracy is expected only whenever cusps develop, this being a local phenomenon. In addition, having control of quadrature errors introduced by (4.11) leads to restrictions on triangle diameters wherever $\|D\Theta^{n-1}\|_{L^\infty(S)}$ exceeds certain tolerance; this is accomplished via (7.20) as well. Since the above interpolation error may accumulate in time, we restrict the number of total mesh changes to $O(\tau^{-1/2})$ so as to preserve an accuracy $O(\tau^{1/2})$.

On the other hand, for all $S \in \mathcal{S}^{n-1}$ intersecting the discrete interface F^{n-1} we have $\{x \in S : 0 \leq U^{n-1}(x) \leq 1\} \neq \emptyset$. For sample problems having a nicely behaved continuous free boundary and verifying a nondegeneracy property, numerical experiments indicate that U^{n-1} may vary from 0 to 1 within one single element. Consequently, even a slight perturbation of triangles S traversed by F^{n-1} would produce an error $\|U^{n-1} - \Pi^n U^{n-1}\|_{L^\infty(S)} = O(1)$ and a subsequent optimal lower bound $\|U^{n-1} - \Pi^n U^{n-1}\|_{L^1(\Omega)} \geq C\tau$, which could be attained provided $\text{length}(F^{n-1}) = O(1)$. This property of F^{n-1} is not enough, however, to ensure the validity of another crucial interpolation estimate, namely $\|\nabla(\Theta^{n-1} - \hat{\Theta}^{n-1})\|_{L^2(\Omega)} \leq C\tau^{1/2}$, unless $\nabla\Theta^{n-1}$ is bounded on F^{n-1}; strong stability would thus break down too. Since such a further constraint on $\nabla\Theta^{n-1}$ rules out the formation of cusps, we should not modify triangles crossed by F^{n-1}. For computational purposes, it is always preferable not to impose this constraint, which is feasible whenever the interface is "smooth"; see [21,22].

2.7.2 Local Mesh Constraints

The key idea hinges upon equidistributing pointwise interpolation errors for Θ^{n-1} in order that the maximum norm error produced by a mesh change be $O(\tau)$. We will now discuss the nature of such an error but first we need some preparatory work. We first remove the superscript n in the various quantities involved and use, instead, the compact notation: $\mathcal{S} := \mathcal{S}^{n-1}$, $\hat{\mathcal{S}} := \mathcal{S}^n$, $\mathcal{E} := \mathcal{E}^{n-1}$, $\mathcal{R} := \mathcal{R}^{n-1}$, $\Pi := \Pi^{n-1}$, $\widehat{\Pi} := \Pi^n$, $U := U^{n-1}$, $\hat{U} := \widehat{\Pi} U$, $\Theta := \Theta^{n-1}$, $F := F^{n-1}$. We then set $\mathcal{S}_F := \{S \in \mathcal{S} : S \cap F \neq \emptyset\}$ and $\mathcal{E}_F := \{e \in \mathcal{E} : e \subset \partial S, S \in \mathcal{S}_F\}$. Set $d_S := |\nabla\Theta|_S|$ for all $S \in \mathcal{S}$ and $D_e := |[\![\nabla\Theta]\!]_e|/h_e$ for all $e \in \mathcal{E}$. Note that these quantities are easy to evaluate in practice. We then introduce the following local parameters

$$\hat{h}_e := \mu_1 \frac{\tau^{1/2}}{D_e^{1/2}}, \text{ for all } e \in \mathcal{E} \backslash \mathcal{E}_F; \quad \hat{h}_S := \mu_2 \frac{\tau^{1/2}}{d_S}, \text{ for all } S \in \mathcal{S} \backslash \mathcal{S}_F; \tag{7.21}$$

$\lambda, \mu_1, \mu_2 > 0$ are arbitrary constants which result from computational considerations [22]. Set $\mathcal{S}_B := \{S \in \mathcal{S} \backslash \mathcal{S}_F : \min_{e \in \mathcal{E} \backslash \mathcal{E}_F : e \subset \partial S}(\hat{h}_S, \hat{h}_e) < \lambda\tau\}$, $\mathcal{E}_B := \{e \in \mathcal{E} \backslash \mathcal{E}_F : e \subset \partial S, S \in \mathcal{S}_B\}$, $\mathcal{B} := \cup_{S \in \mathcal{S}_B} S$, $\mathcal{F} := \cup_{S \in \mathcal{S}_F} S$, $\mathcal{S}_0 := \mathcal{S} \backslash (\mathcal{S}_F \cup \mathcal{S}_B)$, $\mathcal{E}_0 := \mathcal{E} \backslash (\mathcal{E}_F \cup \mathcal{E}_B)$, $\Omega_0 := \Omega \backslash (\mathcal{F} \cup \mathcal{B})$.

In the absence of mush, the remaining local parameter represents the expected local meshsize within the refined region, and so near the discrete interface [21]. For any $S \in \mathcal{S}_F$, we first determine an approximation to the (average) interface velocity V_S [20,21,22,23]. We then define $\mathcal{C}_S$ to be the cone of axis ν (vector normal to $F_S := S \cap F$), vertex at S, opening $\pi/2$ and height $\mu_3 V_S \tau^{1/2}$ as being the region most likely to contain the evolution of F_S for at least $O(\tau^{-1/2})$ time steps. Heuristic arguments, based on balancing the truncation error in time with the interpolation error in space [20,21,22], suggest the following correction

$$\hat{h}_{F_S} := \tau \min\{\max(\lambda, V_S), M\}, \tag{7.22}$$

for the meshsize within $\mathcal{C}_S$. The constants μ_3 and M are arbitrary and result from computational considerations [22]; μ_3 may depend on n. If mushy regions occur we then introduce an extra parameter $\hat{h}_S$ to equidistribute L^1-interpolation errors for enthalpy within the mush, and determine $\hat{h}_{F_S}$ for the nondegenerate mushy boundary only [21,22].

2.7.3 Mesh Selection Algorithm

Various tests are performed on the computed solution Θ to verify admissibility of the current mesh $\mathcal{S}$. The first test consists of checking whether the discrete interface F is within the refined region $\mathcal{R}$ or not. In the event F has escaped from $\mathcal{R}$, we say that the test has failed. Both $\mathcal{S}$ and $\{\Theta, U\}$ are then discarded and the previous computed solution recovered. To prevent the program from performing a useless time step, the boundary of $\mathcal{R}$, called the RED ZONE, alerts that an imminent remeshing must be done; see Figures 2.1 and 2.2. Since the local width of $\mathcal{R}$ is proportional to both $\tau^{1/2}$

and the local interface velocity, we thus expect $\mathcal{S}$ to be admissible for at least $O(\tau^{-1/2})$ time steps, as desired. Rejection of $\mathcal{S}$ is mostly dictated by the failure of this test.

The second test ascertains that interpolation errors are still equidistributed correctly, namely,

$$h_e \leq \mu_1^* \, \hat{h}_e, \text{ for all } e \in \mathcal{E}_0, \quad h_S \leq \mu_2^* \, \hat{h}_S, \text{ for all } S \in \mathcal{S}_0, \tag{7.23}$$

where $\mu_1^*, \mu_2^* > 1$ are suitable constants. This rules out the possibility of an excessive refinement induced by large discrete derivatives.

To guarantee a correct balance between local truncation error and meshsize near F as well as to avoid computational difficulties in specifying the refined region, the third test

$$\mu_3^- \hat{h}_{F_S} \leq h_S \leq \mu_3^+ \hat{h}_{F_S}, \qquad \text{for all } S \in \mathcal{S}_F, \tag{7.24}$$

is enforced; $\mu_3^- < 1 < \mu_3^+$ are suitable constants.

If any one of the above tests fails, then the current mesh $\mathcal{S}$ is rejected as well as the solution $\{\Theta, U\}$, which is overwritten with the previously computed solution. A new *graded* mesh $\hat{\mathcal{S}}$ with the following properties is then generated. To preserve the constraint $h_{\hat{S}} \geq \lambda\tau$, we must keep $\mathcal{S}_B$ fixed because discrete derivatives are badly behaved. In addition, we must not modify $\mathcal{S}_F$, because this would cause an inadmissible interpolation error for U (see Section 2.7.1). Hence,

$$S \in \hat{\mathcal{S}}, \quad \text{for all } S \in \mathcal{S}_B \cup \mathcal{S}_F, \tag{7.25}$$

is the first restriction on $\hat{\mathcal{S}}$. The second one reads

$$\lambda\tau \leq h_{\hat{S}} \leq \min_{\substack{S' \in \mathcal{S}_F : \mathcal{C}_{S'} \cap \hat{S} \neq \emptyset \\ S \in \mathcal{S}_0, e \in \mathcal{E}_0 : S \cap \hat{S}, e \cap \hat{S} \neq \emptyset}} \left(\Lambda \tau^{1/2}, \hat{h}_{F_{S'}}, \hat{h}_e, \hat{h}_S \right), \tag{7.26}$$

for all $\hat{S} \in \hat{\mathcal{S}}$. This accounts for both the equidistribution of pointwise interpolation errors (7.20) and the definition of refined region. In the event of mushy regions, $\hat{h}_{F_{S'}}$ is computed for the nondegenerate mushy boundary only (see [21,22]).

We resort to the automatic mesh generator of [30], whose computational complexity $O(J \log J)$ is quasi-optimal, to minimize the CPU time spent in mesh generation; J is the number of nodes. The three parameters defined in (7.21) and (7.22) are suitably postprocessed for $\hat{\mathcal{S}}$ to satisfy (7.26) (see [21,22]).

2.7.4 Interpolation Theory

Our present concern is interpolation estimates for *noncompatible meshes.* We stick with the notation of Section 2.7.2. Given $W \subset \overline{\Omega}$, we set $\mathcal{S}_W := \{S \in \mathcal{S}_0 : S \cap W \neq \phi\}$, $\mathcal{E}_W := \{e \in \mathcal{E}_0 : e \cap W \neq \phi\}$, and $\tilde{W} := \cup_{S \in \mathcal{S}_W} S$. We point out that there exists a constant $0 < a < 1$, depending only on the regularity of $\mathcal{S}$, such that $\mathrm{dist}(x, S) \geq a h_S$ for all $x \in \bar{\Omega}_0 \backslash \tilde{S}$. We can then distinguish between two opposite (and mutually exclusive) situations in terms of the relative size of triangles in both meshes $\mathcal{S}$ and $\hat{\mathcal{S}}$. Set $\hat{\mathcal{S}}_0 := \hat{\mathcal{S}} \backslash (\mathcal{S}_F \cup \mathcal{S}_B)$ and define the *derefinement* case to be

$$\text{given } \hat{S} \in \hat{\mathcal{S}}_0, \quad a h_S < h_{\hat{S}} \quad \text{for all } S \in \mathcal{S}_{\hat{S}}. \tag{7.27}$$

By contrast, the *refinement* situation reads as follows

$$\text{given } \hat{S} \in \hat{\mathcal{S}}_0, \quad \text{there exists } S \in \mathcal{S}_{\hat{S}} \text{ such that } a h_S \geq h_{\hat{S}}, \tag{7.28}$$

which in turn yields $\hat{S} \subset \tilde{S}$. Let $\hat{\mathcal{S}}_1$ (resp. $\hat{\mathcal{S}}_2$) indicate the set of all $\hat{S}$'s satisfying (7.27) (resp. (7.28)). We then have the following two crucial L^p-interpolation estimates without assumptions on the relative size or location of new and old triangles provided β is piecewise linear. See [21] for the general case.

Lemma 7.1. *Let* $\hat{S} \in \hat{\mathcal{S}}_1$. *Then*

$$\|U-\hat{U}\|_{L^p(\hat{S})} + h_{\hat{S}} \|\nabla(U-\hat{U})\|_{L^p(\hat{S})} \leq C h_{\hat{S}}^2 \begin{cases} (\sum_{e \in \mathcal{E}_{\hat{S}}} h_e^2 D_e^p)^{1/p}, & 1 \leq p < \infty \\ \max_{e \in \mathcal{E}_{\hat{S}}} D_e, & p = \infty. \end{cases} \tag{7.29}$$

The need of control on interpolation errors even for the refinement situation (7.28) arises from the noncompatibility of $\mathcal{S}$ and $\hat{\mathcal{S}}$.

Lemma 7.2. *Let* $\hat{S} \in \hat{\mathcal{S}}_2$. *Then*

$$\|U-\hat{U}\|_{L^p(\hat{S})} + h_{\hat{S}} \|\nabla(U-\hat{U})\|_{L^p(\hat{S})} \leq \begin{cases} C h_{\hat{S}}^{1+2/p} (\sum_{e \in \mathcal{E}_{\hat{S}}} h_e^p D_e^p)^{1/p}, & 1 \leq p < \infty \\ C h_{\hat{S}} \max_{e \in \mathcal{E}_{\hat{S}}} (h_e D_e), & p = \infty. \end{cases} \tag{7.30}$$

It is worth noting that both (7.28) and (7.29) may be rewritten in terms of Θ, rather than U, because they simply coincide or differ by 1 for all $\hat{S} \in \hat{\mathcal{S}}_0 (= \hat{\mathcal{S}} \backslash (\mathcal{S}_F \cup \mathcal{S}_B))$. This assertion holds even for nonlinear β [21]. The following estimates are a consequence of the mesh selection algorithm and the two preceding results [20,21,23].

Theorem 7.3. *The following interpolation error estimates are valid*

$$\|U - \hat{U}\|_{L^\infty(\Omega)} \leq C\tau, \qquad \|\nabla(U - \hat{U})\|_{L^2(\Omega)} \leq C\tau^{1/2}. \tag{7.31}$$

These estimates are sharp.

Proof. In view of (7.25), we only have to consider $\hat{S} \in \hat{\mathcal{S}}_0(= \hat{\mathcal{S}}_1 \cup \hat{\mathcal{S}}_2)$. It turns out that $h_{\hat{S}} \leq \hat{h}_e = C_1(\tau/D_e)^{1/2}$ for all $e \in \mathcal{E}_{\hat{S}}$, as results from (7.26). We first deal with the *derefinement* case which corresponds to the set $\hat{\mathcal{S}}_1$. The estimate (7.29) thus yields

$$\|U - \hat{U}\|_{L^\infty(\hat{S})} \leq C \max_{e \in \mathcal{E}_{\hat{S}}} (h_{\hat{S}}^2 D_e) \leq C\tau,$$

and, using (7.15),

$$\begin{aligned}
\sum_{\hat{S} \in \hat{\mathcal{S}}_1} \|\nabla(U - \hat{U})\|_{L^2(\hat{S})}^2 &\leq C \sum_{\hat{S} \in \hat{\mathcal{S}}_1} \sum_{e \in \mathcal{E}_{\hat{S}}} (h_{\hat{S}}^2 D_e) h_e^2 D_e \\
&\leq C\tau \sum_{e \in \mathcal{E}_0} h_e^2 D_e \\
&\leq C\tau.
\end{aligned}$$

Note that we could replace doubled summations by single ones because, as a consequence of (5.1), card $\{\hat{S} \in \hat{\mathcal{S}}_1 : e \cap \hat{S} \neq \emptyset\} = O(1)$ for all $e \in \mathcal{E}_0$. On the other hand, for all $\hat{S} \in \hat{\mathcal{S}}_2$ (*refinement* case) we use the leftmost constraint in (7.23), namely $h_e \leq \mu_1^* \hat{h}_e$ for all $e \in \mathcal{E}_{\hat{S}}$. Indeed, by virtue of (7.30), we arrive at

$$\|U - \hat{U}\|_{L^\infty(\hat{S})} \leq C \max_{e \in \mathcal{E}_{\hat{S}}} (h_{\hat{S}} h_e D_e) \leq C\tau,$$

and, using (7.15),

$$\begin{aligned}
\sum_{\hat{S} \in \hat{\mathcal{S}}_2} \|\nabla(U - \hat{U})\|_{L^2(\hat{S})}^2 &\leq C \sum_{\hat{S} \in \hat{\mathcal{S}}_2} \sum_{e \in \mathcal{E}_{\hat{S}}} (h_{\hat{S}} h_e D_e)^2 \\
&\leq C\tau \sum_{\hat{S} \in \hat{\mathcal{S}}_2} \sum_{e \in \mathcal{E}_{\hat{S}}} (h_{\hat{S}} h_e D_e) \\
&\leq C\tau \sum_{e \in \mathcal{E}_0} h_e D_e \sum_{\hat{S} \in \hat{\mathcal{S}}_2 : e \cap \hat{S} \neq \phi} h_{\hat{S}} \\
&\leq C\tau \sum_{e \in \mathcal{E}_0} h_e^2 D_e \\
&\leq C\tau.
\end{aligned}$$

In addition, (7.31) is obviously sharp according to the discrete regularity being dealt with, which concludes the proof. ∎

Since meshes $\mathcal{S}$ and $\hat{\mathcal{S}}$ are noncompatible, we cannot expect a pointwise error estimate for ∇U to hold. In fact, consider the refinement case (i.e. $\hat{S} \in \hat{\mathcal{S}}_2$) for which $\|\nabla(U-\hat{U})\|_{L^\infty(\hat{S})} \leq C \max_{e\in\mathcal{E}_{\hat{S}}} h_e D_e$. This yields $\|\nabla(U-\hat{U})\|_{L^\infty(\hat{S})} = O(1)$ provided $D_e = O(h_e^{-1})$, as is expected to happen near the interface F.

2.7.5 Stability

As a consequence of results in Section 2.4, we have the following natural *a priori estimates* for the discrete problem, which resemble those in Section 2.5.1 (see [21]):

$$\max_{1\leq n\leq N} \|U^n\|_{L^\infty(\Omega)} + \max_{1\leq n\leq N} \|\Theta^n\|_{L^\infty(\Omega)} \leq C. \tag{7.32}$$

$$\sum_{n=1}^{N} \|U^n - \hat{U}^{n-1}\|^2_{L^2(\Omega)} + \sum_{n=1}^{N} \tau \|\nabla\Theta^n\|^2_{L^2(\Omega)} \leq C. \tag{7.33}$$

$$\sum_{n=1}^{N} \tau^{-1} \|U^n - \hat{U}^{n-1}\|^2_{L^2(\Omega\setminus\mathcal{R}^n)} + \max_{1\leq n\leq N} \|\nabla\Theta^n\|_{L^2(\Omega)} \leq C, \tag{7.34}$$

$$\max_{1\leq n\leq N} \|U^n - \hat{U}^{n-1}\|_{L^1(\Omega)} \leq C\tau. \tag{7.35}$$

The leftmost term in (7.33) is a discrete-time $H^{1/2}$ estimate and accounts for the global behavior of U^n which, in the limit, is discontinuous. The corresponding term in (7.34) states, instead, a discrete-time H^1 regularity away from the interface F^n, where U^n and Θ^n are equivalent variables. The a priori estimate (7.35) is a discrete analogue of $u_t \in L^\infty(0,T;M(\Omega))$, but is still a bit weaker than the structural assumption $\sum_{e\in\mathcal{E}^n} h_e^2 D_e \leq C$. We stress the need for stability estimates in nonenergy spaces such as (7.32) and (7.35) as they play a relevant role in the error analysis.

Exercises

7.1 Prove the a priori estimates (7.32) and (7.33). Hint: Proceed as in Lemmas 5.2 and 5.6.

7.2 Prove the following property

$$\|\nabla\hat{\Theta}\|^2_{L^2(\Omega)} \leq (1 + C\tau)\|\nabla\Theta\|^2_{L^2(\Omega)} + C\tau \sum_{e\in\mathcal{E}_0} h_e^2 D_e,$$

and use it to derive (7.34). Hint: argue by induction and use the assumption (7.15).

7.3 Prove (7.35). Hint: take the difference between two consecutive equations (7.3) that correspond to the same mesh and select $\varphi = \Pi^n[\psi_\varepsilon(\partial\Theta^n)] \in \boldsymbol{V}^n$ as a test function in the resulting expression. Here, ψ_ε is a suitable regularization of sgn. Use that (7.15) is valid for the mesh changes. This technique is a discrete analogue to that in Exercise 2.8.

2.7.6 Error Estimates

We denote by G^n the discrete Green's operator associated with the mesh $\mathcal{S}^n$ which satisfies, in addition to (4.20),

$$\|(G - G^n)\psi\|_{L^\infty(\Omega)} \leq Ch_n^2|\log h_n|^7\|\psi\|_{L^\infty(\Omega)}, \qquad \text{for all } \psi \in L^\infty(\Omega). \tag{7.36}$$

Here, $h_n := \max_{S\in\mathcal{S}^n} h_S$ and $\rho_n := \min_{S\in\mathcal{S}^n} h_S$ are such that $\lambda\tau \leq \rho_n < h_n \leq \Lambda\tau^{1/2}$. The estimate (7.36) requires the extra assumption $h_n \leq C\rho_n^\gamma$ for $0 < \gamma \leq 1$; in the present case $\gamma = 1/2$. The proof can be found in [21].

We now introduce the *error equation* that is similar to (6.4):

$$\begin{aligned}\langle e_u^n - e_u^{n-1}, \phi\rangle + \langle \int_{I^n} \nabla e_\theta(t)dt, \nabla\phi\rangle + \tau\langle\nabla\Theta^n, \nabla(\phi-\varphi)\rangle \\ = \langle U^n - \hat{U}^{n-1}, \varphi\rangle^n - \langle U^n - \hat{U}^{n-1}, \varphi\rangle \\ + \langle \hat{U}^{n-1} - U^n, \phi - \varphi\rangle + \langle U^{n-1} - \hat{U}^{n-1}, \phi\rangle\end{aligned} \tag{7.37}$$

for all $\phi \in H_0^1(\Omega)$, $\varphi \in \boldsymbol{V}^n$.

Theorem 7.4. *Let (7.15) hold and the total number of mesh changes be limited to $O(\tau^{-1/2})$. Then*

$$\|e_u\|_{L^\infty(0,T;H^{-1}(\Omega))} + \|e_\theta\|_{L^2(0,T;L^2(\Omega))} \leq C\tau^{1/2}|\log\tau|^{7/2}. \tag{7.38}$$

For the practical range of time steps τ, the logarithm above plays no significant role. The restriction on the number of mesh changes accounts for the accumulation of interpolation error $U^{n-1} - \hat{U}^{n-1}$ which is $O(\tau)$. The major novelty in Theorem 7.4 is to be interpreted in terms of properly distributed spatial degrees of freedom (DOF): for well-behaved discrete interfaces, only DOF$=O(\tau^{-3/2})$ are necessary for an $O(\tau^{1/2})$ global accuracy, as opposed to quasi-uniform meshes that require DOF$=O(\tau^{-2})$ because $h_S = O(\tau)$ [9,24,26]; see Theorem 6.1.

Proof. We proceed as in Theorem 6.1. Take first $\phi := Ge_u^n \in H_0^1(\Omega)$ and $\varphi := G^n e_u^n \in \boldsymbol{V}^n$ in (7.37), and next sum over n from 1 to $m \leq N$ to arrive at

$$\begin{aligned}
\sum_{n=1}^{m}\langle e_u^n - e_u^{n-1}, Ge_u^n\rangle + \sum_{n=1}^{m}\langle \int_{I^n} e_\theta(t)dt, e_u^n\rangle &=: I + II \\
= \sum_{n=1}^{m}\Big[\langle U^n - \hat{U}^{n-1}, G^n e_u^n\rangle^n - \langle U^n - \hat{U}^{n-1}, G^n e_u^n\rangle\Big] & \\
+ \sum_{n=1}^{m}\langle \hat{U}^{n-1} - U^n, (G - G^n)e_u^n\rangle & \\
+ \sum_{n=1}^{m}\langle U^{n-1} - \hat{U}^{n-1}, Ge_u^n\rangle & \\
=: III + IV + V. &
\end{aligned} \tag{7.39}$$

The analysis of I goes exactly as in Theorem 6.1. The same happens with II with the only exception of II_3. We intend to apply (6.7) with $\alpha = \beta$, $p = 1$ and $\varphi = U^n$. We first consider $S \in \mathcal{S}_F^n$, for which $h_S = O(\tau)$ and so

$$\|\beta(U^n) - \Pi^n\beta(U^n)\|_{L^1(S)} \le C\tau\|\nabla\Theta^n\|_{L^1(S)} \qquad \text{for all } S \in \mathcal{S}_F^n.$$

If $S \in \mathcal{S}^n\backslash\mathcal{S}_F^n$, then we can exploit the regularity of β to deduce

$$\|\beta(U^n) - \Pi^n[\beta(U^n)]\|_{L^1(S)} \le Ch_S^2\|D^2U^n\|_{L^1(S)} \le C\tau\|\nabla\Theta^n\|_{L^2(S)}^2.$$

Hence, in view of Exercise 2.7 and (7.32), we have

$$|II_3| \le \sum_{n=1}^{m}\|\beta(U^n) - \Pi^n\beta(U^n)\|_{L^1(\Omega)}\int_{I^n}\|e_u(t)\|_{L^\infty(\Omega)}dt \le C\tau.$$

In analyzing term III, we decompose the underlying integrals over all triangles $S \in \mathcal{S}_R^n := \{S \in \mathcal{S}^n : S \subset \mathcal{R}^n\}$, where $h_S = O(\tau)$, and $S \in \mathcal{S}^n\backslash\mathcal{S}_R^n$, where $h_S \le \Lambda\tau^{1/2}$. The inequality (4.13) then yields

$$\begin{aligned}
|III| &\le \sum_{n=1}^{m}\sum_{S\in\mathcal{S}^n}\int_S\Big|(U^n - \hat{U}^{n-1})G^n e_u^n - \Pi^n|_S[(U^n - \hat{U}^{n-1})G^n e_u^n]\Big|\,dx \\
&\le C\sum_{n=1}^{m}\Big(\tau\|U^n - \hat{U}^{n-1}\|_{L^2(\Omega)} \\
&\qquad + \tau^{1/2}\|U^n - \hat{U}^{n-1}\|_{L^2(\Omega\backslash\mathcal{R}^n)}\Big)\|\nabla G^n e_u^n\|_{L^2(\Omega)}.
\end{aligned}$$

Hence, (7.33) and (7.34) lead to

$$\begin{aligned}|III| &\leq \eta A_m + C\eta^{-1}\tau\left(\sum_{n=1}^{m}\|U^n - \hat{U}^{n-1}\|^2_{L^2(\Omega)}\right.\\ &\qquad\left.+\sum_{n=1}^{m}\tau^{-1}\|U^n - \hat{U}^{n-1}\|^2_{L^2(\Omega\setminus\mathcal{R}^n)}\right)\\ &\leq \eta A_m + C\eta^{-1}\tau,\end{aligned}$$

where $\eta > 0$ is to be selected. The contribution due to IV can be handled by means of (7.32), (7.35), (7.37) and Exercise 2.7 as follows:

$$|IV| \leq \sum_{n=1}^{m}\|U^n - \hat{U}^{n-1}\|_{L^1(\Omega)}\|(G-G^n)e_u^n\|_{L^\infty(\Omega)} \leq C\tau|\log\tau|^7.$$

In dealing with term V we resort to the crucial estimate (7.31) and the assumption on the total number of degrees of freedom. In fact, we have

$$\begin{aligned}|V| &\leq \sum_{n=1}^{m}\|U^{n-1} - \hat{U}^{n-1}\|_{L^2(\Omega)}\|Ge_u^n\|_{L^2(\Omega)}\\ &\leq A_m^{1/2}\sum_{n=1}^{m}\|U^{n-1} - \hat{U}^{n-1}\|_{L^2(\Omega)} \leq C\tau^{1/2}A_m^{1/2} \leq \eta A_m + C\eta^{-1}\tau.\end{aligned}$$

A proper choice of η finally allows A_m to be absorbed into the left-hand side of (7.39). Therefore, for all $1 \leq m \leq N$, we have obtained the estimate

$$\max_{1\leq n\leq m}\|e_u^n\|^2_{H^{-1}(\Omega)} + \|e_\theta\|^2_{L^2(0,t^m;L^2(\Omega))} \leq C\tau|\log\tau|^7.$$

The proof then concludes as that of Theorem 6.1 but without an estimate for the flux. ■

The information between consecutive meshes, necessary to advance the algorithm in time, is transfered via a simple interpolation process. Apart from its nontrivial implementation, which requires *quadtree data structures* to reach a nearly optimal computational complexity, it leads to a nonuniform distribution of the spatial degrees of freedom (DOF), $J = O(\tau^{-3/2})$. In fact, most of the DOF are concentrated near discrete interfaces for the so-called *refined region* to be a strip $O(\tau^{1/2})$-wide. This in turn comes from restricting the number of mesh changes to $O(\tau^{-1/2})$ as a consequence of accuracy considerations. The local meshsize within the refined region is $O(\tau)$ and discrete interfaces cannot escape from it without causing mesh modification. As numerical evidence indicates, most of the CPU time is spent in solving the ensuing strongly nonlinear algebraic systems. Such a task is extremely sensitive to J because its total computational complexity is roughly $O(J^{4/3}\tau^{-1})$ [22]. On the other hand, each mesh generation takes about an optimal number of operations, namely $O(J \log J)$ [22,30]. It would thus be preferable to have a narrower refined region, say $O(\tau)$-wide, at the expense of changing the mesh more frequently, say every $O(1)$ time steps. Hence, the resulting meshes would have $J = O(\tau^{-1})$, which appears to be optimal as corresponds to the expected value for the (linear) heat equation.

In the following Exercises we discuss briefly how to eliminate the above constraint on the number of mesh changes, thus allowing $O(\tau^{-1})$ remeshings without compromising accuracy. The idea consists of replacing the elementary interpolation process between consecutive meshes by a more delicate elementwise L^2-projection. This amounts, for every triangle of the new mesh, to computing exactly the integral of a piecewise quadratic function on the old mesh. The use of quadtree data structures is again essential to reach a nearly optimal computational complexity $O(J \log J)$, but the process is intrinsically more expensive than a simple interpolation.

Exercises

7.4 Let $\boldsymbol{W}^n$ be the space of discontinuous piecewise linear functions over $\mathcal{S}^n$. Let $P^n : L^2(\Omega) \to \boldsymbol{W}^n$ be the (local) L^2-projection operator defined, for all $v \in L^2(\Omega)$, by

$$\langle P^n v, \varphi \rangle := \langle v, \varphi \rangle = \sum_{S \in \mathcal{S}^n} \langle v, \chi \rangle_S, \text{ for all } \varphi \in \boldsymbol{W}^n. \tag{7.40}$$

Since no continuity requirements are imposed on $\boldsymbol{W}^n$, $P^n v$ can be computed elementwise by inverting a 3×3 linear system. Set $U^0 := P^1 u_0$. Given a mesh $\mathcal{S}^n$ and a discrete enthalpy $U^{n-1} \in \boldsymbol{V}^n$ for any $1 \le n \le N$, define $\check{U}^{n-1} \in \boldsymbol{V}^n$ by

$$\check{U}^{n-1}(x_j^n) := \sum_{S \cap x_j^n \neq \emptyset} \frac{\text{meas}\,(S)}{\text{meas}(\text{supp}\phi_j^n)} [P^n U^{n-1}]|_S(x_j^n), \tag{7.41}$$

for all $1 \le j \le J^n$. The discrete problem then is similar to (7.1)–(7.3) with $\hat{U}^{n-1}$ and $\hat{\Theta}^{n-1}$ replaced by $\check{U}^{n-1}$ and

$$\check{\Theta}^{n-1} := \Pi^n[\beta(\check{U}^{n-1})],$$

respectively. Prove the equality

$$\langle \check{U}^{n-1}, \varphi \rangle^n = \langle P^n U^{n-1}, \chi \rangle^n, \qquad \text{for all } \varphi \in \boldsymbol{V}^n, \tag{7.42}$$

and the superconvergence error estimate [23]

$$\|U^{n-1} - P^n U^{n-1}\|_{H^{-1}(\Omega)} \le C\tau^{3/2}. \tag{7.43}$$

N	J	S_R	S_P	C	E^2_θ	E^2_u	E^∞_θ	E^∞_I	CPU
40	319	496	148	4(1)	16.1	35.4	11.3	5.37	58
60	593	957	233	4(1)	10.5	29.1	6.65	4.44	132
80	875	1430	328	4	7.54	24.2	5.23	3.45	224
120	1413	2320	510	5	5.34	22.7	3.93	2.60	546
160	2089	3504	679	6(1)	4.07	19.6	3.14	2.08	1111
240	3620	6201	1042	8(1)	2.82	17.7	2.41	1.65	2998

Table 2.1. Adaptive Method.

N	J	S	E^2_θ	E^2_u	E^∞_θ	E^∞_I	CPU
50	448	942	15.9	39.7	20.8	13.8	37
75	1017	2104	10.5	30.4	15.3	9.40	113
100	1812	3718	7.81	26.8	12.4	6.88	292
150	4107	8356	5.57	22.3	7.78	5.65	919
200	7361	14912	4.52	18.6	6.31	3.53	2264

Table 2.2. Fixed Mesh Method.

7.5 Prove Theorem 7.4 without restrictions on the number of mesh changes [23]. Hint: exploit the superconvergence rate (7.43).

Since mesh changes can now be performed every $O(1)$ time steps, the refined region may just be a strip $O(\tau)$-wide in order to avoid failure of the first test in Section 2.7.3. This leads to a quasi-uniform distribution of spatial DOF, as desired.

2.7.7 Numerical Experiments

To illustrate the superior performance of the Adaptive Method (AM) with respect to the Fixed Mesh Method (FMM) in (4.24)–(4.26), we conclude with the severe test below. It is a classical two-phase Stefan problem with an interface that moves up and down. The exact temperature is given by the following expression:

$$\theta(x,y,t) := \begin{cases} 0.75(r^2-1), & r<1, \\ (1.5-\alpha'(t)\sin\varphi)(r-1), & r\geq 1, \end{cases} \tag{7.44}$$

where $r := (x^2+(y-\alpha(t))^2)^{1/2}$, $\alpha(t) := 0.5+\sin(1.25t)$, $\sin\varphi := (y-\alpha(t))/r$, $\Omega := (0,5)\times(0,5)$ and $T := \pi/1.25$. Dirichlet boundary conditions are imposed at $y=0$, $y=5$ and $x=5$, and a homogeneous Neumann condition is prescribed at $x=0$. Since the exact interface $I(t)$ is a circle with center $(0,\alpha(t))$ and radius 1, the velocity $V(x,y,t)$ normal to $I(t)$

at (x, y) exhibits a significant variation along the front, which makes this example an extremely difficult test for our numerical method. Moreover, since $V(x, y, t)$ vanishes at both $(x, y, T/2)$ and $(1, \alpha(t), t)$, and is thus very small nearby, this test constitutes a fair measure of robustness under degenerate situations.

Several numerical experiments were performed with both AM and FMM. For the latter, the constant of proportionality between τ and the (uniform) meshsize was chosen so as to minimize $\|e_\theta\|_{L^2(Q)}$. The various constants introduced in sections 2.7.2 and 2.7.3 are as follows: $\lambda = 1.5$, $\Lambda = 5$, $M = 5$, $\mu_1 = 2.22$, $\mu_2 = 5.59$, average value of $\mu_3 \approx 1.2$, $\mu_1^\star = \mu_2^\star = 3.5$, $\mu_3^- = 0.5$, $\mu_3^+ = 2$.

Results are reported in Tables 2.1 and 2.2, where we have employed the following notation: $\boldsymbol{N}$:= number of (uniform) time steps, $\boldsymbol{J}$:= average number of nodes, $\boldsymbol{S_R}$:= average number of triangles within the refined region, $\boldsymbol{S_P}$:= average number of triangles in the rest of Ω, $\boldsymbol{S}$:= total number of triangles (uniform mesh), $\boldsymbol{C}$:= number of mesh changes (and number of computed solutions rejected, if any), $\boldsymbol{CPU}$:= CPU time in seconds (on a VAX 8530, VMS 4.6) and $\boldsymbol{E}_\theta^2 := \|e_\theta\|_{L^2(Q)}$, $\boldsymbol{E}_u^2 := \|e_u\|_{L^2(Q)}$, $\boldsymbol{E}_\theta^\infty := \|e_\theta\|_{L^\infty(Q)}$, $\boldsymbol{E}_I^\infty := \max_{1 \le n \le \boldsymbol{N}} \mathrm{dist}(I(n\tau), F^n)$, where the errors are scaled by 10^2. Moreover, $\|\theta\|_{L^2(Q)} \approx 33.16$, $\|\theta\|_{L^\infty(Q)} \approx 13.38$, $\|u\|_{L^2(Q)} \approx 39.81$, with an error of one unit in the fourth digit.

In light of these (partial) results, we can certainly claim a superior performance of the Adaptive Method in that it requires less computational labor, say CPU, for a desired global accuracy. The L^2-error for temperature $\boldsymbol{E}_\theta^2$ behaves linearly in τ, thus much better than predicted. We also have a (linear) pointwise error $\boldsymbol{E}_\theta^\infty$ that is far from being theoretically explainable. The same happens with $\boldsymbol{E}_u^2$ and $\boldsymbol{E}_I^\infty$. The improvement gained in L^∞ is clearly more pronounced than that in L^2. The free boundary is located within one single element, thus providing the best possible approximation. This is clearly depicted in Figures 2.2 and 2.3. Since approximability and nondegeneracy are tied together [17], it is worth noting that the nondegeneracy property is not uniform in the present case. The black triangles in Figures 2.1 and 2.2 constitute the boundary of the Refined Region, the so-called RED ZONE.

$\mathbf{N} = 80$, mesh II, $\boldsymbol{S}_R = 1557$, $\boldsymbol{S}_P = 299$, $n = 13 : 38$.

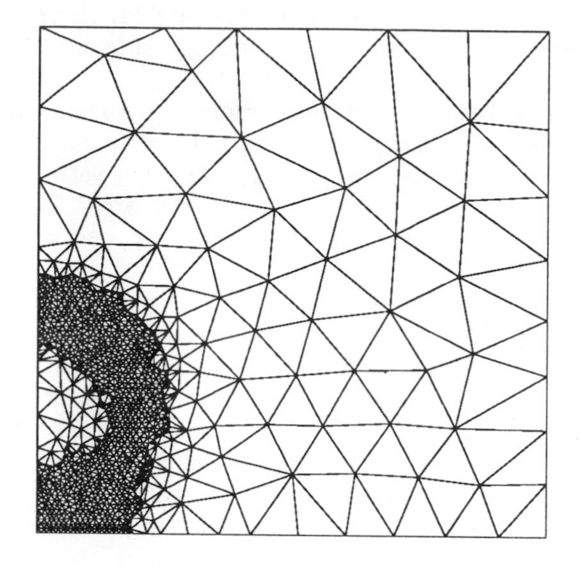

$\mathbf{N} = 80$, mesh III, $\boldsymbol{S}_R = 1058$, $\boldsymbol{S}_P = 339$, $n = 39 : 54$.

Fig. 2.1. AM, $\mathbf{N} = 80$: Two consecutive meshes.

N = 80, interfaces (zoom)

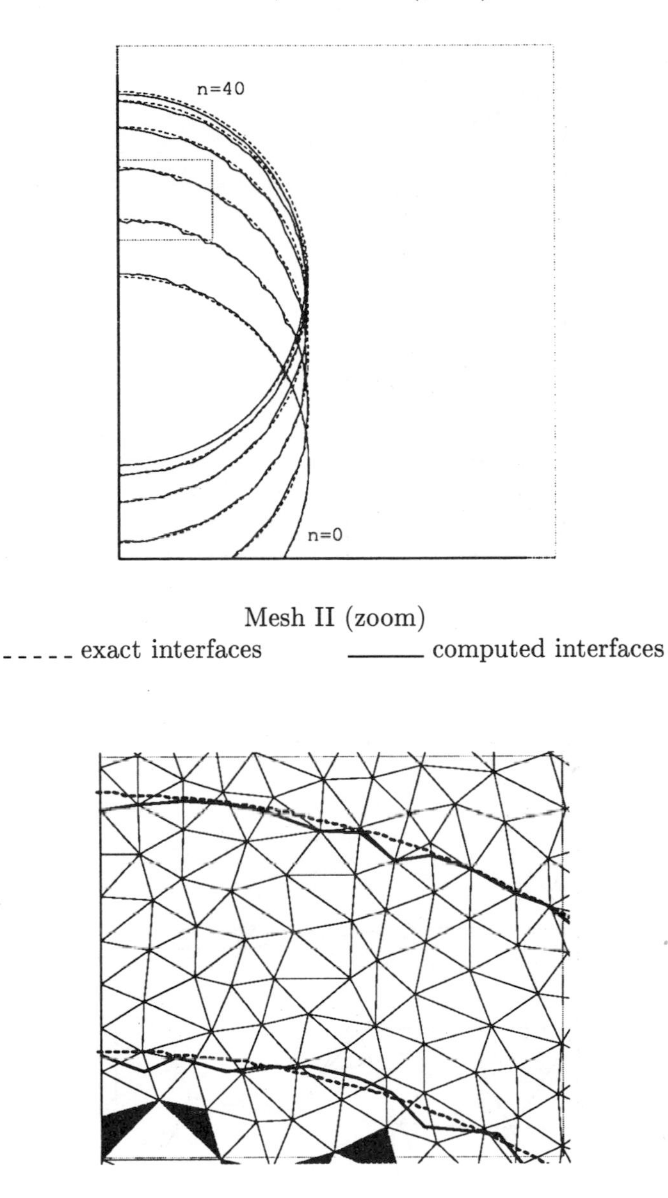

Fig. 2.2. AM, **N** = 80: Interfaces at $n = 8k$ $(0 \leq k \leq 5)$; zoom of Mesh II.

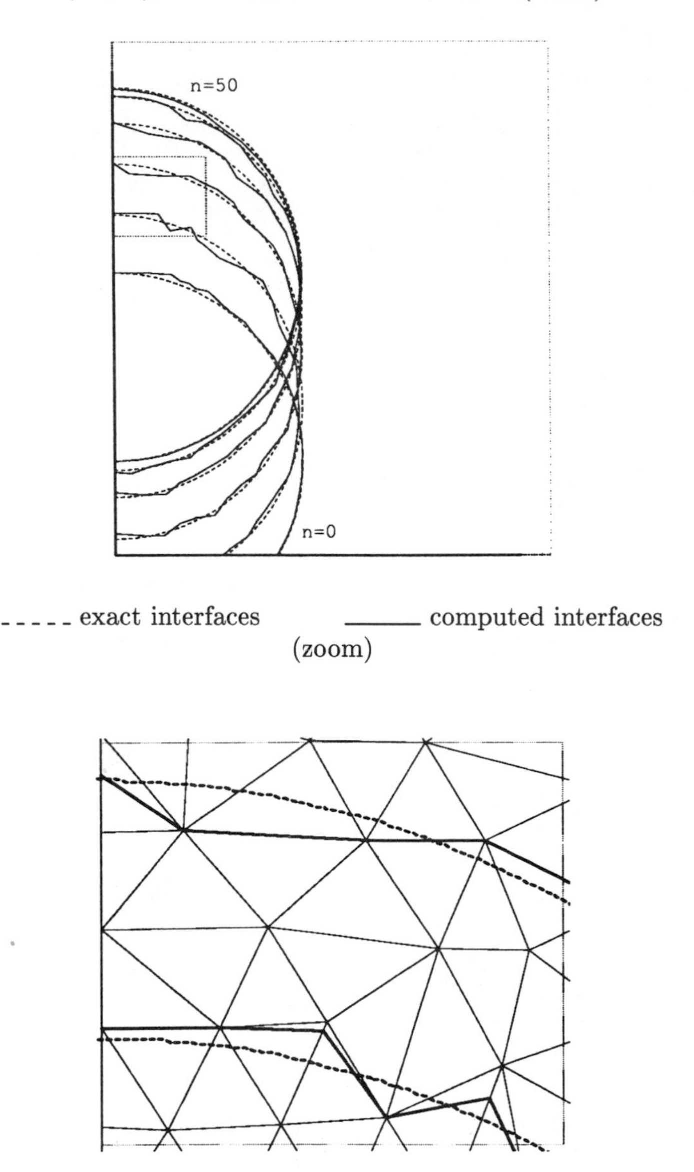

Fig. 2.3. FMM, $\mathbf{N} = 100$: Interfaces at $n = 10k$ $(0 \leq k \leq 5)$; zoom of the mesh.

References

[1] O. Axelsson and V.A. Barker, *Finite element solution of boundary value problems: theory and applications*, Academic Press, Orlando(Florida), 1984

[2] I. Babuška, O.C. Zienkiewicz, J. Gago and E.R. de A. Oliveira, *Accuracy estimates and adaptive refinements in finite element computations*, John Wiley and Sons, Chichester, 1986.

[3] A.E. Berger, H. Brezis and J.C.W. Rogers, *A numerical method for solving the problem* $u_t - \Delta f(u) = 0$, RAIRO Anal. Numer. **13**, 1979, pp. 297–312.

[4] H. Brezis and A. Pazy, *Convergence and approximation of semigroups of nonlinear operators in Banach spaces*, J. Funct. Anal., **9**, 1972, pp. 63–74.

[5] P.G. Ciarlet, *The finite element method for elliptic problems*, North Holland, Amsterdam, 1978.

[6] P.G. Ciarlet and P.A. Raviart, *Maximum principle and uniform convergence for the finite element method*, Comput. Methods Appl. Mech. Engrg., **2**, 1973, pp. 17–31.

[7] J.F. Ciavaldini, *Analyse numérique d'un problème de Stefan à deux phases par une méthode d'éléments finis*, SIAM J. Numer. Anal. **12**, 1975, pp. 464–487.

[8] J. Douglas Jr. , T. Dupont and R. Ewing, *Incomplete iteration for time-stepping a Galerkin method for a quasilinear parabolic problem*, SIAM J. Numer. Anal., **16**, 1979, pp. 503–522.

[9] C.M. Elliott, *Error analysis of the enthalpy method for the Stefan problem*, IMA J. Numer. Anal., **7**, 1987, pp. 61–71.

[10] K. Eriksson and C. Johnson, *Adaptive finite element methods for parabolic problems I: a linear problem*, Preprint University of Göteborg, **31**, 1988.

[11] A. Friedman, *The Stefan problem in several space variables*, Trans. Amer. Math. Soc., **133**, 1968, pp. 51–87.

[12] J.W. Jerome and M. Rose, *Error estimates for the multidimensional two-phase Stefan problem*, Math. Comp. **39**, 1982, pp. 377–414.

[13] E. Magenes, *Problemi di Stefan bifase in più variabili spaziali*, Le Matematiche **36**, 1981, pp. 65–108.

[14] E. Magenes, R.H. Nochetto and C. Verdi, *Energy error estimates for a linear scheme to approximate nonlinear parabolic problems* , RAIRO Model. Math. Anal. Numer., 1987, pp. 655–678.

[15] R.H. Nochetto, *Error estimates for two-phase Stefan problems in several space variables, I: linear boundary conditions*, Calcolo **22**, 1985, pp. 457–499.

[16] R.H. Nochetto, *Error estimates for multidimensional singular parabolic problems*, Japan J. Appl. Math. **4**, 1987, pp. 111–138.

[17] R.H. Nochetto, *A note on the approximation of free boundaries by finite element methods*, RAIRO Model. Math. Anal. Numer. **20**, 1986, pp. 355–368.

[18] R.H. Nochetto, *A class of non-degenerate two-phase Stefan problems in several space variables*, Comm. Partial Differential Equations **12**, 1987, pp. 21–45.

[19] R. Nochetto, *A stable extrapolation method for multidimensional degenerate parabolic problems*, Math. Comp. **53**, 1989, pp. 455–470.

[20] R.H. Nochetto, M. Paolini and C. Verdi, *Selfadaptive mesh modification for parabolic FBPs: theory and computation,* in *Free boundary value problems with special respect to their numerical treatment and optimal control*, Internat. Series Num. Math., 95, Birkhäuser, 1990, 181-206.

[21] R.H. Nochetto, M. Paolini and C. Verdi, *An adaptive finite element method for two-phase Stefan problems in two space dimensions. Part I: stability and error estimates*, Math. Comp., to appear.

[22] R.H. Nochetto, M. Paolini and C. Verdi, *An adaptive finite element method for two-phase Stefan problems in two space dimensions. Part II: Implementation and numerical experiments*, SIAM J. Sci. Statist. Comput., to appear.

[23] R.H. Nochetto, M. Paolini and C. Verdi, *Quasi-optimal mesh adaptation for two-phase Stefan problems in 2D,* in *Computational Mathematics and Applications*, Istituto di Analisi Numerica del C.N.R., 730, Pavia, Italy, 1989, pp. 313–326.

[24] R.H. Nochetto and C. Verdi, *Approximation of degenerate parabolic problems using numerical integration*, SIAM J. Numer. Anal., **25**, 1988, pp. 784–814.

[25] R.H. Nochetto and C. Verdi, *An efficient linear scheme to approximate parabolic free boundary problems: error estimates and implementation*, Math. Comp. **51**, 1988, pp. 27–53.

[26] R.H. Nochetto and C. Verdi, *The combined use of a nonlinear Chernoff formula with a regularization procedure for two-phase Stefan problems*, Numer. Funct. Anal. Optim. **9**, 1987-88, pp. 1177–1192.

[27] J.M. Ortega and W.C. Rheinboldt, *Iterative solution of nonlinear equations in several variables* , Academic Press, New York, 1970.

[28] P.A. Raviart, *The use of numerical integration in finite element methods for solving parabolic equations,* in *Topics in Numerical Analysis,* Academic Press, London, 1973, pp. 233–264.

[29] M. Paolini, G. Sacchi and C. Verdi, *Finite element approximations of singular parabolic problems*, Int. J. Numer. Meth. Eng. **26**, 1988, pp. 1989–2007.

[30] M. Paolini and C. Verdi, *An automatic triangular mesh generator for planar domains*, Rivista di Informatica, to appear.

[31] V. Thomee, *Galerkin finite element methods for parabolic problems,* Lectures Notes in Mathematics 1054, Springer-Verlag, Berlin, 1984.

[32] C. Verdi, *On the numerical approach to a two-phase Stefan problem with nonlinear flux*, Calcolo **22**, 1985, pp. 351–381.

[33] C. Verdi, *Optimal error estimates for an approximation of degenerate parabolic problems*, Numer. Funct. Anal. Optimiz. **9**, 1987, pp. 657–670.

[34] C. Verdi and A. Visintin, *Error estimates for a semiexplicit numerical scheme for Stefan-type problems*, Numer. Math. **52**, 1988, pp. 165–185.

[35] A. Visintin, *Stefan problem with phase relaxation*, IMA J. Appl. Math. **34**, 1985, pp. 225–245.

[36] M. Zlamal, *A finite element solution of the nonlinear heat equation*, RAIRO Anal. Numer. **14**, 1980, pp. 203–216.

Professor R. H. Nochetto
Department of Mathematics
Mathematics Building 084
University of Maryland
College Park
Maryland, 20742
U.S.A.

3

An Introduction to Spectral Methods for Partial Differential Equations

A. Quarteroni

3.1 Introduction

This lecture is an elementary introduction to spectral methods for the numerical approximation of boundary value problems. The basic properties of Fourier and Chebyshev orthogonal expansions are revisited. The discrete Fourier and Chebyshev transforms are introduced, and it is shown how to use them in the framework of spectral collocation approximations. Several examples of applications are shown, including elliptic boundary value problems, advection-diffusion equations and Navier-Stokes equations. Both periodic and non-periodic boundary conditions are considered.

The chapter is concluded with a discussion on domain decomposition methods which are used to enlarge the area of applicability of spectral methods (including non-Cartesian geometries) as well as to match effectively the features of multiprocessor systems.

A comprehensive presentation of spectral methods is given in reference [8]. A modern overview of applications of spectral methods can be found in [10].

3.2 The Fourier Approximation

In this section we review some elementary results on Fourier series, then we introduce the Fourier interpolation and the discrete Fourier transform. These tools are of primary interest in the framework of both Galerkin and collocation approximation of periodic boundary value problems by Fourier methods.

3.2.1 Convergence of the Fourier series

Let N be a positive integer, and set:

$$S_N = \text{span}\,\{\varphi_n, -N \leq n \leq N-1\}, \quad \varphi_n(\theta) = (1/\sqrt{2\pi})\, e^{in\theta}. \tag{2.1}$$

The operator $P_N : L^2(0, 2\pi) \to S_N$, where

$$P_N u = \sum_{n=-N}^{N-1} u_n\ \varphi_n(\theta) \text{ with } u_n = \int_0^{2\pi} u\overline{\varphi}_n, \tag{2.2}$$

is the orthogonal projection operator upon S_N. Hence,

$$||u - P_N u|| \to 0 \text{ as } N \to \infty, \text{ for all } u \in L^2(0, 2\pi). \tag{2.3}$$

Remark 1.1 Instead of (2.1) one could define as well

$$S_N = \text{span}\{\varphi_n(\theta), -N \leq n \leq N\},$$

and define the relative projection; the forthcoming approximation results would not change in their essence. The choice (2.1) is motivated by technical reasons only (the even number of unknowns will make it possible to use the fast Fourier transform).

Result (2.3) means that

$$\int_0^{2\pi} |u(\theta) - P_N u(\theta)|^2\, d\theta \to 0 \quad \text{as } N \to \infty,$$

or equivalently

$$\sum_{|n| \geq N} |u_n|^2 \to 0 \quad \text{as } N \to \infty, \tag{2.4}$$

where, for convenience of notation :

$$\sum_{\underset{\sim}{|n| \leq N}} a_n \equiv \sum_{n=-N}^{N-1} a_n \tag{2.5}$$

$$\sum_{\underset{\sim}{|n| \geq N}} a_n \equiv \sum_{n=-\infty}^{\infty} a_n - \sum_{\underset{\sim}{|n| \leq N}} a_n = \sum_{n=-\infty}^{-N-1} a_n + \sum_{n=N}^{\infty} a_n. \tag{2.6}$$

(a) We recall that the L^2-convergence of a sequence does not imply its convergence *almost everywhere*. However, this happens for the truncated sequence $\{P_N u\}$. That is, for all $u \in L^2(0, 2\pi)$,

$$P_N u(\theta) \to u(\theta) \quad \text{as } N \to \infty \text{ for almost every } \theta \in (0, 2\pi). \tag{2.7}$$

(b) A full characterization of the class of functions u for which the sequence $\{P_N u\}$ *converges uniformly* to u, i.e.,

$$\max_{0\le\theta\le 2\pi} |u(\theta) - P_N u(\theta)| \to 0 \quad \text{as } N \to \infty, \tag{2.8}$$

is not known. However, some partial results are available. First of all, we say that u is a periodic function if $u(0^+)$ and $u(2\pi^-)$ exist and are equal. Since $P_N u$ is periodic, a necessary condition for uniform convergence is that u is a periodic function.

(b1) If u is continous and periodic, its Fourier series does not necessarily converge uniformly.

(b2) If u has bounded variation in $[0, 2\pi]$, then

$$P_N u(\theta) \to (u(\theta^+) + u(\theta^-))/2 \quad \text{for all } \theta \in [0, 2\pi], \tag{2.9}$$

where $u(0^-) = u(2\pi^-)$.

(b3) If u is continous, periodic and has bounded variation in $[0, 2\pi]$, then $P_N u(\theta) \to u(\theta)$ as $N \to \infty$, for all $\theta \in [0, 2\pi]$.

3.2.2 Behaviour of the Fourier Coefficients

We have the following result (see, for example [8]). Let u be m-times differentiable almost everywhere in $(0, 2\pi)$, $m \ge 0$. Under the following hypotheses:

(i) $u^{(m)}$ has bounded variation on $[0, 2\pi]$

(ii) $u^{(j)}$ is periodic, $0 \le j \le m-1$, if $m \ge 1$,

we have that

$$u_k = O(1/k^{m+1}), \quad k = \pm 1, \pm 2, \ldots. \tag{2.10}$$

As a matter of fact, by the definition of u_k and integration by parts we have

$$\begin{aligned} u_k &= \frac{1}{\sqrt{2\pi}} \int_0^{2\pi} u(\theta) e^{-ik\theta} d\theta \\ &= \frac{1}{\sqrt{2\pi}} \left\{ \frac{1}{ik} \int_0^{2\pi} e^{-ik\theta} du(\theta) - \frac{1}{ik} [u(2\pi) - u(0+)] \right\}. \end{aligned} \tag{2.11}$$

If u has total variation V, then the Riemann-Stieltjes integral on the right-hand side can be bounded by $|V|$. Hence

$$u_k = O(1/k). \tag{2.12}$$

If u' exists and has bounded variation, then the integral on the right-hand side of (2.11) is equal to $\sqrt{2\pi}(u')_k$. Hence, it decays like $1/k$. Thus $u_k = O(1/k^2)$ if and only if $u(2\pi^-) = u(0^+)$, i.e., u is periodic. This proves (2.10) for $m = 1$. Iterating this argument the result can be proven for all m.

An important consequence is that whenever u is infinitely differentiable and periodic with all its derivatives on $(0, 2\pi)$, then u_k decays faster than any negative power of k.

From the previous proof it follows that if u is m times differentiable, and its derivatives $u^{(j)}$, $0 \leq j \leq m-1$ are continuous and periodic, then

$$u_k = \frac{1}{(ik)^m} \frac{1}{\sqrt{2\pi}} \int_0^{2\pi} u^{(m)}(\theta) e^{-ik\theta}\, d\theta.$$

Hence

$$|u_k| \leq \frac{1}{|k|^m} M(u^{(m)}), \tag{2.13}$$

where

$$M(v) = \frac{1}{\sqrt{2\pi}} \int_0^{2\pi} |v|\, d\theta$$

is the mean value of $|v|$ (provided it exists). Noting that

$$\left| \sum_{k=-\infty}^{\infty} u_k\, \varphi_k(\theta) \right| \leq \sum_{k=-\infty}^{\infty} |u_k|,$$

it follows from (2.13), that for $m = 2$ the series *converges absolutely and uniformly* to a function U. Since the uniform convergence implies L^2-convergence (for a bounded domain), it follows that $U \equiv u$.

3.2.3 Error Estimates for Projection and Interpolation Operators

Let $\Omega = (0, 2\pi)^d$, $d \geq 1$. We denote by $C_{2\pi}^{\infty}$ the space of those functions which are infinitely differentiable, and whose derivatives of any order are periodic with period 2π. The Sobolev space $H_{2\pi}^k(\Omega)$ is the completion of $C_{2\pi}^{\infty}$ with respect to the norm

$$\|u\|_{H_{2\pi}^k(\Omega)} := \left\{ \sum_{|j| \leq k} |D^j u|^2 \right\}^{1/2}, \quad k \in \mathbb{N}. \tag{2.14}$$

Here $|v|$ denotes the $L^2(\Omega)$-norm of the function v. The L^2-projection operator P_N satisfies the following error inequality :

$$\|u - P_N u\|_{H_{2\pi}^k(\Omega)} \leq C(k,m) N^{k-m} \|u\|_{H_{2\pi}^m(\Omega)}, \quad 0 \leq k \leq m. \tag{2.15}$$

If the space dimension d is greater than one, then the space S_N is defined as follows:

$$S_N = \text{span}\{\varphi_n : \; -N \leq n_j \leq N, \; j = 1, ..., d\}, \tag{2.16}$$

where $\varphi_n(\theta) = (1/\sqrt{2\pi})^d e^{in\theta}$, $n = (n_1, \ldots, n_d)$ is a multi-integer, $\theta = (\theta_1, \ldots, \theta_d)$ and $n\theta = n_1\theta_1 + \ldots + n_d\theta_d$. In the domain Ω we introduce the nodes

$$\theta_j = j\pi/N, \; j \in J = \{(j_1, \ldots, j_d) : j_k = 0, \ldots, 2N-1 \text{ for } k = 1, \ldots, d.\} \tag{2.17}$$

The interpolation operator at these points is $I_N : C^0(\Omega) \to S_N$, where

$$I_N u(\theta_{\mathbf{j}}) = u(\theta_{\mathbf{j}}), \quad \text{for all } j \in J. \tag{2.18}$$

If we define the discrete inner product

$$(u, v)_N = (\pi/N)^d \sum_{j \in J} u(\theta_j)\overline{v(\theta_j)}, \tag{2.19}$$

then we have

$$I_N u(\theta) = \sum_n (u, \varphi_n)_N \varphi_n(\theta)) = \sum_n u_n^* \varphi_n(\theta). \tag{2.20}$$

The coefficients u_n^* are called the *discrete* Fourier coefficients of u. If $u \in S_N$, then $u_n^* = u_n$. The interpolation error can be estimated as follows

$$\|I_N u - u\|_{H^k_{2\pi}(\Omega)} \leq C(k,m) N^{k-m} \|u\|_{H^m_{2\pi}(\Omega)}, \quad 0 \leq k \leq m \text{ and } m > d/2. \tag{2.21}$$

The linear transformation yielding the discrete Fourier coefficients in terms of the values $\{u(\theta_j)\}$ at the nodes (2.17) is the *discrete Fourier transform*. It can be carried out efficiently by the Fast Fourier Transform (F.F.T.) algorithm, which is reviewed in the forthcoming subsection.

3.2.4 The Fast Fourier Transform Algorithm (F.F.T.)

Let f be a function of the space S_N defined in (2.1). Then

$$f(\theta) = \sum_{k=0}^{M-1} a_k \varphi_{k-N}(\theta), \quad M = 2N, \tag{2.22}$$

where

$$a_k = (f, \varphi_{k-N})_N = (\pi/N) \sum_{j=0}^{M-1} f(\theta_j)\varphi_{k-N}(\theta_j),$$

and

$$\theta_j = \pi j/N = 2\pi j/M, \quad 0 \le j \le M-1. \tag{2.23}$$

The linear transformation which gives the Fourier coefficients $\{a_k\}$ of f in terms of the values $\{f(\theta_j)\}$ is the *discrete Fourier transform*. Since

$$\overline{\varphi_{k-N}(\theta_j)} = (-1)^j\overline{\varphi_k(\theta_j)} = [(-1)^j/\sqrt{2\pi}][e^{-2\pi i/M}]^{kj},$$

one has

$$a_k = \sum_{j=0}^{M-1} \sqrt{\pi/2}(1/N)(-1)^j f(\theta_j)[e^{-2\pi i/M}]^{kj},$$

or, equivalently, setting $c_j = \sqrt{\pi/2}(1/N)(-1)^j\ f(\theta_j)$, $w = e^{-2\pi i/M}$,

$$a_k = \sum_{j=0}^{M-1} c_j\ w^{kj}, \quad \text{with } w^M = 1. \tag{2.24}$$

Note that $a_k = \sum_{j=0}^{M-1} c_j x^j$, $x = w^k$, i.e., a_k is the value attained by the polynomial $\sum_{j=0}^{M-1} c_j x^j$ at the point $x = w^k$. Thus Horner's scheme for polynomials can be used to compute each a_k by M operations (here an operation means one complex multiplication plus one complex addition). Hence, the computation of the coefficients a_k, $0 \le k \le M-1$ would require M^2 operations.

We will assume hereafter that M is a power of 2, i.e.,

$$M = 2^m \text{ for some } m \in \mathbb{N}. \tag{2.25}$$

We want to show that by the F.F.T. the computation of the a_k will require

$$2M\log_2 M = 2mM \text{ operations} \tag{2.26}$$

only. We set

$$j = \begin{cases} 2j_1 & \text{if } j \text{ is even} \\ 2j_1+1 & \text{if } j \text{ is odd} \end{cases}, \quad 0 \le j_1 \le (1/2)M - 1. \tag{2.27}$$

Using (2.24) we get

$$\begin{aligned} a_k &= \sum_{j_1=0}^{N-1} c_{2_{j_1}} w^{k2j_1} + \sum_{j_1=0}^{N-1} c_{2_{j_1+1}} w^{k2j_1+1} \\ &= \sum_{j_1=0}^{N-1} c_{2_{j_1}} (w^2)^{kj_1} + \sum_{j_1=0}^{N-1} c_{2_{j_1}} (w^2)^{kj_1} w_k. \end{aligned} \tag{2.28}$$

Let us divide k by N, and denote by α and k_1 respectively the quotient and the remainder:

$$k = \alpha N + k_1.$$

Since $w^N = 1$ we have

$$(w^2)^{kj_1} = (w^2)^{\alpha N j_1}(w^2)^{k_1 j_1} = (w^2)^{k_1 j_1}.$$

Then from (2.28) we deduce

$$a_k = \sum_{j_1=0}^{N-1} c_{2_{j_1}} (w^2)^{k_1 j_1} + w^k \sum_{j_1=0}^{N-1} c_{2_{j_1}} (w^2)^{k_1 j_1},$$

that is,

$$a_k = \varphi(k_1) + w^k \psi(k_1), \quad k = 0, 1, \dots M-1, \tag{2.29}$$

where

$$\varphi(k_1) = \sum_{j_1=0}^{N-1} d_{j_1} \omega^{k_1 j_1}, \quad \psi(k_1) = \sum_{j_1=0}^{N-1} e_{j_1} \omega^{k_1 j_1}, \quad \omega^N = 1, \tag{2.30}$$

and

$$d_{j_1} = c_{2_{j_1}}, \quad e_{j_1} = c_{2_{j_1+1}}, \quad \omega = w^2.$$

The structure of the two discrete transforms (2.30) is exactly the same as that of the discrete transform (2.24). The difference is that the order of (2.30) is $N = M/2$, i.e., half the order of (2.24). Thus, we have replaced a discrete transform of order $M = 2^m$ with two discrete transforms of order $N = M/2 = 2^{m-1}$.

Applying the same splitting procedure to the two discrete transforms (2.30), we will get four discrete transforms, each of them with 2^{m-2} terms. These can be further divided into eight discrete transforms with 2^{m-3} terms, and so on. Once $\varphi(k_1)$ and $\psi(k_1)$ have been computed, the number of operations required to compute the a_k, $0 \leq k \leq M-1$, using (2.29), is at most $2.2^m (= 2M)$.

Denoting by $\mathcal{N}(m)$ the total number of operations required to compute the coefficients when $M = 2^m$ and reasoning as above, we have

$$\begin{aligned} N(m) &\leq 2\mathcal{N}(m-1) + 2.2^m \\ &\leq 2.2^m + 2[2\mathcal{N}(m-2) + 2.2^{m-1}] \\ &= 2.2^m + 2.2^m + 2^2\mathcal{N}(m-2) \\ &= \dots = 2m.2^m, \end{aligned}$$

that is,

$$\mathcal{N}(m) \leq 2M \log_2 M.$$

The *inverse* of the discrete Fourier transform can be computed efficiently by the same algorithm. Indeed, we have from (2.22),

$$f(\theta_j) = \sum_{k=0}^{M-1} a_k \varphi_{k-N}(\theta_j), \quad 0 \leq j \leq M-1.$$

However,

$$\varphi_{k-N}(\theta_j) = (-1)^j \varphi_k(\theta_j) = (-1)^j/\sqrt{2\pi}[e^{2\pi i/M}]^{kj}.$$

Then, setting $f_j = (-1)^j f(\theta_j)$, $b_k = (1/\sqrt{2\pi})a_k$, and $z = e^{2\pi i/M}$, we have:

$$f_j = \sum_{k=0}^{M-1} b_k \, z^{kj}, \quad \text{and } z^M = 1. \tag{2.31}$$

This formula has exactly the same structure as (2.24). Hence the same arguments used there can be applied here.

3.2.5 The Fourier Pseudo-Spectral Derivative

If u is a continous function, its Fourier pseudo-spectral derivative is by definition the derivative of the trigonometric polynomial $I_N u \in S_N$ interpolating u at the nodes θ_j (see (2.17)). In the one-dimensional case, one has

$$(I_N u)'(\theta) = \sum_{k=-N}^{N-1} iku_k^* e^{ik\theta}, \tag{2.32}$$

where $\{u_k^*\}$ are the discrete Fourier coefficients of u. The coefficients of the expansion (2.32) can be computed from the values of u at the nodes $\{\theta_j\}$ using the discrete Fourier transform. For smooth functions, the pseudo-spectral derivative provides a very good approximation of the exact derivative. Actually, using the results of Section 3.1.3 one can easily prove that

$$\max_{0\leq\theta\leq 2\pi} |[u' - (I_N u)'](\theta)| \leq C(m) N^{3/2-m} \|u\|_{H^m_{2\pi}(0,2\pi)}, \; m > 3/2. \tag{2.33}$$

The drawback of using (2.32) is that $(I_N u)'$ is complex even if u is real. An alternative definition is often used in applications. Since

$$e^{iN\theta_j} = e^{-iN\theta_j}, \quad 0 \leq j \leq 2N-1,$$

if we define

$$J_N u(\theta) = \sum_{k=-N}^{N} v_k^* e^{ik\theta}, \tag{2.34}$$

with $v_k^* = u_k^*$ if $|k| \leq N-1$, and $v_N^* = v_{-N}^* = (1/2)u_{-N}^*$, one has

$$J_N u(\theta_j) = I_N u(\theta_j) = u(\theta_j), \quad j = 0, ..., 2N-1.$$

Therefore, J_N is an alternative interpolation operator at the nodes $\{\theta_j\}$, and it still enjoys the approximation property (2.33). Moreover, $(J_N u)'$ is real if u is real. The set of the values $\{(J_N u)'(\theta_j)\}$ are achievable from the values of $\{u(\theta_k)\}$ through the following matrix

$$D_{jk} = \begin{cases} \frac{1}{2}(-1)^{j+k} \cot \frac{\theta_j - \theta_k}{2}, & j \neq k \\ 0 & j = k, \end{cases} \tag{2.35}$$

for $j, k = 0, ..., 2N-1$. The matrix D is called the matrix of the Fourier pseudo-spectral derivative.

3.3 Sturm-Liouville Problems

Spectral methods for non-periodic boundary value problems are based upon finite expansions of orthogonal polynomials which are eigenfunctions of Sturm-Liouville problems. We briefly review here the elementary theory of both regular and singular Sturm-Liouville problems, then we characterize the family of Jacobi polynomials.

For non-periodic problems, we need to use a basis $\{\varphi_k\}$ other than the complex trigonometric basis $1/\sqrt{2\pi}e^{ikx}$ though maintaining the properties of spectral accuracy and, possibily, the fast transform method to 'travel' between physical and frequency spaces. The eigenfunctions of a singular Sturm-Liouville boundary value problem achieve these goals.

Let w be a real, Lebesgue measurable function such that $w(x) > 0$ for $-1 < x < 1$, and

$$\int w(x)\, dx < \infty \quad \text{(properly or improperly)}. \tag{3.1}$$

All integrals in these section are extended to the interval $(-1, 1)$. The function w is called weight function. We will also assume that

$$\int \{w(x)\}^{-1}\, dx < \infty. \tag{3.2}$$

Then we consider the problem

$$Ly(x) = \lambda w(x) y(x), \quad -1 < x < 1, \tag{3.3}$$

where

$$Ly(x) = -(p(x)\; y'(x))'. \tag{3.4}$$

Here, λ is a real parameter and $p \in C^1(-1, 1)$. Then (3.3) is a Sturm-Liouville eigenvalue problem.

3.3.1 Regular Sturm-Liouville Boundary Value Problems

If $p(x) \geq p_0 > 0$, then (3.3) must be supplemented by two conditions, which we assume to have the form:

$$\begin{cases} \alpha_1 u(-1) + \beta_1 u'(-1) = 0, & \alpha_1^2 + \beta_1^2 \neq 0, \quad \alpha_1\beta_1 \leq 0 \\ \alpha_2 u(1) + \beta_2 u'(1) = 0, & \alpha_2^2 + \beta_2^2 \neq 0, \quad \alpha_2\beta_2 \geq 0 \end{cases}. \tag{3.5}$$

The problem (3.3) with (3.5) is a regular Sturm-Liouville boundary value problem. We will abbreviate Sturm-Liouville boundary value problem to SL bvp, henceforward. In the case of a regular SL bvp the eigenvalues of (3.3) form an infinite, unbounded sequence of positive numbers

$$0 < \lambda_0 < \lambda_1 < \ldots \lambda_k < \lambda_{k+1} < \ldots \quad \lambda_n \to \infty \text{ as } n \to \infty. \tag{3.6}$$

The corresponding eigenfunctions φ_k, which satisfy the equations

$$L\varphi_k(x) = \lambda_k w(x)\varphi_k(x), \tag{3.7}$$

are mutually orthogonal with respect to $w(x)$, namely,

$$(\phi_k, \phi_m)_w = \int \phi_k \phi_m \, w dx = 0 \quad \text{if } k \neq m.$$

We introduce now the space

$$L_w^2(-1,1) = \{u : (-1,1) \to \mathbb{R} : \|u\|_w^2 = (u,u)_w < \infty\}. \tag{3.8}$$

Assuming $\|\varphi_k\|_w = 1$ (after normalisation), it is known that $\{\varphi_k, k = 0, 1, 2, \ldots\}$ provides a basis for $L_w^2(-1,1)$. The expansion of $u \in L_w^2(-1,1)$ with respect to this basis is

$$u(x) = \sum_{k=0}^{\infty} u_k \varphi_k(x), \quad u_k = (u, \varphi_k)_w. \tag{3.9}$$

Set

$$u_{(0)} = u, \quad u_{(j)} = \frac{1}{w} L u_{(j-1)}, \; j \geq 1, \tag{3.10}$$

and assume that $u_{(m)} \in L_w^2(-1,1)$ for some m. Then it can be established that

$$|u_k| \leq C \|u_{(m)}\|_w \frac{1}{k^{2m}}, \tag{3.11}$$

provided each $u_{(j)}$ satisfies the boundary conditions (3.5). This is an unnatural requirement, which will prevent our getting spectral accuracy for those functions u whose sequence (3.10) does not obey the boundary conditions (3.5) (see [8, chapter 2]).

3.3.2 Singular Sturm-Liouville Boundary Value Problems

We assume that $p(-1) = 0$, $p(1) = 0$. Moreover, instead of (3.5) we require that

$$\lim_{x \to \pm 1} p(x)u'(x) = 0. \tag{3.12}$$

Since $p(\pm 1) = 0$, this is just a regularity hypothesis for u, and *not* a boundary condition. Set

$$u_{(0)} = u, \; u_{(j)} = \frac{1}{w} L u_{(j-1)}, \quad j \geq 1.$$

If for some $m \geq 1$, $u_{(m)} \in L^2_w(-1,1)$, and $u^{(j)}$ satisfies (3.12) for all $j = 0, \ldots, m-1$, then

$$|u_k| \leq C \frac{1}{k^{2m}} \|u_{(m)}\|_w. \tag{3.13}$$

It follows that if u is infinitely smooth then $pu^{(j)'} = pu^{(j+1)} \to 0$ as $x \to \pm 1$, whence (3.13) holds for all m. In this case all coefficients u_k tend to zero with exponential rate as $k \to \infty$.

The following characterization plays an important role in the applications. Let $\{\varphi_k : k = 0, 1, 2, \ldots\}$ be the eigenfunctions of a singular SL bvp. If we require that φ_k is a polynomial of degree k, then necessarily:

$$w(x) = (1-x)^\alpha (1+x)^\beta \tag{3.14}$$

for some α, β in $(-1, 1)$, and that $p(x) = (1-x)^{\alpha+1}(1+x)^{\beta+1}$. Therefore w is a Jacobi weight function, and the corresponding eigenfunctions are the Jacobi polynomials. For the proof see [8, chapter 2].

Special cases of Jacobi polynomials are:

$\alpha = \beta = 0$	Legendre polynomials
$\alpha = \beta = -1/2$	Chebyshev polynomials of the 1st kind
$\alpha = \beta = 1/2$	Chebyshev polynomials of the 2nd kind.

3.4 The Chebyshev Approximation

Chebyshev polynomials are those most widely used in spectral approximation of boundary value problems. Here we recall some of their properties, then we introduce the discrete Chebyshev transform as well as the pseudospectral Chebyshev derivative. Finally we analyze Chebyshev Gaussian integration as well as projection and interpolation errors in Chebyshev weighted norms.

3.4.1 The Chebyshev Polynomials

As shown in Section 3.3, the Chebyshev polynomials (of the first kind) $\{T_k : k = 0, 1, \ldots\}$ are the eigenfunctions of the singular SL bvp (3.3) corresponding to $w(x) = (1 - x^2)^{-1/2}$ and $p(x) = (1 - x^2)^{1/2}$. Hence they satisfy

$$(-\sqrt{1-x^2}T_k'(x))' = \lambda_k \frac{1}{\sqrt{1-x^2}} T_k(x), \quad k = 0, 1, \ldots, \text{with } \lambda_k = k^2. \tag{4.1}$$

Note that T_k is even if k is even, whereas it is odd if k is odd. If we assume that $T_k(1) = 1$, then

$$T_k(x) = \cos k\theta, \quad \theta = \arccos x. \tag{4.2}$$

Using the trigonometric relation: $\cos(k+1)\theta + \cos(k-1)\theta = 2\cos\theta\cos k\theta$ we immediately have the following recursion formula:

$$T_{k+1}(x) = 2xT_k(x) - T_{k-1}(x), \quad k \geq 2. \tag{4.3}$$

Since $T_k \in \mathbb{P}_k$ and $T_k(1) = 1$, then necessarily $T_0(x) \equiv 1$ and $T_1(x) = x$ (T_0 is even, T_1 is odd). Then (4.3) can be used to generate the whole family $\{T_k\}$. From (4.3) it follows that:

$$|T_k(x)| \leq 1, \quad T_k(1) = 1, \quad T_k(-1) = (-1)^k. \tag{4.4}$$

With the change of variable:

$$x = \cos\theta \tag{4.5}$$

and noting that

$$\frac{d\theta}{dx} = \frac{-1}{\sqrt{1-x^2}} = -w(x), \tag{4.6}$$

we have

$$\int_{-1}^{1} f(x)w(x)\,dx = \int_0^{\pi} f(\cos\theta)\,d\theta. \tag{4.7}$$

Then, by (4.7),

$$\int_{-1}^{1} T_n(x)T_m(x)w(x)\,dx = \int_0^{\pi} \cos n\theta \cos m\theta\,d\theta.$$

It follows that

$$\begin{cases} (T_n, T_m)_w = \frac{\pi}{2} c_n \delta_{nm}, & c_0 = 2,\ c_n = 1 \text{ if } n \geq 1 \\ \|T_n\|_w = \sqrt{\frac{\pi}{2} c_n}, & n = 0, 1, 2, \ldots \end{cases}. \tag{4.8}$$

For any $u \in L^2_w(-1,1)$ its Chebyshev expansion is

$$u(x) = \sum_{k=0}^{\infty} u_k \, T_k(x), \quad u_k = \frac{2}{\pi c_k}(u, T_k)_w. \tag{4.9}$$

Defining the periodic function u^* by $u^*(\theta) = u(\cos\theta)$, one has

$$u^*(\theta) = \sum_{k=0}^{\infty} u_k \cos k\theta. \tag{4.10}$$

Hence the Chebyshev series of u corresponds to a cosine series of u^*.

3.4.2 Relation Between the Chebyshev Coefficients of a Function and Those of its Derivative

We explore here the relationship between the Chebyshev coefficients of a function and those of its derivative. The following property holds. Let $u(x) = \sum_{k=0}^{\infty} u_k T_k(x)$ be such that $u'(x) \in L^2_w(-1,1)$. Then

$$u'(x) = \sum_{k=0}^{\infty} u_k^{(1)} T_k(x), \tag{4.11}$$

where

$$u_k^{(1)} = \frac{2}{c_k} \sum_{\substack{p=k+1 \\ p+k \ odd}}^{\infty} p u_p. \tag{4.12}$$

We present the proof for the convenience of the reader. We confine ourselves to the case of polynomials $u(x) = \sum_{k=0}^{N} u_k T_k(x)$. Then (4.12) must hold with the upper bound of the sum equal to N (otherwise $u_p = 0$). We use the trigonometric identity:

$$2\sin\theta\cos k\theta = \sin(k+1)\theta - \sin(k-1)\theta. \tag{4.13}$$

Noting that

$$T_k'(x) = -k\sin k\theta \frac{d\theta}{dx} = \frac{k\sin k\theta}{\sin\theta},$$

one has $\sin k\theta = \sin\theta T_k'(x)/k$. Using this relation in (4.13) gives:

$$2\sin\theta T_k(x) = \sin\theta\left(\frac{T_{k+1}'(x)}{k+1} - \frac{T_{k-1}'(x)}{k-1}\right).$$

Thus

$$\begin{cases} T_0(x) = T_1'(x) \\ T_k(x) = 1/2\left(\frac{T_{k+1}'(x)}{k+1} - \frac{T_{k-1}'(x)}{k-1}\right), \quad k \geq 1. \end{cases} \tag{4.14}$$

Then

$$u'(x) = \sum_{k=0}^{N} u_k^{(1)} T_k(x) = u_0^{(1)} T_1'(x) + \frac{1}{2} \sum_{k=1}^{N} u_k^{(1)} \left(\frac{T_{k+1}'(x)}{k+1} - \frac{T_{k-1}'(x)}{k-1} \right).$$

On the other hand,

$$u'(x) = \left(\sum_{k=0}^{N} u_k T_k(x) \right) = \sum_{k=1}^{N} u_k T_k'(x), \quad (T_0' \equiv 0). \tag{4.15}$$

Equating the coefficients of T_k', for $0 \le k \le N$ in the expansions above gives

$$\begin{cases} u_0^{(1)} - u_2^{(1)}/2 = u_1 \\ \dfrac{u_{k-1}^{(1)} - u_{k+1}^{(1)}}{2k} = u_k, \quad 2 \le k \le N-2 \\ \dfrac{u_{N-2}^{(1)}}{2(N-1)} = u_{N-1} \\ \dfrac{u_{N-1}^{(1)}}{2N} = u_N. \end{cases} \tag{4.16}$$

We note that this set of equations can be written as follows

$$2k u_k = c_{k-1} u_{k-1}^{(1)} - u_{k+1}^{(1)}, \quad k \ge 1. \tag{4.17}$$

Note that, if $u \in \mathbb{P}_N$, then $u_N^{(1)} = u_{N+1}^{(1)} = 0$. From (4.17) we obtain the formula (4.12).

Note that $u \in \mathbb{P}_N$ has been used in (4.15) to derive the finite expansion of u'. However, if both u and u' are in L_w^2, then the formal derivative of the series for u can still be used. Hence the relations (4.17) hold for all $k \ge 1$.

The form (4.17) suggests the following recurrence relation for $u \in \mathbb{P}_N$:

$$\begin{cases} c_k u_k^{(1)} = u_{k+2}^{(1)} + 2(k+1) u_{k+1}, \quad 0 \le k \le N-1 \\ u_{N+1}^{(1)} = u_N^{(1)} = 0. \end{cases} \tag{4.18}$$

Then, if the u_k are known, the coefficients $\{u_k^{(1)}\}$ can be computed in $3N-2$ multiplications or additions. The above relation can be generalized to a derivative of order $q \ge 1$. If $u^{(q)}(x) = \sum u_k^{(q)} T_k(x)$, then:

$$c_k u_k^{(q)} = u_{k+2}^{(q)} + 2(k+1) u_{k+1}^{(q)}. \tag{4.19}$$

In particular, the expansion coefficients of the second derivative of u are given by:

$$u_k^{(2)} = \frac{1}{c_k} \sum_{\substack{p=k+2 \\ p+k \ \text{even}}}^{\infty} p(p^2 - k^2) u_p.$$

3.4.3 Gauss-Type Quadrature Formulae for Orthogonal Polynomials

We denote here by $\{p_k : k = 0, 1, \ldots\}$ the system of orthogonal polynomials with respect to a general weight w so that

$$(p_k, p_m)_w = \int_{-1}^{1} p_k(x) p_m(x) w(x)\, dx = 0 \quad \text{if } m \neq k. \qquad (4.20)$$

Gauss formulae. Let $x_0, \ldots, x_N$ be the roots of the polynomial p_{N+1} (all of them belong to the interval $(-1, 1)$), and let $w_0, \ldots, w_N$ be the solution of the linear system

$$\sum_{j=0}^{N} (x_j)^k w_j = \int_{-1}^{1} x^k w(x)\, dx, \quad 0 \leq k \leq N. \qquad (4.21)$$

Then

$$\sum_{j=0}^{N} p(x_j) w_j = \int_{-1}^{1} p(x) w(x)\, dx \quad \text{for all } p \in \mathbf{P}_{2N+1}. \qquad (4.22)$$

Gauss-Radau formulae. Let $x_0, x_1, \ldots, x_N$ be the $N+1$ roots of the polynomial

$$q(x) = p_{N+1}(x) + \alpha p_N(x),$$

where α is chosen so that either $q(-1) = 0$, whence

$$\alpha = -p_{N+1}(-1)/p_N(-1),$$

or $q(1) = 0$, whence

$$\alpha = -p_{N+1}(1)/p_N(1)).$$

Let $x_0, x_1, \ldots, x_N$ be the $N+1$ roots of the polynomial q (in the first case they will include the point $x = -1$, in the latter case the point $x = 1$). Let $w_0, \ldots, w_N$ be the solutions of the linear system (4.21) with respect to the new x_j. Then

$$\sum_{j=0}^{N} p(x_j) w_j = \int_{-1}^{1} p(x) w(x)\, dx \quad \text{for all } p \in \mathbf{P}_{2N}. \qquad (4.23)$$

Gauss-Lobatto formulae Let $x_0, \ldots, x_N$ be the $N+1$ roots of the polynomial

$$q(x) = p_{N+1}(x) + \alpha p_N(x) + \beta p_{N-1}(x), \qquad (4.24)$$

where α and β are chosen so that $q(-1) = q(1) = 0$. Let w_j be the solution of the linear system (4.21) with respect to the new points x_j. Then

$$\sum_{j=0}^{N} p(x_j)w_j = \int_{-1}^{1} p(x)w(x)\,dx \quad \text{for all } p \in \mathbf{P}_{2N-1}. \tag{4.25}$$

A different characterization of the Gauss-Lobatto nodes for Jacobi's weights. Let $w(x) = (1-x)^\alpha(1+x)^\beta$, $-1 < \alpha, \beta < 1$ be a Jacobi weight function. Then the internal nodes $x_1, \dots, x_{N-1}$ of the Gauss-Lobatto formula are the extrema of p_N, i.e.,

$$p'_N(x_j) = 0, \quad 1 \le j \le N-1. \tag{4.26}$$

3.4.4 Explicit Expressions for Nodes and Weights of the Gauss-Chebyshev Formulae

In the quadrature formula

$$\sum_{j=0}^{N} f(x_j)w_j \sim \int_{-1}^{1} f(x)w(x)\,dx, \tag{4.27}$$

the abscissae x_j are called the *nodes*, and the coefficients w_j are the *weights*.

We will give now the closed expression of nodes and weights corresponding to the Chebyshev weight function $w(x) = (1-x^2)^{-1/2}$

Chebyshev-Gauss.

$$x_j = \cos\left(\frac{2j+1}{2N+2}\pi\right), \quad w_j = \frac{\pi}{N+1}. \tag{4.28}$$

Chebyshev-Gauss-Radau.

$$x_j = \cos\left(\frac{2j}{2N+1}\pi\right), \quad w_0 = \frac{\pi}{2N+1}, \quad w_j = \frac{2\pi}{2N+1}, \ j \ge 1. \tag{4.29}$$

Chebyshev-Gauss-Lobatto.

$$x_j = \cos\left(\frac{j}{N}\pi\right), \quad w_0 = w_N = \frac{\pi}{2N}, \quad w_j = \frac{\pi}{N}, \ 1 \le j \le N-1. \tag{4.30}$$

3.4.5 Chebyshev Interpolation and Discrete Chebyshev Transforms

The most commonly used Chebyshev points are the Gauss-Lobatto ones (4.30) which we consider in detail hereafter. We define the interpolation operator $I_N : C^0([-1,1]) \to \mathbb{P}_N$ by

$$I_N u(x_j) = u(x_j), \quad x_j = \cos\frac{\pi j}{N},\ j = 0,\dots,N. \tag{4.31}$$

We introduce the discrete L^2_w-inner product

$$(u,v)_N = \sum_{j=0}^{N} u(x_j)v(x_j)w_j, \tag{4.32}$$

where the w_j are given in (4.30). Then, using (4.25) we have:

$$(u,v)_N = (u,v)_w \quad \text{if } uv \in \mathbb{P}_{2N-1}. \tag{4.33}$$

From (4.31) we get, setting $I_N u(x) = \sum_{k=0}^{N} u_k^* T_k(x)$,

$$(u,v)_N = (I_N u, v)_N = \sum_{k=0}^{N} u_k^*(T_k, v)_N.$$

Taking $v = T_m(x)$, $0 \le m \le N-1$, and using (4.33) and (4.8) will give

$$(u,T_m)_N = u_m^* \|T_m\|^2 = u_m \pi/2c_m, \quad 0 \le m \le N-1. \tag{4.34}$$

Taking $v = T_N(x)$ gives

$$(u,T_N)_N = u_N^* \|T_N\|_N^2 = 2u_N^* \|T_N\|_w^2 = u_N^* \pi, \tag{4.35}$$

where we have set

$$\|u\|_N = \sqrt{(u,u)_N} = \sqrt{\frac{\pi}{N}\sum u^2(x_j)w_j}, \tag{4.36}$$

and

$$\|T_N\|_N^2 = \sum \cos^2(N\theta_j)w_j = \sum (\cos\pi j)^2 w_j = \sum w_j = \pi.$$

We can summarize (4.34) and (4.35) as follows:

$$\begin{cases} u_m^* = \dfrac{2}{\pi d_m}(u,T_m)_N = \dfrac{2}{\pi d_m}\displaystyle\sum_{j=0}^{N} w_j T_m(x_j)u(x_j) \\ d_0 = d_N = 2, \quad d_j = 1,\ 1 \le j \le N-1. \end{cases} \tag{4.37}$$

Since $w_j = \pi/Nd_j$, we have

$$u_m^* = \frac{2}{Nd_m} \sum_{j=0}^{N} \frac{1}{d_j} \cos\left(\frac{mj\pi}{N}\right) u(x_j). \tag{4.38}$$

Formula (4.37) (or (4.38)) is the *discrete Chebyshev transform*, going from the physical to the spectral space. Its matrix representation is

$$[u_m^*]^{tr} = A[u(x_j)]^{tr}, \quad a_{mj} = \frac{2}{Nd_m d_j} \cos\left(\frac{mj\pi}{N}\right). \tag{4.39}$$

The *inverse discrete Chebyshev transform*, which goes from the spectral to the physical space, has the following form

$$u(x_j) = I_N u(x_j) = \sum_{j=0}^{N} u_m^* T_m(x_j) = \sum_{j=0}^{N} \cos\left(\frac{mj\pi}{N}\right) u_m^*,$$

that is,

$$[u(x_j)]^{tr} = C[u_m^*]^{tr}, \quad c_{jm} = \cos\frac{mj\pi}{N}, \quad 0 \le m, j \le N. \tag{4.40}$$

Remark 3.1 The continuous and discrete Chebyshev norms, $\|\cdot\|_w$ and $\|\cdot\|_N$, are uniformly equivalent for all polynomials of degree $\le N$. Indeed, as we noticed,

$$\|T_m\|_N = \|T_m\|_w, \quad 0 \le m \le N-1, \quad \|T_N\|_N = 2\|T_N\|_w.$$

Hence, for all $u = \sum_{k=0}^N a_k T_k \in \mathbf{P}_N$, we have:

$$\|u\|_N^2 = \sum_{k=0}^{N} a_k^2 \|T_k\|_N^2 = \sum_{k=0}^{N-1} a_k^2 \|T_k\|_w^2 + a_N^2 \|T_N\|_w^2.$$

Since $\|u\|_w^2 = \sum_{k=0}^N a_k^2 \|T_k\|_w^2$ it follows that

$$\|u\|_w \le \|u\|_N \le 2\|u\|_w \quad \text{for all } u \in \mathbb{P}_N. \tag{4.41}$$

3.4.6 The Pseudo-Spectral Chebyshev Derivative

Let u be a continuous function in $(-1, 1)$. We define the pseudo-spectral Chebyshev derivative of u as follows:

$$\partial_N u = (I_N u)', \tag{4.42}$$

where I_N is the interpolant of u at the points given by (4.30). If the values of $u(x_j)$, $0 \le j \le N$ are known, then the values of the polynomial $\partial_N u \in \mathbb{P}_{N-1}$ at the same nodes can be obtained as follows:

(1) Use the discrete Chebyshev Transform (4.39) to compute the coefficients $\{u_m^*\}_{m=0}^N$ of the expansion of $I_N u$;

(2) Use the recursion formulae (4.18) to get the Chebyshev coefficients $u_m^{(1)}$ of $(I_N u)'$;

(3) Use the inverse of the discrete Chebyshev transform (4.40) to compute the gridvalues $(\partial_N u)(x_j) = (I_N u)'(x_j)$.

Both the direct and inverse discrete Chebyshev transform can be computed through the FFT. Hence the whole process (1), (2) and (3) will require $(\mathcal{O}(N \log_2 N)$ operations.

As for the Fourier case, the pseudo-spectral Chebyshev derivative can be represented by a single matrix D. In order to find the entries of D we set

$$I_N u(x) = \sum_{j=0}^{N} u(x_j)\psi_j(x), \quad \text{where } \psi_j \in \mathbf{P}_N \text{ and } \psi_j(x_m) = \delta_{jm}. \tag{4.43}$$

For the Gauss-Lobatto points one has

$$\psi_j(x) = (-1)^{j+1} \frac{(1-x^2)T_N'(x)}{d_j N^2 (x - x_j)}. \tag{4.44}$$

Since $T_N'(x) = -N \sin(N\theta)/\sin\theta$ we have $T_N'(x_m) = 0$ if $m \neq j$, whence $\psi_j(x_m) = 0$ if $m \neq j$. Now

$$(\partial_N u)(x_m) = (I_N u)'(x_m) = \sum_{j=0}^{N} \psi_j'(x_m) u(x_j). \tag{4.45}$$

Hence

$$\begin{cases} [(\partial_N u)(x_m)]^{tr} = D[u(x_j)]^{tr} \\ D_{mj} = \psi_j'(x_m). \end{cases}$$

It can be shown that

$$D_{mj} = \begin{cases} \dfrac{d_m}{d_j} \dfrac{(-1)^{m+j}}{x_m - x_j} & \text{if } m \neq j \\ -\dfrac{x_j}{2(1-x_j^2)} & \text{if } 1 \leq m = j \leq N-1 \\ \dfrac{2N^2+1}{6} & \text{if } m = j = 0 \\ -\dfrac{2N^2+1}{6} & \text{if } m = j = N. \end{cases} \tag{4.46}$$

This matrix is not skew-symmetric, in contrast to the matrix of the Fourier differentiation (see Section 3.1). Its only eigenvalue is $\lambda = 0$, with algebraic multiplicity $N+1$. Of course, the matrix $D^{(2)}$ representing the second derivative in physical space will be $D^{(2)} = DD = (D)^2$, which is no longer symmetric.

3.4.7 Error Estimates for Projection and Interpolation Operators

The natural Sobolev norms to measure approximation errors for the Chebyshev system involve the Chebyshev weight. We set

$$\|u\|_{H^m_w(-1,1)} = \left(\sum_{k=0}^{m} \int_{-1}^{1} |u^{(k)}(x)|^2 w(x)\, dx \right)^{1/2}. \tag{4.47}$$

The Hilbert space associated with this norm is denoted by $H^m_w(-1,1)$. The truncation error $u - P_N\, u$, where now $P_N u = \sum_{k=0}^{N} u_k T_k$ is the truncated Chebyshev series of u, satisfies the inequality

$$\|u - P_N u\|_{L^2_w(-1,1)} \le C N^{-m} \|u\|_{H^m_w(-1,1)}, \tag{4.48}$$

for all $u \in H^m_w(-1,1)$, with $m \ge 0$. The truncation error in higher order Sobolev norms is estimated by the inequality

$$\|u - P_N u\|_{H^k_w(-1,1)} \le C N^{-1/2} N^{2k-m} \|u\|_{H^m_w(-1,1)} \tag{4.49}$$

for $u \in H^m_w(-1,1)$, with $1 \le k \le m$. Thus, the asymptotic behaviour of the Chebyshev truncation error is non-optimal with respect to the exponent of N. In order to define the polynomial of best approximation in $H^1_w(-1,1)$ we introduce the inner product

$$((u,v))_w = \int_{-1}^{1} (u'v' + uv) w\, dx, \quad \text{for all } u, v \in H^1_w(-1,1), \tag{4.50}$$

and define the related orthogonal projection on $\mathbb{P}_N$ to be the polynomial $P^1_N u \in \mathbb{P}_N$ such that

$$((P^1_N u, \phi))_w = ((u, \phi))_w, \quad \text{for all } \phi \in \mathbb{P}_N. \tag{4.51}$$

The corresponding estimate is

$$\|u - P^1_N u\|_{H^k_w(-1,1)} \le C N^{k-m} \|u\|_{H^m_w(-1,1)} \tag{4.52}$$

for all $u \in H^m_w(-1,1)$ with $m \ge 1$, and $k = 0, 1$.

These estimates extend to functions satisfying prescribed boundary data. For instance, one can find a polynomial $u_N \in \mathbb{P}_N$, $u_N(\pm 1) = 0$, whose distance from u decays in an optimal way both in the H^1_w-norm and in the L^2_w-norm, i.e.,

$$\|u - u^N\|_{H^k_w(-1,1)} \le C N^{k-m} \|u\|_{H^m_w(-1,1)}, \quad k = 0, 1. \tag{4.53}$$

We consider now the interpolation error. Let $I_N u \in \mathbb{P}_N$ denote the interpolation of u at any of the three families of Chebyshev-Gauss points (4.28), (4.29) or (4.30). Then the following estimate holds:

$$\|u - I_N u\|_{L^2_w(-1,1)} \leq CN^{-m}\|u\|_{H^m_w(-1,1)}, \tag{4.54}$$

provided $u \in H^m_w(-1,1)$ for some $m \geq 1$. In higher order Sobolev norms one has

$$\|u - I_N u\|_{H^k_w(-1,1)} \leq CN^{2k-m}\|u\|_{H^m_w(-1,1)}, \tag{4.55}$$

for $0 \leq k \leq m$. As a consequence, we have

$$\|u' - (I_N u)'\|_{L^2_w(-1,1)} \leq CN^{2-m}\|u\|_{H^m_w(-1,1)}.$$

When the function u is analytic, the error $u' - (I_N u)'$ decays exponentially in N. Precisely, if u is analytic in $[-1,1]$ and has a regularity ellipse whose sum of semi-axes equals $e^{\eta 0} > 1$, then

$$\|u' - (I_N u)'\|_{L^2_w(-1,1)} \leq C(\eta)N^2 e^{-N\eta}, \tag{4.56}$$

for all η^0, $0 < \eta < \eta^0$. The interpolation error in the maximum norm decays as follows:

$$\|u - I_N u\|_{L^\infty(-1,1)} \leq CN^{1/2-m}\|u\|_{H^m_w(-1,1)},$$

under the same assumptions required for (4.54).

3.5 Elliptic and Parabolic Boundary Value Problems

We review in this Section some basic aspects of finite element and spectral methods for elliptic boundary value problems in a plane domain. We consider as a model example the second order elliptic boundary value problem

$$-\Delta u + \alpha u = f \quad \text{in } \Omega, \tag{5.1}$$

with $u = 0$ on $\partial\Omega$, where Ω is an open bounded domain of $\mathbb{R}^2, \partial\Omega$ is its boundary (a piecewise C^1-curve), $\alpha \geq 0$, f is a given function of $L^2(\Omega)$ and $\Delta := \partial^2/\partial x_1^2 + \partial^2/\partial x_2^2$ is the Laplace operator. The variational formulation of (5.1) is

$$\text{Find } u \in H^1_0(\Omega) : \int_\Omega (\nabla u \cdot \nabla v + \alpha uv)\, dx = \int_\Omega fv\, dx \text{ for all } v \in H^1_0(\Omega), \tag{5.2}$$

where $H_0^1(\Omega)$ is the space of the functions in $L^2(\Omega)$, whose (distributional) derivatives of first order are in $L^2(\Omega)$, and which vanish on $\partial\Omega$. Problem (5.2) has been formally obtained from (5.1) using the Green's formula

$$-\int_\Omega (\Delta u)v\,dx = \int_\Omega \nabla u\cdot\nabla v\,dx - \int_{\partial\Omega} \partial u/\partial n v\,d\sigma, \tag{5.3}$$

where $\nabla := (\partial/\partial x_1, \partial/\partial x_2)^{tr}$ is the gradient operator, and $\partial/\partial n$ is the outward normal derivative on $\partial\Omega$. Due to the Poincaré inequality, $\|v\| := (\int_\Omega |\nabla v|^2\,dx)^{1/2}$ is a norm for $H_0^1(\Omega)$. Then, setting

$$a(u,v) := \int_\Omega (\nabla u\cdot\nabla v + \alpha uv)\,dx, \text{ for all } u,v\in H_0^1(\Omega), \tag{5.4}$$

we note that there exist two positive constants k_1 and k_2 such that

$$a(v,v)\geq k_1\|v\|^2,\ |a(u,v)|\leq k_2\|u\|\,\|v\|, \text{ for all } u,v\in H_0^1(\Omega). \tag{5.5}$$

Since (5.2) can now be written as

$$u\in H_0^1(\Omega): a(u,v) = \int_\Omega fv\,dx, \quad \text{for all } v\in H_0^1(\Omega), \tag{5.6}$$

in view of the Lax-Milgram theorem it follows that this problem has a unique solution, and $\|u\|\leq \|f\|_*/k_1$, where $\|\cdot\|_*$ is the norm of $H^{-1}(\Omega)$, the dual of the space $H_0^1(\Omega)$.

Remark 4.1. Since the bilinear form $a(\cdot,\cdot)$ is symmetric, u is the solution to the minimisation problem

$$J(u) = \min\{J(v)\,|\,v\in H_0^1(\Omega)\},$$

where $J(v) = \{\frac{1}{2}a(v,v) - \int_\Omega fv\,dx\}$ is a quadratic functional.

We introduce now a family of finite dimensional subspaces V_h of $H_0^1(\Omega)$, whose dimension $I(h)$ will tend to infinity as $h\to 0$. The Galerkin approximation to (5.6) is defined as follows:

$$u_h\in V_h: a(u_h,v_h) = \int_\Omega fv_h\,dx \text{ for all } v_h\in V_h. \tag{5.7}$$

The Lax-Milgram theorem can still be applied to deduce that for each h (5.7) has a unique solution which satisfies the stability estimate: $\|u_h\|\leq 1/k_1\|f\|_*$.

A more constructive proof can be provided. Let $\{\varphi_1, \ldots, \varphi_{I(h)}\}$ be a basis of V_h, and set

$$u_h(x) = \sum_{j=1}^{I(h)} u_j \varphi_j(x). \tag{5.8}$$

Then (5.7) is equivalent to the linear system

$$A\mathbf{u} = \mathbf{f}, \text{ with } A_{ij} = a(\varphi_j, \varphi_i),\ f_i = \int_\Omega f \varphi_i \, dx, \tag{5.9}$$

and $\mathbf{u} = \{u_j\}$ is the unknown vector. Owing to (5.5), A is symmetric. Moreover, for any $\mathbf{v} \in \mathbb{R}^{I(h)}$ one has $\mathbf{v}^* A \mathbf{v} = a(v, v)$, where $v(x) = \sum_{j=1}^{I(h)} v_j \varphi_j(x)$. Thus A is positive definite. It follows that A is non-singular, and therefore the system (5.9) has a unique solution.

Concerning now the approximation properties of u_h to u, we note that, subtracting (5.7) from (5.6) we obtain the *consistency* estimate

$$a(u - u_h, v_h) = 0 \quad \text{for all } v_h \in V_h,$$

whence, due to (5.5), we deduce the inequality

$$\|u - u_h\| \leq (k_2/k_1) \inf_{v_h \in V_h} \|u - v_h\|. \tag{5.10}$$

This basic estimate states that the error $u - u_h$ behaves like the best approximation error in the norm of $H_0^1(\Omega)$.

3.5.1 The Finite Element Method

The *finite element approximation* relies upon a particular strategy for choosing the subspace V_h. We give here a short description of the method; for a thorough discussion we refer to [9] or [21]. Let the domain Ω be partitioned into a finite union $\mathcal{T}_h$ of polygons T (typically, triangles and/or quadrilaterals, possibly with curved boundaries), called finite elements, such that h is their maximum diameter, $\overline{\Omega} = \cup \overline{T}$, and if $i \neq j$ and $T_i \cap T_j$ is not empty, then $T_i \cap T_j$ is either a vertex or a side shared by T_i and T_j. Thus, a partition like that in Figure 3.1(b) will not be allowed.

We denote by $\Sigma = \{a_j\}_{j=1}^N$ a set of distinct points of T, and by P a N-dimensional vector space of functions (typically, polynomials). The set Σ is said to be *P-unisolvent* if, for any given set $\{\alpha_j, j = 1, \ldots, N\}$ of real values, there exists a unique function $p \in P$ such that $p(a_j) = \alpha_j$, $j = 1, \ldots, N$. In this case, the triplet (T, P, Σ) is called a *finite element.* If (T, P, Σ) is a finite element, the functions $p_i \in P$ such that $p_i(a_j) = \delta_{ij}$ are called (Lagrangian) *basis functions.* We can now define the *Lagrange finite element interpolant*

$$\pi_T : C^0(T) \to P,\ \pi_T v(x) = \sum_{j=1}^{N} v(a_j) p_j(x) \quad \text{for all } v \in C^0(T). \tag{5.11}$$

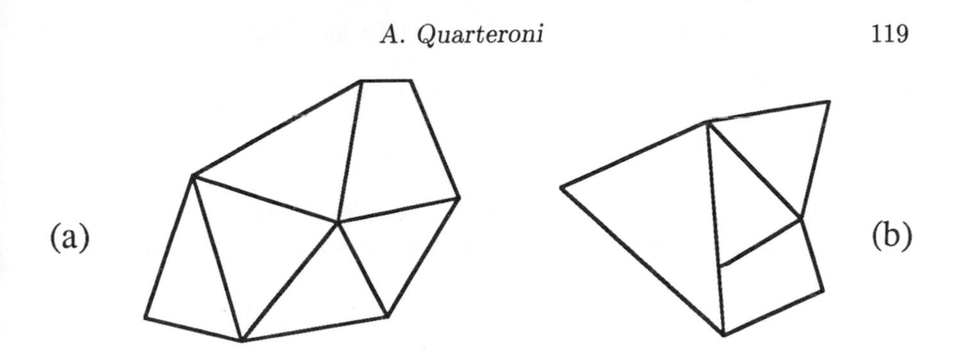

Fig. 3.1. Correct and unacceptable finite element decompositions

For any integer $k \geq 0$, we denote by $\mathbf{P}_k$ the space of polynomials of degree not greater than k, i.e.,

$$\mathbf{P}_k = \left\{ p(x) = \sum_{\substack{i,j=0 \\ i+j\leq k}}^{k} a_{ij} x_1^i x_2^j,\ a_{ij} \in \mathbb{R} \right\},\ \dim(\mathbf{P}_k) = \frac{(k+1)(k+2)}{2}, \tag{5.12}$$

and by Q_k the space of polynomials of degree $\leq k$ in each variable, i.e.,

$$\mathbf{Q}_k = \left\{ q(x) = \sum_{i,j=0}^{k} a_{ij} x_1^i x_2^j,\ a_{ij} \in \mathbb{R} \right\}, \dim(\mathbf{Q}_k) = (k+1)^2. \tag{5.13}$$

With any given decomposition $\mathcal{T}_h$ of Ω we associate the finite element spaces (we assume hereafter that Ω is a polygonal domain, so that all the T are polygons):

$$X_h^r := \{v \in C^0(\Omega) \,|\, v_{|T} \in \mathbf{P}_r \text{ (or } \mathbf{Q}_r^*), \text{ for all } T \in \mathcal{T}_h\} \subset H^1(\Omega) \tag{5.14}$$

$$V_h^r := \{v \in X_h \,|\, v = 0 \text{ on } \partial\Omega\} \subset H_0^1(\Omega), \tag{5.15}$$

where $r \geq 0$ is a positive integer. If T is a quadrilateral then $Q_r^* = Q_r \circ \Psi_T$, where Ψ_T is the function mapping T into the reference square $(0,1)^2$.

We can now define the global *finite element interpolant* $\Pi_{h,r} : C^0(\Omega) \to X_h^r$ as follows. Let $(T, \Sigma_r(T), P_r(T))$ be the *triangular* finite element of order r, and set $\Sigma_h = \cup\{\Sigma_r(T), T \in \mathcal{T}_h\}$. Then Σ_h is the set of the *nodes* of $\mathcal{T}_h$, and we set $\Sigma_h = \{a_j, j = 1, \ldots, I\}$. Now $\Pi_{h,r}v$ is the only function of X_h^r such that $\Pi_{h,r}v(a_j) = v(a_j)$, $j = 1, \ldots, I$. It is easy to see that $(\Pi_{h,r}v)_{|T} = \pi_T(v_{|T})$ for all $T \in \mathcal{T}_h$, and

$$\Pi_{h,r}v(x) = \sum_{j=1}^{I} v(a_j)\varphi_j(x), \tag{5.16}$$

where $\{\varphi_j\}$ is the Lagrange basis of X_h^r, i.e.,

$$\varphi_j \in X_h^r : \varphi_j(a_i) = \delta_{ij}, \quad i, j = 1, \dots, I. \tag{5.17}$$

The values $\{v(a_j),\ j = 1, \dots, I\}$ are called degrees of freedom of the function v. For the finite element interpolant the following error estimate holds

$$\|v - \Pi_{h,r}v\|_{H^k(\Omega)} \leq Ch^{m-k}\|v\|_{H^m(\Omega)}, \quad k = 0, 1,\ m > 1. \tag{5.18}$$

Here $H^k(\Omega)$ is the Hilbert space of those functions $v \in L^2(\Omega)$ whose partial (distributional) derivatives of order at most k are represented by functions of $L^2(\Omega)$. Hereafter, C denotes a generic positive constant, independent of h.

In view of the convergence estimate (5.10) that was stated for a general Galerkin approximation, it is now easy to conclude that the solution of the finite element Galerkin approximation (5.7), with V_h given by (5.15), converges to the solution of (5.6) as follows:

$$\|u - u_h\|_{H^1(\Omega)} \leq C\frac{k_2}{k_1}h^{m-1}\|u\|_{H^m(\Omega)}, \quad m > 1. \tag{5.19}$$

Actually, assuming that $u \in H^m(\Omega) \cap H_0^1(\Omega)$, with $m > 1$, one has $\Pi_{h,r}\ u \in V_h^r$. Hence it is enough to take $v_h = \Pi_{h,r}u$ in (5.10) and use (5.18) with $v = u$ and $k = 1$.

We make now a short comment on the algebraic aspects of the problem (5.7) with the finite element space (5.15). We note that from the Lagrange basis $\{\varphi_j\}$ of X_h^r defined in (5.17), we can easily get a Lagrange basis for V_h^r by simply disregarding those functions φ_j for which the node a_j belongs to the boundary of Ω. If we denote by $\{\varphi_i, i = 1, \dots, I(h)\}$ this basis, and we express the finite element solution in the form (5.8), it follows that $u_j = u_h(a_j)$, $j = 1, \dots, I(h)$. Moreover, the associated matrix A defined in (5.9) is sparse and banded. Its bandwith depends on the numbering of the vertices of the triangles, and especially on the maximum difference $|i - j|$ of two contiguous vertices a_i and a_j. Efficient methods to solve the finite element system include the Choleski factorization method (for banded matrices), the frontal method, the preconditioned conjugate-gradient method and the multigrid method. The interested reader can refer to [13], for example.

3.5.2 Spectral Methods

Spectral methods form a highly accurate class of techniques for the solution of partial differential equations. Especially when used in a collocation form these methods achieve most of their effectiveness owing to fast transform methods, such as the fast Fourier and Chebyshev transforms. In these cases, the fundamental unknowns are the solution values at the collocation points ('physical unknowns'). Otherwise, when implemented in a Galerkin form, the basic unknowns are the expansion coefficients of the solution with respect to a basis of orthogonal polynomials ('spectral unknowns').

The key step is to use globally smooth functions to approximate the solution in order to achieve exponential type accuracy. With this purpose, taking S_N as the space of either trigonometric (for periodic solutions) or algebraic polynomials of degree at most N (in all other cases) one has (see Sections 3.1 and 3.4)

$$\inf_{v_N \in S_N} \|u - v_N\|_k \leq C(k,m) N^{k-m} \|u\|_m, \quad 0 \leq k \leq m. \tag{5.20}$$

Here $\|\cdot\|_k$ denotes the norm of the Sobolev space of order k, where $k \geq 0$. (See (2.14) and (4.47).) It is clear from Sections 3.1 and 3.4 that any polynomial, either algebraic or trigonometric, can be represented either by its expansion coefficients ('spectral representation') or by the values attained at selected Gaussian points ('physical representation'). Since the number of coefficients in the Fourier (or polynomial) expansion equals the number of Gaussian points, one can go from physical to spectral representation (and back) by linear, invertible mappings. These are precisely the discrete transforms, which can be carried out by $\mathcal{O}(N \log_2 N)$ operations (rather than $\mathcal{O}(N^2)$) for both families of Fourier and Chebyshev Gaussian points. According to the method that is used, one of the above representations will be the primary one. However, the concurrence of the two representations is a common feature of most spectral methods. As we said before, in collocation spectral methods the primary unknowns are the solution values at collocation points (physical unknowns). However, it is customary to use the expansion coefficients (spectral or frequency unknowns) in order to calculate effectively the derivatives with the help of recurrence relations, and then to go back to the physical unknowns. Symmetrically, for those nonlinear problems whose primary unknowns are the spectral ones, the nonlinear terms give rise to convolution sums whose direct calculations might be prohibitive. On the contrary, one can go to the space of physical unknowns, computing the nonlinear terms by simple multiplications, then go back to the spectral representation.

Generally speaking, choosing the primary unknowns amounts to choose the type of spectral method to be used for the solution of a given problem. Both Galerkin methods (mostly used for Fourier approximations) and Tau methods (used for approximating problems with prescribed boundary conditions by Chebyshev or Legendre polynomials) use the frequency unknowns. If physical unknowns are adopted as the primary ones, the method that has to be used is the collocation method. In such case, the differential equation is satisfied exactly at those Gaussian knots which are internal to the physical domain, while the boundary knots are used to enforce the prescribed boundary conditions. Collocation methods are used for either periodic or non-periodic problems. Generally speaking, the use of collocation spectral methods is mandatory unless the problem is quite simple (linear, with constant coefficients).

We present now some examples of spectral methods, using either Fourier or Chebyshev polynomial expansions.

Fourier Spectral Methods

Consider the Helmholtz equation (5.1) with $\alpha > 0$, $\Omega = (0, 2\pi)^2$ and u periodic with period 2π in each direction. We use the Fourier-Galerkin method for its approximation. For any given integer N, we look for $u_N \in S_N$, with

$$S_N = \text{span}\{\varphi_{km}(x) = e^{i(kx_1+mx_2)},\ -N \le k, m \le N-1\} \tag{5.21}$$

satisfying

$$(-\Delta u_N + \alpha u_N, v) = (f, v), \quad \text{for all } v \in S_N, \tag{5.22}$$

where $(u, v) = \int_\Omega u(x)\overline{v(x)}\,dx$ is the $L^2(\Omega)$-scalar product for complex valued functions. In the current situation the abstract convergence result (5.10) yields

$$\|u - u_N\|_{H^1_{2\pi}(\Omega)} \le C \inf_{v_N \in S_N} \|u - v_N\|_{H^1_{2\pi}(\Omega)}. \tag{5.23}$$

Denoting the $L^2(\Omega)$-projection of u upon S_N by $P_N u$, i.e.,

$$P_N : L^2(\Omega) \to S_N,\ (P_N u, v) = (u, v), \quad \text{for all } v \in S_N, \tag{5.24}$$

one has, assuming that $u \in H^m_{2\pi}(\Omega)$ for some $m \ge 2$ (see (2.15)):

$$\|u - P_N u\|_{H^1_{2\pi}(\Omega)} \le C N^{1-m} \|u\|_{H^m_{2\pi}(\Omega)}. \tag{5.25}$$

Thus from (5.23) and (5.1) the following error bound holds

$$\|u - u_N\|_{H^1_{2\pi}(\Omega)} \le C N^{1-m} \|f\|_{H^{m-2}_{2\pi}(\Omega)}. \tag{5.26}$$

The order of convergence is only bounded by the smoothness degree of the data, as opposed to the finite element case where the rate of convergence is governed by the degree of the piecewise polynomials used. Actually, if f is an analytic function, then (5.26) can be improved to give

$$\|u - u_N\|_{H^1_{2\pi}(\Omega)} \le C(f) \exp(-\gamma N), \quad \gamma > 0. \tag{5.27}$$

This explains why spectral methods are said to converge with exponential accuracy. The solution of problem (5.22) is straightforward using the spectral representation

$$u_N(x) = \sum_{k,m} u_{km}\varphi_{km}(x).$$

Actually, the orthogonality of the $\{\varphi_{km}\}$ under the inner product $(\cdot,\cdot)$ yields

$$u_{km}(\alpha + k^2 + m^2) = f_{km}, \quad -N \le k, m \le N-1, \tag{5.28}$$

where $f_{km} = (f, \varphi_{km})$. Thus the Fourier coefficients of u_N are achievable by $4N^2$ divisions only, provided the coefficients f_{km} are available. If this is not the case, we can use in the right hand side of (5.29) the discrete Fourier coefficients of f, which in turn can be computed by the discrete Fourier transform (see Section 3.1). It is not difficult to see that, in the latter case, one actually gets the *collocation* Fourier solution at the collocation nodes $x_{ij} = (\pi i/N, \pi j/N)$, $i, j = 0, \dots, 2N-1$.

Chebyshev Spectral Methods

To approximate the problem (5.1) with Dirichlet boundary conditions $u = 0$ on $\partial\Omega$, Chebyshev (or Legendre) rather than Fourier expansions should be used. This amounts to looking for an approximation of the form

$$u_N \in V_N : a_w(u_N, v_N) = \int_\Omega f v_N w \, dx, \quad \text{for all } v_N \in V_N, \tag{5.29}$$

where V_N is the set of polynomials of degree less than or equal to N in each variable that vanish on $\partial\Omega$, and

$$a_w(u, v) = \int_\Omega \nabla u \nabla(vw) \, dx,$$

where $w(x) = [(1 - x_1^2)(1 - x_2^2)]^{-1/2}$ is the Chebyshev weight function (we are assuming here that $\Omega = (-1, 1)^2$). On the other hand, the collocation Chebyshev method consists of looking for $u_N \in V_N$ such that

$$-\Delta u_N + \alpha u_N = f$$

at each Chebyshev-Lobatto point $x_{ij} = \left(\cos\frac{\pi i}{N}, \cos\frac{\pi j}{N}\right)$, for $1 \le i, j \le N-1$. In both cases, the spectral solution is stable and the following error estimate can be obtained:

$$\|u - u_N\|_{H^1_w(\Omega)} \le C N^{1-m} \|u\|_{H^m_w(\Omega)}, \quad n \ge 1.$$

Spectral Approximations to other Boundary Value Problems

We consider further examples of approximation by spectral methods to problems that are essentially driven by the Laplace operator. Let us consider now a Fourier approximation to a fully periodic, incompressible, 3 dimensional Navier-Stokes equation.

If we denote by Re the Reynolds number, the time independent Navier-Stokes equations read as follows

$$\begin{cases} -(1/Re)\,\Delta\mathbf{u} + (\mathbf{u}\cdot\nabla)\mathbf{u} + \nabla p = \mathbf{f} & \text{in } \Omega = (0,2\pi)^3 \\ \operatorname{div}\mathbf{u} = 0 & \text{in } \Omega. \end{cases} \tag{5.30}$$

Let us set

$$S_N = \operatorname{span}\{\varphi_k(x) = \exp(i\mathbf{k}\cdot\mathbf{x})\,;\mathbf{k}\in Z^3,\ -N \le k_\alpha \le N-1,\ \alpha = 1,2,3\},$$

then look for

$$\mathbf{u}_N^\alpha(x) = \sum_{\mathbf{k}} u_k^\alpha \varphi_k(x) \in S_N, \quad \alpha = 1,2,3,$$

and project the momentum equation onto S_N. Note that, with $\psi_k = \nabla\varphi_k$ as test function, one has

$$\begin{aligned} (\Delta\mathbf{u}_N, \psi_k) &= 0, \\ (\nabla p_N, \psi_k) &= -((\mathbf{u}_N\cdot\nabla)\mathbf{u}_N, \nabla\varphi_k) + (\mathbf{f}, \psi_k). \end{aligned}$$

Thus p_N is eliminated in favour of $\mathbf{u}_N$ and $\mathbf{f}$, and we find:

$$\frac{|\mathbf{k}|^2}{Re}u_k^\alpha = -\left\{[(\mathbf{u}_N\cdot\nabla)u_N^\alpha]_k + \frac{ik_\alpha}{|\mathbf{k}|^2}[\operatorname{div}(\mathbf{u}_N\cdot\nabla)\mathbf{u}_N]_k\right\} - ik_\alpha f_k^\alpha,$$

for $-N \le k_\alpha \le N-1$, $\alpha = 1,2,3$ and

$$[\operatorname{div}(\mathbf{u}_N\cdot\nabla)\mathbf{u}_N]_{\mathbf{k}} = -\sum_{\beta=1}^{3}(ik_\beta)[(\mathbf{u}_N\cdot\nabla)u_N^\beta]_{\mathbf{k}}.$$

As usual, $[v]_{\mathbf{k}}$ denotes the $\mathbf{k}-^{th}$ Fourier coefficient of the function v. Analysis by an implicit function theorem yields the following result (see [17]):

$$\|\mathbf{u}(Re) - \mathbf{u}_N(Re)\|_{H^1_{2\pi}(\Omega)} \le C(Re)N^{1-s}\|\mathbf{u}(Re)\|_{H^s_{2\pi}(\Omega)}, \quad s \ge 1.$$

A direct calculation of the convective terms involves the evaluation of convolution sums. Actually, one has:

$$[u^\alpha \partial u^\beta/\partial x_\alpha]_{\mathbf{n}} = \sum_{\mathbf{k}} i/2\pi(n_\alpha - k_\alpha)u_{\mathbf{k}}^\alpha u_{\mathbf{n}-\mathbf{k}}^\beta$$

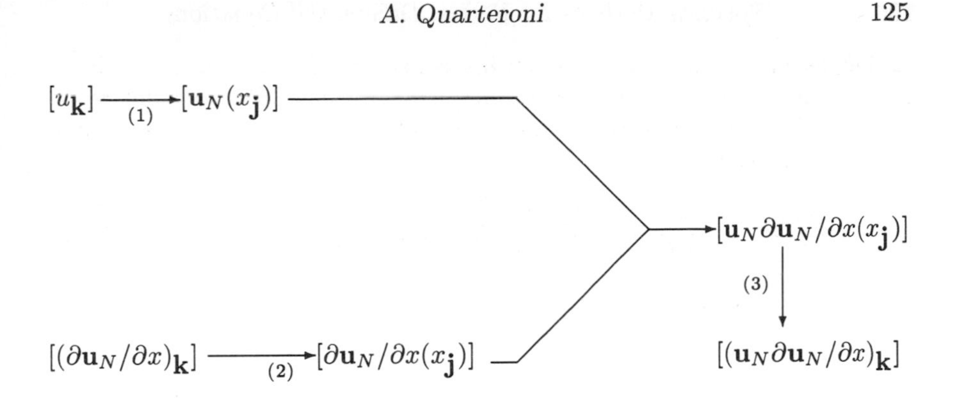

Fig. 3.2. Flow Diagram for Fourier calculation of convective terms

yielding $\mathcal{O}(N^6)$ operations. For example, if $N = 16$, one has $(2N)^3 = 33792$ unknowns, and more than 2×10^9 operations are needed to compute the convective term $(\mathbf{u}_N \cdot \nabla)\mathbf{u}_N$ through the above convolution sums.

The use of discrete Fourier transforms (D.F.T.) is made possible by introducing the gridpoints

$$x_{\mathbf{j}} = \pi \mathbf{j}/N, \quad 0 \leq j_\alpha \leq 2N-1, \quad \alpha = 1,2,3.$$

The procedure is schematically represented by the flow diagram in Figure 3.2.

Using the Fast Fourier Transform to carry out the D.F.T.s (1), (2) and (3), only $\mathcal{O}((2N)^3 \log_2 N)$ operations suffice to compute the nonlinear convective terms.

We illustrate now the Chebyshev collocation method on the following *viscous Burgers equation.* We consider the initial-boundary-value problem

$$\begin{cases} \frac{\partial u}{\partial t} - \nu \frac{\partial^2 u}{\partial x^2} + u\frac{\partial u}{\partial x} = 0, & -1 < x < 1,\ t > 0 \\ u(\pm 1, t) = 0 & t > 0 \\ u(x,0) = u_0(x) & -1 < x < 1. \end{cases}$$

The analysis of this equation can be carried out by the Cole-Hopf transform: $u = 2\nu\psi_x/\psi$ (note that ψ solves the linear heat equation $\psi_t = 2\nu\psi_{xx}$). By this approach one can prove global existence, and that $u(t,x) \to 0$ uniformly as $t \to \infty$. Then by standard monotonicity arguments it is straightforward to obtain that if $\max_x |u_0(x)| \leq M$ then $\max_{x,t} |u(x,t)| \leq M$.

Let us now introduce the following semidiscrete (continuous in time) approximation by the *Chebyshev collocation method.* We look for $u_N : (0,T) \to \mathbb{P}_N$ (algebraic polynomials in x of degree at most N such that for all $t < T$) such that

$$\frac{\partial u_N}{\partial t} - \nu \frac{\partial^2 u_N}{\partial x^2} + \frac{\partial_N(u_N^2)}{2} = 0 \text{ at } x_j,\ 1 \le j \le N-1, \tag{5.31}$$

and $u_N(\pm 1) = 0$. Here

$$x_j = \cos \pi j/N, \quad 0 \le j \le N,$$

are the Chebyshev-Lobatto points and $\partial_N \varphi := \partial/\partial x(I_N \varphi)$ is the Chebyshev pseudo-spectral derivative (see Section 3.4). We recall that

$$I_N \varphi(x) = \sum_{k=0}^{N} \varphi_k^* T_k(x) \in \mathbb{P}_N$$

denotes the interpolant of φ at the Chebyshev nodes $\{x_j,\ 0 \le j \le N\}$. Since $T_k(x_j) = \cos k\theta_j$ the discrete Chebyshev transforms (D.C.T.)

$$\{\varphi(x_j)\} \ \leftrightarrow\ \{\varphi_k^*\}$$

can be computed in a fast way by the F.F.T. (see Section 3.4). For each $t \le T$, using the approximation results concerning the Chebyshev interpolation and projection operators we can prove that for all $t > 0$,

$$\|(u - u_N)(t)\|_{L^2_w(-1,1)} \le C N^{1-s} \|u_0\|_{H^s_w(-1,1)}, \quad s \ge 1. \tag{5.32}$$

(For the notation, see Section 3.4).

Problem (5.31) is a system of ordinary differential equations whose associated matrix has a condition number growing like $\mathcal{O}(N^4)$. Explicit schemes to advance in time are therefore discouraged, for they would introduce a C.F.L. stability limit on the time step Δt of the form $\Delta t \le CN^{-4}$. Instead, it is advisable to use a *semi-implicit scheme*, i.e., a method dealing with the viscous terms implicitly and with the convective terms explicitly. The advantage is twofold: the severe C.F.L. condition is weakened; moreover, the nonlinear terms can be effectively computed through D.C.T. at previously known time levels. The same strategy is used in other cases, such as pseudoparabolic equations, the Korteweg-de Vries equation and the Navier-Stokes equations.

An example of such a method for the problem (5.31) is provided by the first order scheme:

$$\frac{u_N^{k+1} - u_N^k}{\Delta t} - \nu \frac{\partial^2 u_N^{k+1}}{\partial x^2} = -\partial_N((u_N^k)^2/2) \quad \text{at } x_j,\ 1 \le j \le N-1,$$

where u_N^k is the spectral solution at the time-level $t^k = k\Delta t$. The analysis shows that this method is unconditionally stable for any fixed time interval, viz.,

$$\|u_N^k\|_{L^2_w(-1,1)} \leq C(T)\|u_N^0\|_{L^2_w(-1,1)},$$

for all k such that $t^k \leq T$, and for all N, Δt. It is also spectrally convergent, i.e.,

$$\|u_N^k - u(k\Delta t)\|_{L^2_w(-1,1)} \leq C(T)\|u^0\|_{H^s_w(-1,1)}(N^{1-s} + \Delta t).$$

From the algorithmic point of view, for each k we have to solve the linear system (with fixed matrix):

$$(\frac{1}{\Delta t} - \nu D^2)\mathbf{V}^{k+1} = \frac{1}{\Delta t}\mathbf{V}^k - D(\mathbf{V}^k)^2/2, \tag{5.33}$$

where $\mathbf{V}^k = [u_N^k(x_j)]^{tr}$ and $D : [\varphi(x_j)]^{tr} \to [\partial_N\varphi(x_j)]^{tr}$ is the matrix of the pseudo-spectral Chebyshev derivative (see Section 3.4). We can write (5.33) as

$$L_{sp}\mathbf{X} = \mathbf{F} \quad \text{with} \quad L_{sp} = 1/\Delta t - \nu D^2. \tag{5.34}$$

The drawback is that the matrix L_{sp} associated with (5.33) is 'full' (precisely, $2N^3$ out of N^4 entries are non-zero), and severely ill-conditioned. Actually, if $\lambda_{\max}$ and $\lambda_{\min}$ denote its maximum and minimum eigenvalues (which are real), one has:

$$c(L_{sp}) := \lambda_{\max}/\lambda_{\min} = \mathcal{O}(N^4).$$

The remedy is, for instance, to solve the pseudo-evolutionary problem

$$\partial/\partial t(A\mathbf{Y}) + L_{sp}\mathbf{Y} = \mathbf{F}, \quad (\Rightarrow \mathbf{X} = \lim_{t\to+\infty} \mathbf{Y}(t))$$

by the explicit forward difference scheme. This yields the *Richardson* scheme

$$A(\mathbf{Y}^{m+1} - \mathbf{Y}^m) = \theta(\mathbf{F} - L_{sp}\mathbf{Y}^m),$$

where $\theta > 0$ is an acceleration parameter (dimensionally, θ is a time-step), and A is a suitable preconditioning matrix. Taking $\theta = \theta_{opt}$ (such an optimal value is achieved for $2/(\Gamma_{min} + \Gamma_{max})$, where $\{\Gamma\}$ are the eigenvalues of $A^{-1}L_{sp}$, provided they are real), the error reduction factor per iteration is

$$\rho = \frac{c(A^{-1}L_{sp}) - 1}{c(A^{-1}L_{sp}) + 1}.$$

Taking A as the finite difference (or finite element) approximation to the operator $1/\Delta t - \nu\partial^2_{xx}$ yields:

(i) A is banded (hence the system is easy to solve)

(ii) $Re\Gamma > 0$. (Thus the scheme is convergent.)
(iii) $c(A^{-1}L_{sp})$ is independent of N (therefore ρ is independent of N).

The convergence is extremely fast, especially if θ is chosen dynamically according to a minimal residual strategy. Other methods can be efficiently adopted to solve spectral systems of the form (5.34). Among them, we mention the preconditioned conjugate gradient method, the multigrid method and the alternating direction method (for problems in more than one dimension). An extensive review is presented in [8, Chapter 5].

3.6 Incompressible Navier-Stokes Equations

We consider in this Section the Navier-Stokes equations for incompressible viscous flows in a connected, bounded domain Ω of $\mathbb{R}^2$:

$$\begin{cases} -\nu\Delta\mathbf{u} + \mathbf{u}\cdot\nabla\mathbf{u} + \nabla p = \mathbf{f} & \text{in } \Omega \\ \operatorname{div}\mathbf{u} = 0 & \text{in } \Omega \\ \mathbf{u} = \mathbf{g} & \text{on } \partial\Omega. \end{cases} \tag{6.1}$$

Here $\nu > 0$ denotes the kinematic viscosity, $\mathbf{f}$ and $\mathbf{g}$ are two given vector functions, and $\mathbf{g}$ satisfies the compatibility condition $\int_{\partial\Omega}\mathbf{g}\cdot\mathbf{n} = 0$, where $\mathbf{n}$ is the outward normal direction to $\partial\Omega$; $\mathbf{u}$ is the velocity field and p is the pressure. An extensive mathematical analysis of the Navier-Stokes equations can be found in [22].

In (6.1) one looks for a solution $\mathbf{u}$ in the subspace V^* of $[H^1(\Omega)]^2$ of the vector fields satisfying the prescribed boundary conditions, and for a pressure $p \in Q = \{q \in L^2(\Omega) : \int_\Omega q dx = 0\}$. Associated with the full Navier-Stokes problem (6.1) is the (reduced) linear Stokes problem:

$$\begin{cases} -\nu\Delta\mathbf{u} + \nabla p = \mathbf{f} & \text{in } \Omega \\ \operatorname{div}\mathbf{u} = 0 & \text{in } \Omega \\ \mathbf{u} = \mathbf{g} & \text{on } \partial\Omega. \end{cases} \tag{6.2}$$

The analysis of any numerical approximation to the problem (6.1) can be reduced to analysing the corresponding Stokes problem (6.2). Actually, taking $\mathbf{g} = \mathbf{0}$ for simplicity, so that V^* reduces to the space $V := [H_0^1(\Omega)]^2$, let us define the Stokes solution operator $T : V' \to V \times Q$ as $T\mathbf{f} = (\mathbf{u}, p)$. Here $V' = [H^{-1}(\Omega)]^2$ is the dual of the space V. Now define the nonlinear operator $G : V \times Q \to V$ as $G(\mathbf{u}, p) = f - \mathbf{u}\cdot\nabla\mathbf{u}$. If we set $\mathbf{x} = (\mathbf{u}, p)$ and $F(\mathbf{x}) = \mathbf{x} - TG(\mathbf{x})$, the solution $\mathbf{x}$ to (6.1) is characterized by the equation:

$$\mathbf{x} \in V \times Q : F(\mathbf{x}) = 0. \tag{6.3}$$

Consider a finite dimensional approximation to the full Navier-Stokes problem (6.1), in which the velocity $\mathbf{u}_h$ is sought in $V_h \subset V$ and the pressure

p_h in $Q_h \subset Q$, V_h and Q_h being, of course, some suitable finite dimensional spaces. The operator T is approximated by $T_h : V' \to V_h \times Q_h$, and G by $G_h : V_h \times Q_h \to V'$. Setting $\mathbf{x}_h = (\mathbf{u}_h, p_h)$ and $F_h(\mathbf{x}_h) = \mathbf{x}_h - T_h G_h(\mathbf{x}_h)$, as before the finite dimensional problem can be written as

$$\mathbf{x}_h \in V_h \times Q_h : F_h(\mathbf{x}_h) = 0.$$

At this point one can invoke abstract theorems concerning the approximation of a nonlinear problem of the form (6.3) by the family of finite dimensional problems defined above. With these theorems, which are based on an implicit function theorem, (see, e.g., [5,6,7]) one exploits the approximation properties of T_h to T and of G_h to G. Typically, G_h is obtained from G using either numerical integration or numerical interpolation. Hence its convergence to G relies upon classical approximation results. On the other hand, the convergence of T_h to T amounts to the convergence of the solution to a discrete Stokes problem to (6.2).

From now on we will focus only on problem (6.2), and we will investigate its finite dimensional approximations using finite dimensional methods.

3.6.1 Abstract Stability and Convergence Results for the Stokes Problem

With the aim of stating a suitable framework for the analysis, it is convenient to write (6.2) in a variational form. We define

$$a(\mathbf{u}, \mathbf{v}) := \nu \int_\Omega \nabla \mathbf{u} \cdot \nabla \mathbf{v}\, dx, \quad b(\mathbf{v}, q) := -\int_\Omega q \operatorname{div} \mathbf{v}\, dx.$$

As usual, we will denote the norms of $L^2(\Omega)$ and $[H^1(\Omega)]^2$ by

$$|q| = \left\{ \int_\Omega |q^2(x)|\, dx \right\}^{1/2} \quad \text{and} \quad \|\mathbf{v}\| = \{|v|^2 + |\nabla v|^2\}^{1/2},$$

respectively. If we denote by $\mathbf{u}_0$ a vector function of $[H_0^1(\Omega)]^2$ such that $\operatorname{div}\mathbf{u}_0 = 0$, $\mathbf{u}_0 = \mathbf{g}$ on $\partial\Omega$, and we set $\mathbf{w} = \mathbf{u} - \mathbf{u}_0$, we can reformulate (6.2) as follows:

$$\begin{cases} \text{find } (\mathbf{w}, p) \in V \times Q \text{ such that} & \\ a(\mathbf{w}, \mathbf{v}) + b(\mathbf{v}, p) = \langle \mathbf{L}, \mathbf{v} \rangle & \text{for all } \mathbf{v} \in V \\ b(\mathbf{w}, q) = 0 & \text{for all } q \in Q. \end{cases} \tag{6.4}$$

We have set: $L \in V'$, $\langle \mathbf{L}, \mathbf{v} \rangle := \langle \mathbf{f}, \mathbf{v} \rangle - a(\mathbf{u}_0, \mathbf{v})$. The forms $a : V \times V \to \mathbb{R}$, and $b : V \times Q \to \mathbb{R}$, are bilinear and continuous. Moreover a is V-elliptic, i.e., there exists a constant $\alpha > 0$ for which

$$a(\mathbf{v}, \mathbf{v}) \geq \alpha \|\mathbf{v}\|^2 \quad \text{for all } \mathbf{v} \in V, \tag{6.5}$$

and b satisfies the following 'inf-sup' condition: there exists a constant $\beta > 0$ such that

$$\sup_{\mathbf{v}\in V} b(\mathbf{v}, q)/\|\mathbf{v}\| \geq \beta|q| \quad \text{for all } q \in Q. \tag{6.6}$$

Throughout this section $\|\cdot\|$ denotes the norm in V, and $|\cdot|$ the norm in Q. Due to (6.5) and (6.6), one can conclude that if $\mathbf{f} \in [H^{-1}(\Omega)]^2$, and $\mathbf{g} \in [H^{1/2}(\partial\Omega)]^2$ (see [15]), then there exists a unique solution of (6.4).

We introduce now a finite dimensional approximation to the problem (6.4). Let $h > 0$ be the parameter of the discretization, and let V_h and Q_h be two finite dimensional subspaces of V and Q, respectively. We consider the following approximation to (6.4):

$$\begin{cases} \text{find } (\mathbf{w}_h, p_h) \in V_h \times Q_h \text{ such that} & \\ a(\mathbf{w}_h, \mathbf{v}) + b(\mathbf{v}, p_h) = \langle \mathbf{L}, \mathbf{v}\rangle & \text{for all } \mathbf{v} \in V_h \\ b(\mathbf{w}_h, q) = 0 & \text{for all } q \in Q_h. \end{cases} \tag{6.7}$$

We define

$$V_h^b := \{\mathbf{v} \in V_h : b(\mathbf{v}, q) = 0 \text{ for all } q \in Q_h\}, \tag{6.8}$$

and we note that the unknown $\mathbf{w}_h$ of (6.7) satisfies:

$$\mathbf{w}_h \in V_h^b : a(\mathbf{w}_h, \mathbf{v}) = \langle \mathbf{L}, \mathbf{v}\rangle, \quad \text{for all } \mathbf{v} \in V_h^b. \tag{6.9}$$

If we set $V^b := \{\mathbf{v} \in V : b(\mathbf{v}, q) = 0 \text{ for all } q \in Q\}$, then (6.9) can be viewed as a direct, external (since $V_h^b \not\subset V^b$ in general) approximation to the problem

$$\mathbf{w} \in V^b : a(\mathbf{w}, \mathbf{v}) = \langle \mathbf{L}, \mathbf{v}\rangle \quad \text{for all } \mathbf{v} \in V^b, \tag{6.10}$$

which is satisfied by the solution $\mathbf{w}$ of (6.4). The following theorem states a basic result concerning the problem (6.7). For its proof we refer to [4] or [12].

Theorem 6.1. *Assume that there exists a constant $\alpha^* > 0$ such that:*

$$a(\mathbf{v}, \mathbf{v}) \geq \alpha^* \|\mathbf{v}\|^2 \quad \textit{for all } \mathbf{v} \in V_h^b. \tag{6.11}$$

Then problem (6.9) has a unique solution, and there is a constant C_1 depending only upon α^*, $\|a\|$ and $\|b\|$ (the norms of the bilinear forms a and b) such that the following 'optimal' error bound holds:

$$\|\mathbf{u} - \mathbf{u}_h\| \leq C_1 \left\{ \inf_{v_h \in V_h^b} \|\mathbf{u} - \mathbf{v}_h\| + \inf_{q_h \in Q_h} |p - q_h| \right\}. \tag{6.12}$$

If we assume in addition that there exists a constant $\beta^* > 0$ *such that*

$$\sup_{\mathbf{v}_h \in V_h} b(\mathbf{v}_h, q_h)/\|\mathbf{v}_h\| \geq \beta^* |q_h| \quad \text{for all } q_h \in Q_h, \tag{6.13}$$

then there exists $p_h \in Q_h$ *such that* $(\mathbf{w}_h, p_h)$ *is the only solution of problem (6.7). Furthermore, there exists a constant* C_2 *depending upon* α^*, β^*, $\|a\|$ *and* $\|b\|$ *such that:*

$$\|\mathbf{u} - \mathbf{u}_h\| + |p - p_h| \leq C_2 \left\{ \inf_{v_h \in V_h} \|\mathbf{u} - \mathbf{v}_h\| + \inf_{q_h \in Q_h} |p - q_h| \right\}. \tag{6.14}$$

In particular, C_2 *is proportional to* $1/\beta^*$.

We note that (6.11) is a consequence of the assumption (6.5) whenever V_h is a subspace of V. On the contrary, (6.13) is not a direct consequence of (6.6), but entails a strong compatibility relation between the finite dimensional subspaces V_h and Q_h, that should be fulfilled for the well-posedness of problem (6.7). Clearly, if β^* is independent of h, then (6.14) yields an optimal error bound. The verification of the discrete inf-sup condition (6.13) is therefore a crucial step in the construction and the analysis of finite element approximations to the Stokes equations. The following result (which is due to Fortin) provides a useful criterior for such verification.

Lemma 6.2. *The discrete inf-sup condition (6.13) holds with a constant* $\beta^* > 0$ *independent of* h *provided there exists an operator* $\Pi_h : V \to V_h$ *satisfying*

$$b(\mathbf{v} - \Pi_h \mathbf{v}, q_h) = 0, \quad \text{for all } q_h \in Q_h \text{ and for all } \mathbf{v} \in V, \tag{6.15}$$

and

$$\|\Pi_h \mathbf{v}\| \leq C\|\mathbf{v}\|, \quad \text{for all } \mathbf{v} \in V, \tag{6.16}$$

with a constant $C > 0$ *independent of* h.

Actually, if such an operator exists, owing to (6.15) we have for all $q_h \in Q_h$:

$$\sup_{v_h \in V_h} b(\mathbf{v}_h, q_h)/\|\mathbf{v}_h\| \geq \sup_{v \in V} b(\Pi_h \mathbf{v}, q_h)/\|\Pi_h \mathbf{v}\| = \sup_{v \in V} b(\mathbf{v}, q_h)/\|\Pi_h \mathbf{v}\|.$$

Since $q_h \in Q$, from (6.16) and the 'continuous' inf-sup condition (6.6) it follows that

$$\sup_{v \in V} b(\mathbf{v}, q_h)/\|\Pi_h \mathbf{v}\| \geq C^{-1} \sup_{v \in V} b(\mathbf{v}, q_h)/\|\mathbf{v}\| \geq \beta C^{-1} |q_h|.$$

Whence (6.13) follows with $\beta^* = \beta/C$. Furthermore, one can easily prove that the existence of such an operator is also a necessary condition in order that (6.13) be satisfied.

3.6.2 Approximation to the Stokes Equations by Spectral Methods

Spectral approximations of the Stokes problem (6.2) are based on either Galerkin or collocation methods using finite dimensional subspaces V_N and Q_N of V and Q made by global polynomials. Typically, V_N is the square of the space of polynomials of degree at most N in each variable. Here by polynomial we mean an algebraic polynomial, or even a trigonometric polynomial for those directions (if any) in which the solution $\mathbf{u}$ is periodic. Accordingly, the most natural space for the pressure is the subspace of Q made by polynomials of degree at most N in each variable. Let us denote this space by $\mathcal{Q}_N$. With this choice, however, the compatibility condition (6.13) may not be fulfilled. (From now on, when we refer to the notation and results of Section 3.6.1, the identification $V_h = V_N, Q_h = \mathcal{Q}_N$, will be understood.)

As a matter of fact, assume we use a Galerkin projection method with respect to the inner product

$$(\mathbf{u}, \mathbf{v}) = \int_{\Omega} \mathbf{u}(x)\mathbf{v}(x)\, dx \tag{6.17}$$

to approximate the momentum equation. Then we look for $\mathbf{u}_N \in V_N$ and $p_N \in \mathcal{Q}_N$ such that

$$-\nu(\Delta \mathbf{u}_N, \mathbf{v}) + (\nabla p_N, \mathbf{v}) = (\mathbf{f}, \mathbf{v}) \quad \text{for all } \mathbf{v} \in V_N. \tag{6.18}$$

We are assuming that $\mathbf{f} \in [L^2(\Omega)]^2$, so that the integral on the right hand side of (6.18) makes sense. If there exists a polynomial $p^* \in \mathcal{Q}_N$, $p^* \neq 0$, such that

$$(\nabla p^*, \mathbf{v}) = 0 \quad \text{for all } \mathbf{v} \in V_N, \tag{6.19}$$

then the couple $(\mathbf{u}_N, p_N + p^*)$ would also be a solution of (6.18). In such a circumstance, the spectral algorithm would fail to define a unique pressure. The pressure functions $p^* \neq 0$ satisfying (6.19) are called *spurious modes* or *parasitic modes*. They form a linear subspace X_N of $\mathcal{Q}_N$. Clearly, the inf-sup condition (6.13) is not satisfied if the pressure space shares some elements with X_N.

In the next subsection the space of parastic modes X_N will be characterized for spectral approximations of Stokes equations with different kinds of boundary conditions. We will consider fully periodic solutions, then solutions which are periodic in one direction and submitted to a homogeneous Dirichlet condition in the others, and finally the case of fully homogeneous Dirichlet conditions.

Fully Periodic Problems

We assume that $\Omega = (0, 2\pi)^2$, and that $\mathbf{f}$ is periodic of period 2π in each direction, so that $\mathbf{u}$ and p are also 2π-periodic. We denote by S_N the space of trigonometric polynomials of degree at most N in each variable (see Section 3.1). The space of velocities is $V_N = [S_N \cap Q]^2$ so that their average is fixed to zero, and the space of pressures is $\mathcal{Q}_N = S_N \cap Q$. This gives a pure Fourier Galerkin approximation to the Stokes equations, which was presented in section 3.5.2 on the full Navier-Stokes problem. In this way, $\nabla p \in V_N$ for all $p \in \mathcal{Q}_N$. Then choosing $\mathbf{v} = \nabla p^*$ in (6.19) we get $\nabla p^* = 0$, and therefore $p^* = 0$. It follows that the pure Fourier method is free of spurious modes, i.e., X_N is empty.

The spaces $V_N, \mathcal{Q}_N$ are precisely those used in the classical Orszag-Patterson method, which was first analyzed in [17].

Mixed Periodic-Nonperiodic Problems

We assume now that $\Omega = (0, 2\pi) \times (-1, 1)$, and that the solutions to (6.2) are periodic of period 2π with respect to the x-direction, whereas $\mathbf{u}(x, \pm 1) = 0$ for all $x \in (0, 2\pi)$. We take $V_N = [S_N \otimes \mathbb{P}_N^0]^2$, where S_N is now the space of trigonometric polynomials of degree at most N in one variable, and $\mathbb{P}_N^0$ is the subspace of the algebraic polynomials $\mathbb{P}_N$ which vanish at $y = \pm 1$. We look for spurious modes in the pressure space $\mathcal{Q}_N = [S_N \otimes \mathbb{P}_N] \cap Q$.

We write each $p^* \in \mathcal{Q}_N$ as $p^*(x, y) = \sum_{k=-N}^{N-1} \sum_{n=0}^{N} p_{kn}^* e^{ikx} L_n(y)$, where $L_n(y)$ denotes the nth Legendre polynomial (see Section 3.3). (Notice that $p_{00}^* = 0$ as p^* has zero average). The equation (6.19) yields the two relations:

$$\int_\Omega p^* \partial v / \partial y = 0, \quad \text{for all } v \in S_N \otimes \mathbb{P}_N^0, \tag{6.20}$$

$$\int_\Omega p^* \partial v / \partial x = 0, \quad \text{for all } v \in S_N \otimes \mathbb{P}_N^0. \tag{6.21}$$

The following set of functions is a basis in $S_N \otimes \mathbb{P}_N^0$:

$$\{q_{km}(x, y) = e^{ikx}(1 - y^2)L_m'(y) : -N \leq k \leq N - 1,\ 1 \leq m \leq N - 1\}. \tag{6.22}$$

We recall that the Legendre polynomials satisfy the Sturm-Liouville equation

$$((1 - y^2)L_m'(y))' + m(m + 1)L_m(y) = 0. \tag{6.23}$$

Taking $v = q_{km}$ in (6.20), and using (6.23) yields the set of equations

$$p_{km}^* = 0, \quad -N \leq k \leq N - 1,\ 1 \leq n \leq N - 1.$$

An alternative basis for $S_N \otimes \mathbb{P}_N^0$ is given by

$$\{\sigma_{km}(x,y) = e^{ikx}[L_m(y) - L_\alpha(y)],\ -N \le k \le N-1,\ 2 \le m \le N\},\quad (6.24)$$

where $\alpha = 0$ if m is even, whereas $\alpha = 1$ if m is odd. Using $v = \sigma_{km}$ in (6.21) and exploiting the orthogonality of the Legendre polynomials yields the new set of equations

$$p^*_{km} \int_{-1}^{1} [L_m(y)]^2\, dy = p^*_{k\alpha} \int_{-1}^{1} [L_\alpha(y)]^2\, dy,\ -N \le k \le N-1,\ 2 \le m \le N. \quad (6.25)$$

Since $\int_{-1}^{1} (L_i)^2 dy = (i + 1/2)^{-1}$, as a consequence of (6.24) and (6.25), we deduce that

$$p^*_{km} = 0, \quad \text{for all } k,\ m, \text{ except } p^*_{0N}.$$

Thus the only spurious mode is the one corresponding to p^*_{0N}, i.e., $X_N = \text{span}\{L_N(y)\}$ and $\dim X_N = 1$.

Fully Nonperiodic Problems

Consider now the case in which $\mathbf{u} = \mathbf{0}$ on $\partial\Omega$; V_N is the space $[\mathbb{P}_N^0 \otimes \mathbb{P}_N^0]^2$, and $Q_N = [\mathbb{P}_N \otimes \mathbb{P}_N] \cap Q$. If we take $(1 - s^2)L'_m(s)$ as basis functions in the direction of differentiation, and $L_m(s) - L_\alpha(s)$ in the other direction (again $\alpha = 0$ if m is even, and $\alpha = 1$ if m is odd), we find as non-trivial solutions to (6.19) the four modes:

$$L_i(x)L_j(y), \quad i, j = 0 \text{ or } N,$$

and also four other functions p^*_{ij} $(i, j = 0$ or $1)$, which are suitable combinations of the remaining modes with the same parity:

$$p^*_{ij}(x,y) = \sum_{k=i(\text{mod}2)} \sum_{m=j(\text{mod}2)} b_{ijkm} L_k(x) L_m(y).$$

Since we assumed from the beginning that $p^*_{00} = 0$, we conclude that in this case $\dim X_N = 7$. If a Chebyshev (rather than Legendre) Galerkin method is used, the same results hold provided each L_k is replaced by T_k, the kth Chebyshev polynomial.

Finally, the collocation (rather than Galerkin) method is used to approximate the Stokes equations, with the Chebyshev knots

$$x_{ij} = (\cos \pi i/N, \cos \pi j/N), \quad i, j = 0, \ldots, N,$$

then the spurious modes are spanned by the seven functions

$$T_i(x)T_j(y), \quad i, j = 0 \text{ or } N\ (ij \ne 0), \text{ and } T'_N(x)T'_N(y)(1 \pm x)(1 \pm y).$$

In the above discussion we have determined those parasitic modes in the polynomial pressure space Q_N that affect the momentum equation approximated by the Galerkin method. We consider now a full approximation to the Stokes equations by Galerkin methods. We use Galerkin rather than the most practical collocation approach in order to simplify our analysis. The spectral problem is

$$\begin{cases} \text{find } \mathbf{u}_N \in V_N,\ p_N \in Q_N \text{ such that} & \\ -\nu(\Delta \mathbf{u}_N, \mathbf{v}) + (\mathbf{v}, \nabla p_N) = (\mathbf{f}, \mathbf{v}), & \text{for all } \mathbf{v} \in V_N \\ (\operatorname{div} \mathbf{u}_N, q) = 0, & \text{for all } q \in Q_N. \end{cases} \tag{6.26}$$

A collocation method can also be cast in the same framework, provided the L^2-inner product (6.17) is replaced by a discrete inner product at the collocation nodes (see Section 3.4.3). After integration by parts, problem (6.26) can be written in the abstract form

$$\begin{cases} \text{find } \mathbf{u}_N \in V_N,\ p_N \in Q_N \text{ such that} & \\ a(\mathbf{u}_N, \mathbf{v}) + b(\mathbf{v}, p_N) = (\mathbf{f}, \mathbf{v}), & \text{for all } \mathbf{v} \in V_N \\ b(\mathbf{u}_N, q) = 0, & \text{for all } q \in Q_N, \end{cases} \tag{6.27}$$

with the usual definitions of a and b given at the beginning of this Section. This is a problem like (6.7) (with the obvious modification of notation). Therefore the abstract results of theorem 6.1 can be applied to infer its stability and convergence properties.

In the current situation the space defined in (6.8) becomes

$$V_N^b = \{\mathbf{v} \in V_N : b(\mathbf{v}, q) = 0, \text{ for all } q \in Q_N.\}$$

It is therefore a subspace of $V = [H_0^1(\Omega)]^2$. Hence (6.11) holds with $\alpha^* = \alpha$ (independent of N) owing to (6.5). We look now for a pressure space Q_N that satisfies (6.13) (written for V_N and Q_N instead of V_h and Q_h) with a constant β^* possibly independent of N, or at least decaying with N at the lowest possible rate. In the latter case, estimate (6.14) will still imply convergence provided that the infima on the right-hand side decay to zero sufficiently fast to compensate for the growth of the constant C_2.

Recall that we have chosen as V_N the space of polynomials of degree at most N that satisfy the prescribed boundary conditions. As a consequence of the analysis carried out in the previous subsections, it is clear that the space Q_N cannot contain any element of X_N, the space of parasitic modes. Actually, due to (6.19), the functions $p^* \in X_N$ are precisely those which yield zero in the left-hand side of (6.13), while the right-hand side would be positive (unless we are in the trivial case $p^* \equiv 0$). Thus, we are led to choose as Q_N any subspace of $\mathcal{Q}_N$ (the full space of polynomials of degree at most N with zero average) such that:

$$Q_N \cap X_N = \{0\}, \ \dim Q_N + \dim X_N = \dim \mathcal{Q}_N. \tag{6.28}$$

A space Q_N with this property is a *supplementary* space of X_N in $\mathcal{Q}_N$. Any such space satisfies (6.13) with a suitable constant β^*. Actually, if for a given $q \in Q_N$ one has

$$\sup_{\mathbf{v} \in V_N} b(\mathbf{v}, q)/\|\mathbf{v}\| = 0, \tag{6.29}$$

then $b(\mathbf{v}, q) = 0$ for all $\mathbf{v} \in V_N$. Hence $q \in X_N$ and therefore $q = 0$ owing to (6.28). It follows that the left-hand side of (6.29) is a norm for Q_N, which must be equivalent to the norm $|q|$ since Q_N has finite dimension.

In conclusion, if V_N is chosen as above, and Q_N satisfies (6.28), the spectral Stokes approximation (6.27) has a unique solution. Among the spaces fulfilling these conditions, it is advisable to use those for which the constant β^* is the largest possible, so that C_2 in (6.14) is the lowest possible. Of course the larger is Q_N, the smaller is C_2. The best choice is to take Q_N as the orthogonal complement of X_N in $\mathcal{Q}_N$ with respect to the inner product (6.17). Other choices however have also been pursued in the applications. For a review we refer the reader to [8, Chapters 7 and 11].

The infima in the right-hand side of the convergence estimate (6.14) can be evaluated accordingly, by resorting to the error estimates involving spectral projection and interpolation operators (see Sections 3.1, 3.3 and 3.4 and [8, Chapter 9]. The results we obtain is

$$\|\mathbf{u} - \mathbf{u}_N\| + |p - p_N| \leq C(\beta^*)^{-1} N^{1-s} \{\|\mathbf{u}\|_{H^s(\Omega)} + \|p\|_{H^{s-1}(\Omega)}\},$$

assuming of course that for some $s \geq 2$ the norms in the right-hand side are bounded.

Before concluding this section, we want to mention that there exist spectral algorithms in which the divergence free condition is not discretized directly as we did in the second equation of (6.26). An alternative approach consists of discretizing a Poisson equation for the pressure. This equation is obtained by taking the divergence of the momentum equation and using the continuity equation to eliminate the velocity from the resulting equation. The proper (unknown) boundary value for p is then determined by an influence operator technique by requiring that the corresponding velocity field be divergence free. In the spectral frame, this method has been proposed and extensively used by Kleiser and Schumann[14]. Other methods are considered in [16].

We conclude this chapter with a short comment on the issue of *time-discretization* of the unsteady Navier-Stokes equations. Several methods have been proposed in recent years for advancing in time the spatially discretized momentum and continuity equations. Among them we mention finite difference schemes, *explicit* or *implicit*. For an analysis of one-step and multistep stable schemes to be used for finite element spatial approximations we refer to [22,12].

Semi-implicit finite difference schemes are also used, especially for spectral approximations in space using Fourier or Chebyshev methods. Actually, if we deal with viscous and pressure terms implicitly and with the convective nonlinear terms explicitly, the latter terms can be effectively computed by fast transform techniques (e.g., the fast Fourier/Chebyshev transform). Furthermore, the CFL stability limit (see [20]) is not too severe, since the higher order viscous term is dealt with implicitly (see Section 3.5.2, [8, Chapter 11] and [16]).

An alternative approach is the use of *fractional-step methods* (or splitting techniques). With such techniques at each time interval an advection-diffusion step is generally decoupled from a pressure step in which the pressure is corrected with the purpose of recovering the divergence-free constraint. Several variants can be adopted, however. For example, the full time-stepping can be the results of three (rather than two) steps, one for the viscous terms, the second for nonlinear advective terms and the last for the pressure correction. In the frame of finite element approximations, fractional-step methods are described in [11]. For spectral approximations in space, a review of the most commonly used fractional-step algorithms is given in [8, Chapter 7].

3.7 Domain Decomposition Methods

Domain decomposition is a very effective technique for large scale solution of partial differential equations. In this Section, we consider domain decompositions in which the interaction between the subdomains is handled by proper iterative methods, in the framework of spectral methods. These iterative domain decomposition methods allow the reduction of a problem set in a complicated geometry into a sequence of problems of similar type but with smaller size in every subdomain. Due to their high degree of parallelism, these methods can be implemented on multiprocessor systems. We present here an approach which relies upon an equivalence principle between single-domain and multi-domain formulation of boundary value problems. A proper iterative procedure among the subdomains yields a sequence of single-domain problems with alternating Dirichlet and Neumann boundary conditions at each subdomain interface. Such an approach is so general that it naturally applies to differential problems of several types.

Among them we mention elliptic equations and Navier-Stokes equations. The rate of convergence of the iterative method proposed here does not decrease as the number of unknowns N in each subdomain increases. From an algebraic point of view, the iteration-by-subdomain method considered here amounts to solving the capacitance system governing the interface unknowns by an iteration procedure using a proper block diagonal preconditioner. Other preconditioners for the capacitance system can be used, and they give rise to different iteration-by-subdomains algorithms. A survey on this domain decomposition approach can be found in [18]. Domain decomposition based on the so-called spectral element method is also of great interest (see, e.g., [16] and the references therein).

3.7.1 Multidomain Formulation of an Elliptic Problem

We consider the boundary-value-problem

$$Lu := -\Delta u + \alpha u = f \text{ in } \Omega,\ u = 0 \text{ on } \partial\Omega, \tag{7.1}$$

where $\alpha \geq 0$ and Ω is a two dimensional domain with boundary $\partial\Omega$. In variational form (7.1) reads

$$u \in H_0^1(\Omega) : a(u,v) = \int_\Omega f v\, dx \quad \text{for all } v \in H_0^1(\Omega). \tag{7.2}$$

(See (5.4) for the definition of $a(\cdot,\cdot)$). Let us consider a partition of Ω by non-intersecting open subdomains Ω_i, $i = 1,\ldots,M$, and denote by Γ_i the common boundary between Ω_i and Ω_{i+1}, $i \leq M-1$ (see Figure 3.3 for an example). Then set $\Gamma = \cup\Gamma_i$, and, for each i, we denote with $\Phi := \prod_{i=1}^{M-1} H_{00}^{1/2}(\Gamma_i)$ the space of functions defined on Γ, which are the traces on Γ of functions of $H_0^1(\Omega)$ (see [15]). Note that these functions vanish at the endpoints of Γ_i for $i = 1,\ldots,M-1$. Let us define

$$a_i(u,v) = \int_{\Omega_i} (\nabla u \nabla v + \alpha_0 uv)\, dx.$$

If we set $u_i = u_{|\Omega_i}$, then it is readily seen that (7.2) is equivalent to:

$$\begin{aligned}
a_i(u_i,v) &= \int_{\Omega_i} f v\, dx \quad \text{for all } v \in H_0^1(\Omega_i), \quad (7.3)\\
&\qquad\qquad 1 \leq i \leq M\\
u_i &= u_{i+1} \quad \text{on } \Gamma_i,\ 1 \leq i \leq M-1, \quad (7.4)\\
u_i &= 0 \quad \text{on } \partial\Omega_i \cap \partial\Omega,\ 1 \leq i \leq M, \quad (7.5)\\
\sum_{\substack{i=1\\ i\ odd}}^{M} \left\{a_i(u_i,\overline{\varphi}) - \int_{\Omega_i} f\overline{\varphi}\, dx\right\} &= -\sum_{\substack{j=2\\ j\ even}}^{M} \left\{a_j(u_j,\overline{\varphi}) - \int_{\Omega_j} f\overline{\varphi}\, dx\right\}, \quad (7.6)\\
&\qquad\qquad \text{for all } \varphi \in \Phi.
\end{aligned}$$

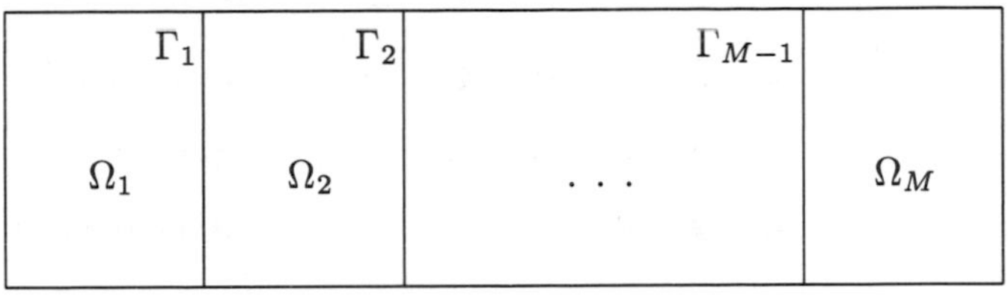

Fig. 3.3. Decomposition by aligned subdomains

Here $\overline{\varphi}$ denotes any continuous extension of φ to $H_0^1(\Omega)$. Clearly, (7.3) amounts to requiring that

$$Lu_i = f \quad \text{in } \Omega_i,\ 1 \le i \le M, \tag{7.7}$$

while (7.6) is equivalent to the *transmission condition*

$$\frac{\partial u_i}{\partial \nu} = \frac{\partial u_{i+1}}{\partial \nu} \quad \text{on } \Gamma_i,\ 1 \le i \le M-1, \tag{7.8}$$

where ν is the outward unit normal to Γ_i.

3.7.2 Spectral Collocation Approximation to the Multidomain Problem

A spectral collocation approximation to the problem (7.3)–(7.6) consists of collocating the differential equations (7.7) at Gaussian collocation points internal to Ω_i, and enforcing the interface conditions at some selected points of Γ. Either Gaussian-Legendre or Gaussian-Chebyshev points can be used.

For the sake of simplicity we confine ourselves to the case of a rectangular domain Ω partitioned into M non-intersecting rectangles Ω_i without internal cross points (see Figure 3.3). Let D be the reference domain $(-1,1)^2$ and let N_x, N_y be two given natural numbers. The Legendre Gauss-Lobatto collocation points in D are the roots $\{\xi_{km},\ 0 \le k \le N_x,\ 0 \le m \le N_y\}$ of the polynomial

$$\frac{\partial}{\partial x} L_{N_x}(x) \frac{\partial}{\partial y} L_{N_y}(y),$$

where $L_k(t)$ is the Legendre polynomial of degree k in $[-1,1]$ (see Section 3.3). In the Chebyshev case, the Gauss Lobatto collocation points in D are given by

$$\xi_{km} = \left(-\cos\frac{\pi k}{N_x}, -\cos\frac{\pi m}{N_y}\right), \quad \text{for } 0 \le k \le N_x,\ 0 \le m \le N_y.$$

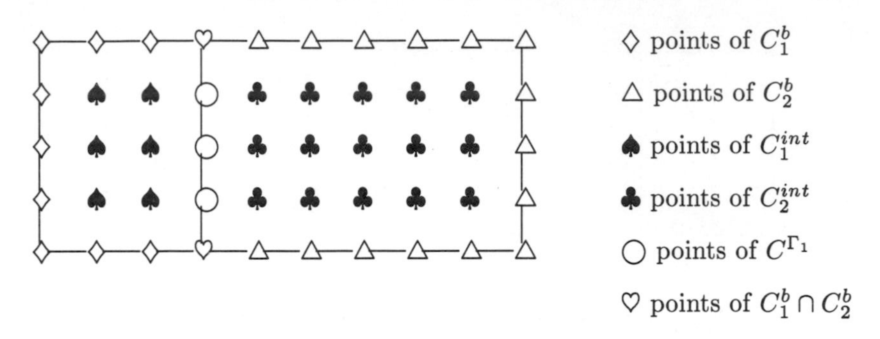

Fig. 3.4. Interior and Boundary Collocation Points

We recall that, in both Legendre and Chebyshev cases, ξ_{km} lies on the boundary of D if either k equals 0 or N_x, or m equals 0 or N_y. Moreover,

$$C_i := \{(x_k^i, y_m),\ 0 \le k \le N_x^i,\ 0 \le m \le N_y\}, \quad \text{for } i = 1, \dots, M,$$

will denote the set of collocation points in the domain Ω_i. Given M natural numbers N_x^i, $i = 1, \dots, M$, we take the points of C_i as the images of the Legendre (or Chebyshev) points in D, with $N_x = N_x^i$, through the affine transformation which maps D into Ω_i. For the convenience of exposition we set

$$C_i^{int} := C_i \cap \Omega_i,\ C_i^b := C_i \cap \partial\Omega,\ C^{\Gamma_i} := C_i \setminus (C_i^{int} \cup C_i^b), \quad \text{for } i = 1, \dots, M.$$

See Figure 3.4 for an example with $M = 2$.

For $i = 1, \dots, M$ we denote by $\mathbb{P}_N(\Omega_i)$ the space of polynomials of degree N_x^i with respect to the x variable and degree N_y with respect to the y variable. Moreover we set

$$\mathbb{P}_N^0(\Omega_i) := \{p \in \mathbb{P}_N(\Omega_i) : p = 0 \text{ on } \partial\Omega_i\}.$$

For the convenience of the reader we recall the Legendre and Chebyshev Gauss-Lobatto formulae.

Legendre Gauss-Lobatto formula.

$$\int_{\Omega_i} g(x, y)\, dxdy \cong \sum_{k=0}^{N_x^i} \sum_{m=0}^{N_y} g(x_k^i, y_m)\omega_k^i \omega_m, \tag{7.9}$$

where

$$\omega_k^i = \frac{a_i}{N_x^i(N_x^i+1)} \cdot \frac{1}{[L_{N_x}(x_k^i)]^2} \quad \text{and} \quad \omega_m = \frac{2}{N_y(N_y+1)} \cdot \frac{1}{[L_{N_y}(y_m)]^2}. \tag{7.10}$$

We have set $a_i = x_i - x_{i-1}$, where x_i denotes the abscissa of Γ_i.

Chebyshev Gauss-Lobatto formula.

$$\int_{\Omega_i} g(x,y)\omega^i(x,y)\,dxdy \cong \sum_{k=0}^{N_x^i}\sum_{m=0}^{N_y} g(x_k^i, y_m)\omega_k^i\omega_m, \tag{7.11}$$

where

$$\begin{cases} \omega^i(x,y) = [(x-x_{i-1})(x_i-x)]^{-1/2}(1-y^2)^{-1/2}, \\ \omega_k^i = \dfrac{a_i\pi}{2N_x^i}, \quad \text{for } k=1,\dots,N_x^i-1, \\ \omega_k^i = \dfrac{a_i\pi}{4N_x^i}, \quad \text{for } k=0 \text{ and } k=N_x^i, \\ \omega_m = \dfrac{\pi}{N_y}, \quad \text{for } m=1,\dots,N_y-1, \\ \omega_m = \dfrac{\pi}{2N_y}, \quad \text{for } m=0 \text{ and } m=N_y. \end{cases} \tag{7.12}$$

The *spectral collocation* solution u_N to the problem (7.1) is such that:

$$u_{i,N} := u_{N|\Omega_i} \in P_N(\Omega_i), \quad i=1,\dots,M, \tag{7.13}$$

$$Lu_{i,N} = f \quad \text{in } C_i^{int},\ i=1,\dots,M, \tag{7.14}$$

$$u_{i,N} = u_{i+1,N} \quad \text{in } C^{\Gamma_i},\ i=1,\dots,M-1, \tag{7.15}$$

$$u_{i,N} = 0 \quad \text{in } C_i^b,\ i=1,\dots,M, \tag{7.16}$$

$$\frac{\partial u_{i,N}}{\partial x} - \frac{\partial u_{i+1,N}}{\partial x} = -(Lu_{i,N}-f)\omega_i^- - (Lu_{i+1,N}-f)\omega_i^+ \quad \text{in } C^{\Gamma_i},\ i=1,\dots,M-1. \tag{7.17}$$

Here we have set $\omega_i^- = \omega_{N_x^i}^i$ and $\omega_i^+ = \omega_0^{i+1}$ for $i=1,\dots,M-1$. Note that at the interface collocation points, the jump of the normal derivative is not asked to vanish (as in (7.8)), but rather to balance a suitable linear combination of the residual of the equation from both sides.

We now introduce an iteration-by-subdomains procedure for the solution of the above collocation problem. Let us suppose, here and in the sequel, that M is an odd number; the case of M even can be studied similarly. For a given $g^0 \in \mathbb{P}_N^0(\Gamma)$ (i.e., $g^0_{|\Gamma_i} \in \mathbb{P}_N^0(\Gamma_i)$ for each $i=1,\dots,M-1$) we look for a sequence $u_{i,N}^n$, $n \geq 1$, satisfying: $u_{i,N}^n \in \mathbb{P}_N(\Omega_i)$, $i = 1, \ldots, M$, and for i odd:

$$Lu_{i,N}^n = f \quad \text{in } C_i^{int}, \tag{7.18}$$

$$u_{i,N}^n = g^{n-1} \text{ in } C^{\Gamma_{i-1}} \cup C^{\Gamma_i} \quad \text{for } i \neq 1, M, \tag{7.19}$$

$$u_{i,N}^n = g^{n-1} \text{ in } C^{\Gamma_i}, \quad u_{M,N}^n = g^{n-1} \text{ in } C^{\Gamma_{M-1}}, \tag{7.20}$$

$$u_{i,N}^n = 0 \quad \text{in } C_i^b. \tag{7.21}$$

For i even we require that

$$Lu_{i,N}^n = f \quad \text{in } C_i^{int}, \tag{7.22}$$

$$\frac{\partial u_{i,N}^n}{\partial x} - (Lu_{i,N}^n - f)\omega_{i-1}^+ = \frac{\partial u_{i-1,N}^n}{\partial x} + (Lu_{i-1,N}^n - f)\omega_{i-1}^- \text{ in } C^{\Gamma_{i-1}}, \tag{7.23}$$

$$\frac{\partial u_{i,N}^n}{\partial x} + (Lu_{i,N}^n - f)\omega_i^- = \frac{\partial u_{i+1,N}^n}{\partial x} - (Lu_{i+1}^n - f)\omega_i^+ \text{ in } C^{\Gamma_i}, \tag{7.24}$$

$$u_{i,N}^n = 0 \quad \text{in } C_i^b. \tag{7.25}$$

Then set:

$$g^n = \begin{cases} \theta_n u_{i-1,N}^n + (1-\theta_n)g^{n-1} & \text{on } C^{\Gamma_{i-1}} \text{ for } i = 3, \ldots, M-1, \, i \text{ odd} \\ \theta_n u_{i+1,N}^n + (1-\theta_n)g^{n-1} & \text{on } C^{\Gamma_i} \text{ for } i = 1, \ldots, M-2, \, i \text{ odd}, \end{cases} \tag{7.26}$$

and restart from (7.18) with $n+1$ instead of n. In (7.26), θ_n is a positive relaxation parameter. According to this method, at each step one solves the $(M+1)/2$ *independent Dirichlet problems* (7.18)–(7.21); then, after computing the (independent) right-hand sides of (7.23)–(7.24), one has to solve the $(M-1)/2$ *independent mixed problems* (7.22)–(7.25).

3.7.3 Convergence of the Iterative Method for the Collocation Multidomain Problem

In this section we confine ourselves to the case of Legendre collocation points, where we will make use of the following notation. We define a discrete inner product, approximating that of $L^2(\Omega_i)$, as follows:

$$(v, z)_{i,N} := \sum_{k=0}^{N_x^i} \sum_{m=0}^{N_y} (vz)(x_k^i, y_m)\omega_k^i \omega_m, \quad i = 1, \ldots, M.$$

We also define the discrete bilinear form

$$a_{i,N}(v, z) := \left(\frac{\partial v}{\partial x}, \frac{\partial z}{\partial x}\right)_{i,N} + \left(\frac{\partial v}{\partial y}, \frac{\partial z}{\partial y}\right)_{i,N} + (\alpha v, z)_{i,N}, \quad i = 1, \ldots, M.$$

It is known (see, for example, [8, Chapter 11]) that there exists β independent of N such that

$$\|w\|^2_{H^1(\Omega_i)} \leq \beta a_{i,N}(w,w), \quad \text{for all } w \in P_N(\Omega_i).$$

We set $\Gamma = \cup\Gamma_i$, and define

$$\mathbb{P}^0_N(\Gamma) := \left\{ \begin{array}{l} p : \Gamma \to \mathbb{R} : p_i := p_{|\Gamma_i} \text{ is a polynomial of degree at most} \\ \quad N_y \text{ on } \Gamma_i \text{ vanishing at the endpoints of } \Gamma_i \end{array} \right\}.$$

For any $\varphi \in \mathbb{P}^0_N(\Gamma)$, we denote with $\varphi^* \in \mathbb{P}_N(\Omega)$ any piecewise polynomial extension of φ to Ω which is determined by the values of φ_{i-1} and φ_i solely. For example, φ^* can be either the discrete harmonic extension or the interpolant extension of φ. We can now prove the following equivalence result.

The collocation multidomain problem (7.13)–(7.17) is equivalent to looking for u_N: $u_{i,N} := u_{N|\Omega_i} \in \mathbb{P}_N(\Omega_i)$, $i = 1, \ldots, M$ which satisfies

$$a_{i,N}(u_{i,N}, v) = (f,v)_{i,N} \text{ for all } v \in \mathbb{P}^0_N(\Omega_i),\ i = 1, ..., M, \tag{7.27}$$

$$u_{i,N} = 0 \quad \text{on } \partial\Omega_i \cap \partial\Omega\ i = 1, \ldots, M, \tag{7.28}$$

$$u_{i,N} = u_{i+1,N} \quad \text{on } \Gamma_i,\ i = 1, \ldots, M-1, \tag{7.29}$$

$$\sum_{\substack{i=1 \\ i \text{ even}}}^{M} a_{i,N}(u_{i,N}, \varphi^*) = -\sum_{\substack{j=1 \\ j \text{ odd}}}^{M} a_{j,N}(u_{j,N}, \varphi^*) + \sum_{k=1}^{M} (f, \varphi^*)_{k,N}, \quad \text{for all } \varphi \in \mathbb{P}^0_N(\Gamma). \tag{7.30}$$

Therefore the iterative procedure (7.18)–(7.26) admits the following equivalent variational formulation:

For all i odd ($i = 1, \ldots, M$) solve

$$a_{i,N}(u^n_{i,N}, v) = (f,v)_{i,N} \quad \text{for all } v \in \mathbb{P}^0_N(\Omega_i),\ i = 1, \ldots, M, \tag{7.31}$$

$$u^n_{i,N} = 0 \quad \text{on } \partial\Omega \cap \partial\Omega_i,\ i = 1, \ldots, M, \tag{7.32}$$

$$u^n_{i,N} = \left\{ \begin{array}{l} \theta_{n-1} u^{n-1}_{i-1,N} + (1-\theta_{n-1}) u^{n-1}_{i,N} \\ \qquad \text{on } \Gamma_{i-1},\ i = 3, \ldots, M-1 \\ \theta_{n-1} u^{n-1}_{i+1,N} + (1-\theta_{n-1}) u^{n-1}_{i,N} \\ \qquad \text{on } \Gamma_i,\ i = 1, \ldots, M-2. \end{array} \right. \tag{7.33}$$

For all i even, ($i = 2, \ldots, M-1$) solve:

$$a_{i,N}(u^n_{i,N}, v) = (f,v)_{i,N} \quad \text{for all } v \in \mathbb{P}^0_N(\Omega_i), \tag{7.34}$$

$$\begin{aligned}
u_{i,N}^n &= 0 \quad \text{on } \partial\Omega \cap \partial\Omega_i && (7.35)\\
a_{i,N}(u_{i,N}^n, \varphi^*) &= -a_{i-1,N}(u_{i-1,N}^n, \varphi^*) + (f, \varphi^*)_{i,N} + (f, \varphi^*)_{i-1,N} \\
&\qquad \text{for all } \varphi \in \mathbb{P}_N^0(\Gamma)\text{: } \varphi_{|\Gamma_j} \equiv 0 \text{ if } j \neq i-1, && (7.36)\\
a_{i,N}(u_{i,N}^n, \varphi^*) &= -a_{i+1,N}(u_{i+1,N}^n, \varphi^*) + (f, \varphi^*)_{i,N} + (f, \varphi^*)_{i+1,N} \\
&\qquad \text{for all } \varphi \in \mathbb{P}_N^0(\Gamma)\text{: } \varphi_{|\Gamma_j} \equiv 0 \text{ if } j \neq i. && (7.37)
\end{aligned}$$

Let $u_{i,N}$, $i = 1, \ldots, M$, be the spectral collocation solution of (7.27)–(7.30), and let $u_{i,N}^n$, $i = 1, \ldots, M$, be that of (7.31)–(7.37). The following convergence result holds.

Theorem 7.1. *There exist θ', θ'' and κ with $0 < \theta' < \theta''$ and $\kappa < 1$ such that for all $N > 0$ if $\theta_n \in [\theta', \theta'']$ for all $n \geq 1$, one has*

$$\left[\sum_{i=1}^{M} \|u_{i,N}^n - u_{i,N}\|_{H^1(\Omega_i)}^2\right]^{1/2} \leq C\kappa^n. \qquad (7.38)$$

Here, C is a constant independent of N.

Remark 6.1 In (7.33) $\{\theta_n\}$ is a sequence of positive relaxation parameters that are selected dynamically in order to accelerate the convergence of the iterative schemes. These parameters can be automatically evaluated within the iterative procedure and do not require any initial guess. With such a choice the convergence of the procedure is very fast.
Remark 6.2 The approach we have used is based on a sequence of differential problems to be solved in each subdomain. An (a priori) different point of view consists of applying the influence (or capacitance) matrix method. The influence matrix coincide with the Schur complement of the matrix of the system (7.13)–(7.17) with respect to the interface variables. It is precisely the matrix of the system of the interface unknowns, and it is derived from the global system by block Gaussian elimination. It is possible to prove that the iteration-by-subdomain method is equivalent to a preconditioned Richardson iterative method for the resolution of the influence system.

The generalization of the domain decomposition approach presented in the previous sections to more general problems is straightforward. For instance, linear and nonlinear advection-diffusion problems can be examined, and the same iterative procedure among subdomains can be used.

Another important generalization is the one to the Stokes equations for incompressible flows. A multidomain approach is suitable for any finite dimensional approximation to the weak variational formulation of these equations. Then the previous iteration-by-subdomain procedure can be applied in order to reduce the problem to a sequence of finite dimensional Stokes problems on each subdomain. At each subdomain interface, the continuity of the velocity field as well as that of the normal stress are enforced. The convergence still holds with a rate that is independent of the polynomial of degree of the spectral solution. For this and other applications we refer the interested reader to [18].

References

[1] R. Adams, *Sobolev spaces*, Academic Press, New York, 1975.

[2] D. N. Arnold, F. Brezzi and M. Fortin, *A stable finite element for the Stokes equations*, CALCOLO 21 (1984), pp. 337-344.

[3] H. Brezis, *Analyse Fonctionelle: Théorie et Applications*, Masson, Paris, 1983.

[4] F. Brezzi, *On the existence, uniqueness and approximation of saddle point problems arising from Lagrangian multipliers*, RAIRO Num. Anal. 8 (1974), pp. 129-151.

[5] F. Brezzi, J. Rappaz, P. A. Raviart, *Finite dimensional approximation of nonlinear problems*, Part I: *Branches of nonsingular solutions.* Numer. Math. 36, (1980), pp. 1-25.

[6] F. Brezzi, J. Rappaz, P. A. Raviart, *Finite dimensional approximation of nonlinear problems*, Part II: *Limit points.* Numer. Math. 37, (1981), pp. 1-28.

[7] F. Brezzi, J. Rappaz, P. A. Raviart, *Finite dimensional approximation of nonlinear problems*, Part III: *Simple bifurcation points.* Numer. Math. 38, (1981), pp. 1-30.

[8] C. Canuto, M. Y. Hussaini, A. Quarteroni and T. A. Zang, *Spectral Methods in Fluid Dynamics*, Springer, New York, 1988.

[9] P. G. Ciarlet, *The Finite Element Method for Elliptic Problems*, North Holland, Amsterdam, 1978.

[10] C. Canuto and A. Quarteroni, Eds., *Spectral and High Order Methods for Partial Differential Equations*, Elsevier, North-Holland, Amsterdam, 1990.

[11] R. Glowinski, *Numerical Methods for Nonlinear Variational Problems*, Springer, Berlin, 1984.

[12] V. Girault and P. A. Raviart, *Finite Element Approximation of the Navier-Stokes Equations: Theory and Algorithms*, Springer, Berlin, 1986.

[13] C. Johnson, *Numerical Solution of Partial Differential Equations by the Finite Element Method*, Cambridge Univ. Press, Cambridge, 1987.

[14] L. Kleiser, U. Schumann, *Spectral Simulation of the Laminar-Turbulent Transition Process in Plane Poiseuille Flow*, in *Spectral Methods for Partial Differential Equations*, ed. by R. G. Voigt, D. Gottlieb, M. Y. Hussaini (SIAM-CBMS, Philadelphia), 1982, pp. 141-163.

[15] J. L. Lions and E. Magenes, *Nonhomogeneous Boundary Value Problems and Applications*, Vol. I, Springer, Berlin, Heidelberg, New York, 1972.

[16] Y. Maday and A. T. Patera, *Spectral Element Methods for the Incompressible Navier-Stokes Equations*, in *State-of-the-Art Surveys on Computational Mechanics*, A. K. Noor and J. T. Oden, Eds., Book No.H00410, 1989.

[17] Y. Maday and A. Quarteroni, *Spectral and pseudo-spectral approximation of Navier-Stokes equations*, SIAM J. Numer. Anal. 19 (1982), pp. 761-780.

[18] A. Quarteroni, *Domain Decomposition Method for the Numerical Solution of Partial Differential Equations*, Report of the Minnesota Supercomputer Institute, Minneapolis, Minnesota, 1990.

[19] A. Quarteroni and G. Sacchi-Landriani, *Domain decomposition preconditioners for the spectral collocation method*, J. Sci. Comput. 3 (1988), pp. 45-76.

[20] R. O. Richtmyer and R. Morton, *Numerical methods for Initial Value Problems*, J. Wiley, New York, 1966.

[21] P. A. Raviart and J. M. Thomas, *Introduction à l'Analyse Numérique des Equations aux Derivées Partielles*, Masson, Paris, 1983.

[22] R. Temam, *Navier-Stokes equations*, North-Holland, Amsterdam, 1977.

Professor A. Quateroni
Department of Mathematics
Universitá Cattolica del Sacro Cuore
Via Trieste 17
25121 Brescia
Italy

4
Two Topics in Nonlinear Stability

J. M. Sanz-Serna

4.1 Introduction

There are several ways in which the word stability may be understood in a numerical analysis context. First of all, stability is used in expressions like 'stability and consistency imply convergence'. In this first sense, stability refers to dependence of a numerical result on the data, and, as such, applies to most numerical computations, including numerical linear algebra, quadrature, differential equations, etc. This first notion is akin to the idea of 'well posedness'. In fact, in many cases, a numerical procedure is said to be stable in this sense if it is well posed uniformly with respect to the relevant parameters, such as the dimension of the problem, grid-size, etc. In a second use, the term stability applies in connection with the long time behaviour of discretizations of time-dependent problems in ordinary or partial differential equations. The stability of discretizations of nonlinear differential equations, ordinary or partial, is the unifying theme of the present work. In Sections 4.2 to 4.8, we deal with the question of how best to define 'stability' of a nonlinear discretization so that the familiar 'stability and consistency imply convergence' holds. We present a definition of stability, that when combined with an important lemma due to Stetter and with a suitable linearization theorem, has revealed itself to be very helpful in proving the convergence of nonlinear finite-difference, spectral and Galerkin methods. In Sections 4.9–4.11 we are concerned with stability in the second sense. We show how to employ dynamical system results to investigate the stability of numerical methods that in a linear analysis are neutrally stable. Our study leads in a natural way to the consideration of symplectic or canonical integrators, a subject briefly surveyed in the final Section 4.12.

4.2 Discretizations

In order to study the idea of stability, it is advisable to work in an abstract framework, so that the attention may be focused on the key issues and a number of application fields can be covered at once. Here we work within the framework used in Sanz-Serna[29], that, while being general enough to treat most application areas, is not unnecessarily abstract. In this section we summarize some of the basic ideas of Sanz-Serna[29]. The reader is referred to the original paper for a more detailed treatment and for information on other available general frameworks.

We consider a given problem involving a linear or nonlinear differential equation supplemented by suitable side conditions, such as boundary conditions, initial conditions, etc. Let u denote a solution of this problem (well-behaved nonlinear problems may of course possess more than one solution). For the numerical solution we use the notation U_h, where the subscript h reflects the dependence of U_h on a (small) parameter such as a mesh-size, element diameter, reciprocal of number of terms retained when truncating a series, etc. We always assume that h takes values in a set H of positive numbers with $\inf H = 0$. The numerical approximation U_h is reached by solving a *discretized problem*

$$\Phi_h(U_h) = 0, \tag{2.1}$$

where, for each h in H, Φ_h is a mapping with domain D_h and taking values in Y_h. Here Y_h is a vector space and D_h is a subset of a vector space X_h with

$$\dim(X_h) = \dim(Y_h) < \infty. \tag{2.2}$$

It is typical of nonlinear situations that Φ_h cannot be defined everywhere in X_h: the analytic expression of Φ_h may involve logarithms, square roots, etc., which only make sense for vectors in a set D_h smaller that X_h. As h ranges in H, the family of discrete problems (2.1) is called a discretization.

Most discretizations used in practice for stationary and time-dependent problems may readily be cast in the format (2.1). This applies not only to finite-difference techniques, but also to Galerkin, collocation, spectral methods, etc., (more on this later).

When a solution U_h of (2.1) has been obtained, the question arises as to what extent does U_h provide a good approximation to u. A first difficulty in answering this question stems from the fact that U_h can be completely dissimilar to u. Typically, in finite differences, U_h is a vector with, say, d entries, while u is a function of one or several continuous variables. This difficulty is circumvented as follows. Since any solution U_h of (2.1) is bound to be an element in D_h, we first make up our minds as to which element u_h in D_h should be regarded as the most desirable numerical result. Typically in finite differences u_h contains d nodal values of u. Once u_h has been chosen, the vector $e_h = u_h - U_h$ is defined to be the *(global) error* in U_h. In order to measure the size of e_h, we introduce, for each h in H, a norm $\|\cdot\|$ in X_h. (Norms in different spaces will simply be denoted by $\|\cdot\|$ without mention of the space.) We say that the discretization (2.1) is *convergent* if there exists $h^* > 0$, such that for h in H, $h < h^*$, (2.1) possesses a solution U_h such that $\lim \|u_h - U_h\| = 0$, $h \to 0$. If, furthermore, $\|e_h\| = O(h^p)$, $h \to 0$, then the convergence is said to be of order p.

The *(local) discretization error* τ_h in u_h is, by definition, the element $\tau_h = \Phi_h(u_h) \in Y_h$, i.e., the residual by which the element u_h fails to satisfy the discrete equations. The measurement of the size of τ_h requires, therefore, the introduction, for each h in H, of a norm $\|\cdot\|$ in Y_h. When these norms have been chosen, (2.1) is said to be consistent (resp. consistent of order p) if, as $h \to 0$, $\|\tau_h\| \to 0$ (resp. $\|\tau_h\| = O(h^p)$).

Before we review how convergence is obtained from consistency and stability, it is convenient to illustrate the set-up described above with an example. In the interest of clarity, the example refers to a simple finite-difference scheme. More interesting finite-difference discretizations have been treated within our framework in López-Marcos and Sanz-Serna[20], Frutos and Sanz-Serna[10], Ortega and Sanz-Serna[23]. For examples of the application to Galerkin methods see López-Marcos and Sanz-Serna [21], Süli[44], Murdoch and Budd[22] and for spectral and pseudospectral techniques see Frutos and Sanz-Serna[9], Frutos, Ortega and Sanz-Serna[11], Abia and Sanz-Serna[1].

Example A. Consider the following periodic initial-value reaction-diffusion problem

$$u_t = u_{xx} + f(u), \quad -\infty < x < \infty,\ 0 \leq t \leq T < \infty, \tag{2.3}$$

$$u(x+1,t) = u(x,t), \quad -\infty < x < \infty,\ 0 \leq t \leq T < \infty, \tag{2.4}$$

$$u(x,0) = u^0(x), \quad -\infty < x < \infty. \tag{2.5}$$

In (2.3), f is a smooth real function of the real variable u, $-\infty < u < \infty$. In (2.5), u^0 is a given real 1-periodic function and it is assumed that f, T and u^0 are such that (2.3)–(2.5) possesses a unique smooth solution up to $t = T$.

To set up the numerical scheme, choose a positive constant r (the mesh-ratio) and an integer $J > 2$. Set $h = 1/J$ and consider the grid-points $x_j = jh$, j integer, and the time levels $t_n = nk$, $k = rh^2$, $n = 0, \ldots, N = [T/k]$. For $j = 1, \ldots, J$ and $n = 0, \ldots, N-1$ set

$$\frac{U_j^{n+1} - U_j^n}{k} - \frac{U_{j-1}^n - 2U_j^n + U_{j+1}^n}{h^2} - f(U_j^n) = 0, \tag{2.6}$$

where it is obviously understood that $U_0^n = U_J^n$ and $U_{J+1}^n = U_1^n$. For $j = 1, \ldots, J$ set

$$U_j^0 - u_0(x_j) = 0. \tag{2.7}$$

Formulae (2.6)–(2.7) are cast in the format (2.1) as follows. Let Z_h denote the vector space of grid functions $\mathbf{U} = [U_1, \ldots, U_J]$ defined on $\{x_j : 1 \leq j \leq J\}$. For each n, all the numerical approximations U_j^n associated with the time level t_n form a vector $\mathbf{U}^n$ in Z_h. Thus (2.6)–(2.7) may be rewritten

$$\mathbf{U}^0 - \mathbf{u}^0 = \mathbf{0}, \quad \mathbf{u}^0 = [u^0(x_1), \ldots, u^0(x_J)], \tag{2.8}$$

$$\frac{\mathbf{U}^{n+1} - \mathbf{U}^n}{k} - D^2\mathbf{U}^n - \mathbf{f}(\mathbf{U}^n) = \mathbf{0}, \quad n = 0, \ldots, N-1, \tag{2.9}$$

where D^2 is the standard matrix replacement of the second derivative operator with periodic boundary conditions and the notation $\mathbf{f}(\mathbf{U}^n)$ is self-explanatory. Next choose $X_h = D_h = Y_h$ equal to the product of $N+1$ copies $Z_h \times \ldots \times Z_h$. Thus $U_h := [\mathbf{U}^0, \ldots, \mathbf{U}^N]$ is a vector in X_h and (2.8)–(2.9) are clearly of the form (2.1) for a suitable choice of Φ_h. For u_h the obvious choice is given by the vector of grid restrictions $[\mathbf{u}_0, \ldots, \mathbf{u}_N]$ of u. In Z_h we use the maximum norm and in X_h we use the $L\infty(L\infty)$ norm

$$\|[\mathbf{V}^0, \ldots, \mathbf{V}^N]\| = \max_n \|\mathbf{V}_n\|, \quad [\mathbf{V}^0, \ldots, \mathbf{V}^N] \in X_h. \tag{2.10}$$

With this norm, convergence in the sense of the abstract framework means uniform convergence in time of the maximum spatial norm. The local discretization error τ_h is of the form $[\boldsymbol{\tau}^0, \ldots, \boldsymbol{\tau}^N]$, where, according to (2.8), $\boldsymbol{\tau}^0 = \mathbf{0}$, while for $n = 0, \ldots, N-1$, $\boldsymbol{\tau}^{n+1}$ contains the familiar truncation errors of the formula (2.6), i.e., the value of the left hand side of (2.6) after replacing each 'numerical' U_j^n by its 'theoretical' counterpart $u(x_j, t_n)$. Standard Taylor expansions show that each component of $\boldsymbol{\tau}^n$ is $O(h^2 + k)$. If in Y_h we use the $L_1(L_\infty)$ norm

$$\|[\boldsymbol{\rho}^0, \ldots, \boldsymbol{\rho}^N]\| = \|\boldsymbol{\rho}^0\| + \sum_{n=1}^{N} k\|\boldsymbol{\rho}^n\|, \quad [\boldsymbol{\rho}^0, \ldots, \boldsymbol{\rho}^N] \in Y_h, \tag{2.11}$$

the $O(h^2 + k)$ behaviour of the components of $\boldsymbol{\tau}^n$ yields

$$\|\tau_h\| = O(h^2 + k) = O(h^2 + rh^2) = O(h^2),$$

i.e., second order of consistency.

4.3 Stability in the linear case

The general idea of stability will first be presented in the context of linear discretizations. Let us suppose that (2.1) takes the linear form

$$\Phi_h(U_h) \equiv \Psi_h U_h - g_h = 0, \tag{3.1}$$

i.e.

$$\Psi_h U_h = g_h, \tag{3.2}$$

where, for each h in H, Ψ_h is a linear operator (matrix) mapping X_h into Y_h and g_h is a fixed vector in Y_h. The discretization (3.1)–(3.2) is stable with respect to the chosen norms in X_h and Y_h, if there exist positive constants S (the stability constant) and h_0 such that for each h in H, $h \leq h_0$, and for each V_h in X_h

$$\|V_h\| \leq S\|\Psi_h V_h\|. \tag{3.3}$$

In other words stability represents an *a priori* bound for the solutions of (3.2) with a constant S that must be independent of h. The bound (3.3) can be used in two ways:

(i) For h fixed, $h \leq h_0$, (3.3) implies that the kernel of Ψ_h is trivial. This combined with (2.2) reveals that the solution U_h of (3.1)–(3.2) exists and is unique. The unique existence of the numerical solution is obvious in some cases, e.g., explicit algorithms in initial value problems, but of course cannot in general be taken for granted. We emphasize that, in the linear case, existence and uniqueness of U_h *follow* from stability.

(ii) For $h \leq h_0$, we may write

$$\|u_h - U_h\| \leq S\|\Psi_h(u_h - U_h)\| = S\|\Phi_h(u_h) - \Phi_h(U_h)\|, \tag{3.4}$$

which according to (2.1) and the definition of local discretization error, implies

$$\|e_h\| = \|u_h - U_h\| \leq S\|\Phi_h(u_h) - \Phi_h(U_h)\| = S\|\tau_h\|. \tag{3.5}$$

This bounds the global error in terms of local discretization error and shows that "consistent (of order p) + stable $\Longrightarrow$ convergent (of order p)".

Example A (revisited). Consider (2.3)–(2.7) with $f \equiv 0$ and let us check whether (3.3) holds. If $V_h = [\mathbf{V}^0, \ldots, \mathbf{V}^N] \in X_h$ and $\Phi_h(V_h) = [\boldsymbol{\rho}^0, \ldots, \boldsymbol{\rho}^N] \in Y_h$, then

$$\mathbf{V}^0 = \boldsymbol{\rho}^0, \tag{3.6}$$

$$\frac{\mathbf{V}^{n+1} - \mathbf{V}^n}{k} - D^2\mathbf{V}^n = \boldsymbol{\rho}^{n+1}, \quad n = 0, \ldots, N-1. \tag{3.7}$$

It is expedient to rewrite (3.7) in the form

$$\mathbf{V}^{n+1} = C_h\mathbf{V}^n + k\boldsymbol{\rho}^{n+1}, \tag{3.8}$$

where

$$C_h = I + kD^2, \tag{3.9}$$

is the 'transition matrix', i.e. the matrix that in (2.9) with $f \equiv 0$ maps $\mathbf{U}^n$ into $\mathbf{U}^{n+1}$. Recursion in (3.8) shows that for $n = 0, \ldots, N$,

$$\mathbf{V}^n = C_h^n \mathbf{V}^0 + kC_h^{n-1}\boldsymbol{\rho}^1 + kC_h^{n-2}\boldsymbol{\rho}^2 + \ldots + k\boldsymbol{\rho}^n,$$

so that, by (2.10)–(2.11) and (3.6)

$$\|V_h\| \leq \{\max_{0\leq n\leq N} \|C_h^n\|\}\|\Psi_h(V_h)\|, \tag{3.10}$$

where $\|C_h^n\|$ is the operator norm for linear transformations in Z_h. It is easy to check that the constant in brackets in the bound (3.10) is the best possible. Therefore stability in the sense of the framework with (best) stability constant S is equivalent to the following Lax-stability requirement (boundedness of powers of the transition matrix)

$$S := \max\{\|C_h^n\| : h \in H,\ h \leq h_0,\ 0 \leq n \leq N\}, \tag{3.11}$$

(cf. Palencia and Sanz-Serna[24]). It is well known that (3.11) holds if and only if $r \leq 1/2$, and then $\|C_h^n\| = 1$ for all h and n. Note that the equivalence between stability in the abstract sense and Lax stability is quite general in that it holds for (3.8) independently of the specific nature of Z_h and C_h. (The equivalence hinges on the $L_\infty(Z_h)$ and $L_1(Z_h)$ choice of norms for X_h and Y_h, see Sanz-Serna and Palencia[35].)

4.4 Nonlinear stability

When trying to decide how to modify the linear stability definition to cater for nonlinear cases, the key observation is of course that we would like (3.5) to be valid, i.e., in the linear case (3.3) is applied to *differences* of vectors in X_h. Therefore, it is natural to define (2.1) to be stable if positive constants h_0 and S exist such that for each h in H, $h \leq h_0$, and each pair V_h, W_h of vectors in D_h

$$\|V_h - W_h\| \leq S\|\Phi_h(V_h) - \Phi_h(W_h)\|. \tag{4.1}$$

Clearly, in linear cases, this is equivalent to (3.3). We shall say that discretizations stable in the sense of this definition are N-stable (N for natural or for naive, according to your preferences). If the existence of U_h is obvious or has been proved in some way, then "consistent (of order p) + N-stable $\Longrightarrow$ convergent (of order p)". The trivial proof is again given by (3.5).

Example A (revisited). Consider (2.3)–(2.7) with $r \leq 1/2$ and f globally Lipschitz, i.e.

$$|f(v) - f(w)| \leq L|v - w|, \tag{4.2}$$

for all $v, w \in \mathbb{R}$. Then, if $V_h = [\mathbf{V}^0, \ldots, \mathbf{V}^N]$, $W_h = [\mathbf{W}^0, \ldots, \mathbf{W}^N]$, $\Phi_h(V_h) = [\boldsymbol{\rho}^0, \ldots, \boldsymbol{\rho}^N]$, $\Phi_h(W_h) = [\boldsymbol{\sigma}^0, \ldots, \boldsymbol{\sigma}^N]$,

$$\mathbf{V}^0 - \mathbf{u}^0 = \boldsymbol{\rho}^0, \tag{4.3}$$

$$\frac{\mathbf{V}^{n+1} - \mathbf{V}^n}{k} - D^2\mathbf{V}^n - \mathbf{f}(\mathbf{V}^n) = \boldsymbol{\rho}^{n+1}, \quad n = 0, \ldots, N-1. \tag{4.4}$$

$$\mathbf{W}_0 - \mathbf{u}^0 = \boldsymbol{\sigma}^0, \tag{4.5}$$

$$\frac{\mathbf{W}^{n+1} - \mathbf{W}^n}{k} - D^2\mathbf{W}^n - \mathbf{f}(\mathbf{W}^n) = \boldsymbol{\sigma}^{n+1}, \quad n = 0, \ldots, N-1. \tag{4.6}$$

Subtract (4.6) from (4.4) and use (3.9), to obtain, for $n = 0, \ldots, N-1$,

$$\mathbf{V}^{n+1} - \mathbf{W}^{n+1} = C_h(\mathbf{V}^n - \mathbf{W}^n) + k[\mathbf{f}(\mathbf{V}^n) - \mathbf{f}(\mathbf{W}^n)] + k[\boldsymbol{\rho}^{n+1} - \boldsymbol{\sigma}^{n+1}], \tag{4.7}$$

(cf. (3.8)). Note that (4.2) implies

$$\|\mathbf{f}(\mathbf{V}^n) - \mathbf{f}(\mathbf{W}^n)\| \leq L\|\mathbf{V}^n - \mathbf{W}^n\|, \tag{4.8}$$

so that, since $\|C_h\| = 1$, (4.7) yields

$$\|\mathbf{V}^{n+1} - \mathbf{W}^{n+1}\| \leq (1 + kL)\|\mathbf{V}^n - \mathbf{W}^n\| + k\|\boldsymbol{\rho}^{n+1} - \boldsymbol{\sigma}^{n+1}\|, \tag{4.9}$$

and a standard recursion leads to (4.1) with $S = \exp(LT)$. Thus, for $r \leq 1/2$, and f globally Lipschitz, the scheme is N-stable and hence convergent.

The argument used in this example to show N-stability is essentially identical to the argument frequently used to show that one or multistep ordinary differential equation discretizations for $y' = f(y)$ are N-stable if they are N-stable as applied to $y' = 0$ (0-stability see e.g. Hairer et al.[14], Section III.4).

4.5 Stability thresholds

As pointed out before, the abstract N-stability definition (4.1) includes, as particular cases, the notions of Lax-stability of linear initial-value problems in partial differential equations and of 0-stability in numerical ordinary differential equations. In these two application areas the idea of stability is well understood and frequently invoked. However, for nonlinear partial differential equations there is no general notion of stability that is commonly invoked to prove convergence. This may be due to the fact that the theory in Section 4.4 suffers from some drawbacks. Firstly, the question of existence of discrete solutions, which in the linear case is implied by stability, was not addressed. Secondly the scope of application of (4.1) is too restrictive. Let us illustrate this second point in the context of Example A.

Example A (revisited). The global Lipschitz condition used in the proof of N-stability is so demanding that few functions of interest satisfy it. In fact in reaction-diffusion problems $f(u)$ is often a polynomial and (4.2) does not hold. Now it is not difficult to see that, when f is not globally Lipschitz, the scheme (2.8)–(2.9) is not N-stable. In fact, set $f(u) = u^2$; $T = 1$; $\mathbf{V}^n = \mathbf{0}$, $n = 0, \ldots, N$; $\mathbf{W}^0 = [1, \ldots, 1]$ and $\mathbf{W}^{n+1}$, $n = 0, \ldots, N-1$, determined by (4.6) with $\boldsymbol{\sigma}^{n+1} = 0$. Then

$$\|\Phi_h(V_h) - \Phi_h(W_h)| = \|\mathbf{V}^0 - \mathbf{W}^0\| = \|\mathbf{W}^0\| = 1,$$

while

$$\|V_h - W_h\| \geq \|\mathbf{V}^N - \mathbf{W}^N\| = \|\mathbf{W}^N\|,$$

a quantity that, according to Sanz-Serna and Verwer[38], grows like

$$1/(h|\log h|).$$

This behaviour of the difference scheme should not be surprising. The vectors $\{\mathbf{V}^n\}$ and $\{\mathbf{W}^n\}$ are the numerical solutions of the scheme (2.8)–(2.9) when $u^0 \equiv 0$ and $u^0 \equiv 1$, respectively. For these initial conditions the solutions of (2.3)–(2.5) are, respectively, $u \equiv 0$ and $u^* = 1/(1-t)$. Clearly $u(x,t) - u^*(x,t)$ cannot be bounded in terms of the difference in initial condition and it cannot be hoped that, in the numerical scheme, $\|\mathbf{V}^n - \mathbf{W}^n\|$ can be bounded in terms of $\|\mathbf{V}^0 - \mathbf{W}^0\|$.

Since, for f not globally Lipschitz, (2.8)–(2.9) is not N-stable, it follows that the convergence of the scheme cannot be derived by invoking the 'consistency + N-stability' result of Section 4.4. In practice the convergence of (2.8)–(2.9) can be proved by using various well-known ad hoc techniques (tricks). Probably the quickest trick is the following (Shampine and Gordon[39], p.24). Introduce a globally Lipschitz function f^* that coincides with f in a neighbourhood Ω of $\{u(x,t) : 0 \leq x \leq 1,\ 0 \leq t \leq T\}$. The scheme for f^* is convergent by the material in Section 4.4. Now the theoretical solution u^* corresponding to f^* is the solution u corresponding to f. Hence the numerical solution U_h^* corresponding to f^*, for h small, takes values in Ω, which in turn shows that U_h^* is the numerical solution U_h corresponding to f. Therefore, the convergence of U_h^* to u^* actually implies the convergence of U_h to u. This technique is often used for terms of the form $f(u)$, but does not apply to more general nonlinearities such as the common uu_x. A second trick, of a wider applicability, is as follows. Since f is smooth, (4.2) holds when v and w are restricted to belong to the neighbourhood Ω introduced above. Assume *a priori* that the numerical solution takes values in Ω. Then (4.8) holds for the particular choice $\mathbf{V}^n = \mathbf{u}^n$, $\mathbf{W}^n = \mathbf{U}^n$. Consequently, recursion from (4.9), taking into account that $\boldsymbol{\rho}^{n+1} = O(h^2)$, $\boldsymbol{\sigma}^{n+1} = \mathbf{0}$, leads to a $O(h^2)$ estimate for

$\|\mathbf{u}^{n+1} - \mathbf{U}^{n+1}\|$, as long as the vectors $\mathbf{U}^n, \mathbf{U}^{n-1}, \ldots, \mathbf{U}^0$ have their components in Ω. The estimate shows that for h small enough $\mathbf{U}^{n+1}$ will also have its components in Ω. By induction, the estimate holds for n up to N.

The lack of applicability of the N-definition to the case at hand may be attributed to its *global* character: all V_h and W_h are allowed in (4.1). On the other hand the tricks used to prove convergence attract the attention to the point that in error estimation we are only interested in the local behaviour of Φ_h near the theoretical solution. The question arises of whether the definition in (4.1) cannot be made local in some way, so that with the new local definition common convergent schemes like (2.8)–(2.9), are again classified as stable. Local versions of the idea of stability have been introduced by Stetter[42] and H. B. Keller[16]. The advantages and drawbacks of the two versions have been compared by López-Marcos and Sanz-Serna[20]. It turns out that Keller's approach should be favoured. Keller's definition is as follows. The discretization (2.1) is said to be K-stable (K for Keller) if there exist constants $S > 0$, $h_0 > 0$ and R, $0 < R \leq \infty$, such that for each h in H, $h \leq h_0$, the open ball

$$B(u_h, R) = \{V_h \in X_h : \|V_h - u_h\| < R\}$$

is contained in the domain D_h, and, for each V_h and W_h in $B(u_h, R)$, (4.1) holds.

The quantity R is called the ***stability threshold.*** It is easy to show that when (2.1) takes the linear form (3.1), a discretization that is stable with threshold $R < \infty$ is also stable for the threshold $R = \infty$. For this choice (4.1), is asked to hold for all V_h, W_h in X_h and we recover the standard linear definition. Note that K-stability is a local notion that explicitly refers to the 'theoretical' vectors u_h. This is different from the naive situation whether the stability or otherwise of a discretization does not relate to the theoretical solution being approximated.

For each real $R > 0$, the scheme (2.8)–(2.9), with $r \leq 1/2$ and f smooth, is K-stable with threshold R and stability constant $S = \exp(LT)$, where $L = L(R)$ is the Lispchitz constant of f in

$$\Omega = \{v : |v - u(x,t)| < R \text{ for some } (x,t),\, 0 \leq x \leq 1,\, 0 \leq t \leq T\}.$$

This is shown by the argument in (4.3)–(4.9). Note that now (4.1) has only to be proved for V_h and W_h in $B(u_h, R)$ and that such vectors have their components in Ω, so that (4.8) holds for them.

Now that we know that, in the new sense, (2.8)–(2.9) is stable, it is appropriate to ask whether this knowledge is of any help in proving the convergence of that discretization. Actually, it is not evident that, for general discretizations (2.1), consistency and K-stability imply convergence. To begin with, the question of the existence of U_h must be answered. Then, it is doubtful that the argument in (3.5) can be applied as it is not clear that U_h lies in the ball $B(u_h, R)$ where the stability bound (4.1) holds. At first glance it may seem that the general result would read 'consistency + K-stability + existence of U_h + a priori estimate $\|u_h - U_h\| < R'$ imply convergence'. In the context of the scheme (2.8)–(2.9), where the existence of U_h is obvious, we would need, on top of stability and convergence, the *a priori* bounds $|u(x_j, t_n) - U_j^n| < R$. This more or less sends us back to the tricks mentioned above, so that apparently little has been gained from the introduction of the notion of K-stability.

However the considerations just outlined are unduly pessimistic. It is true that consistency and K-stability *on their own* imply convergence. The key ingredient of the proof is the following Lemma due to Stetter[42], whose usefulness in this context was first shown by López-Marcos[18].

Lemma 5.1. *Let Φ be a Y-valued mapping defined and continuous in an open ball $B(v^*, R) = \{v \in X : \|v - v^*\| < R\}$, where X and Y are finite-dimensional normed spaces with $\dim(X) = \dim(Y)$. Assume that a positive constant S exists such that for all v and w in B_R*

$$\|v - w\| \leq S\|\Phi(v) - \Phi(w)\|.$$

Then the inverse mapping Φ^{-1} exists (uniquely) in the open ball of radius R/S centered at $\Phi(v^*)$.

Now assume that in (2.1) Φ_h is a continuous mapping (a hypothesis usually satisfied in the applications) and suppose that (2.1) is consistent and K-stable. The application of the lemma to Φ_h in the ball centered at u_h with radius equal to the stability threshold shows that for h sufficiently small $\Phi^{-1}(0)$ exists in the open ball $B(u_h, R)$, i.e., there is a solution U_h of the discrete equations that satisfies the bound $\|u_h - U_h\| < R$. Then (3.5) can be used to bound the global error in terms of the local discretization error and convergence follows.

The discrete solution U_h is unique in $B(u_h, R)$, so that any other solution U_h^* of (2.1) is away from u_h in the sense that $\|u_h - U_h^*\| \geq R$. However global uniqueness of U_h cannot be expected. On the one hand, the original problem being solved is likely to possess several solutions u. In this situation, often found in nonlinear stationary problems, (2.1) will typically have a solution approximating each possible u_h. Also nonlinear discretizations should be expected to have spurious discrete solutions, i.e., solutions with no 'theoretical counterpart'. As a simple example take the backward Euler discretization $u^{n+1} = u^n + h[-u^{n+1} + (u^{n+1})^2]$, of the well behaved ordinary differential equation Cauchy problem $u' = -u + u^2$, $0 \leq t \leq 1$, $u(0) = 1/2$. The equation to be solved *at each step* is quadratic and, for

u^n near the theoretical solution has two real roots one of which is spurious. Hence the discretized equations (2.1) that embrace the computation of u^n at all time levels t_n, $n = 1, \ldots, [1/h]$ certainly possess many solutions. This multiplicity of solutions is very frequent in real-life nonlinear discretizations (for ordinary differential equation problems see Hairer et al. [13], Iserles[15]). Yet it is found to be suprising by some numerical analysts that have been brought up with the concept of naive stability. (Note that N-stability clearly implies global uniqueness of solutions.)

4.6 h-dependent stability thresholds

As shown above, the notion of stability due to Keller, when combined with Stetter's lemma, provides a very convenient method for the analysis of discretizations. Unfortunately some interesting numerical schemes for partial differential equation problems are not K-stable (see e.g., Frutos and Sanz-Serna[10]). In this section we present a useful extension of Keller's definition. It is expedient to study first the following example.

Example B. Consider the periodic initial-value hyperbolic problem

$$u_t + uu_x = 0, \quad -\infty < x < \infty,\ 0 \le t \le T < \infty, \tag{6.1}$$

$$u(x+1,t) = u(x,t), \quad -\infty < x < \infty, 0 \le t \le T < \infty, \tag{6.2}$$

$$u(x,0) = u^0(x), \quad -\infty < x < \infty. \tag{6.3}$$

In (6.3) u^0 is a given smooth, 1-periodic function and it is assumed T is small enough so that the solution of (6.1)–(6.3) is smooth up to $t = T$, i.e., the first crossing of characteristics occurs after $t = T$ (see e.g., Whitham[47]). The equation (6.1) has often been used in the study of nonlinear stability issues, see e.g., Richtmyer and Morton[26], Fornberg[8], Vadillo and Sanz-Serna[46].

The notation for the numerical scheme is similar to that employed in Example A. Choose a positive constant r (the mesh-ratio) and an integer $J > 2$. Set $h = 1/J$; $x_j = jh$, j integer; $t_n = nk$, $k = rh$, $n = 0, \ldots, N = [T/k]$. For $j = 1, \ldots, J$ and $n = 0, \ldots, N-1$ set

$$\frac{U_j^{n+1} - U_j^n}{k} + U_j^{n+1/2} \frac{U_{j+1}^{n+1/2} - U_{j-1}^{n+1/2}}{2h} = 0, \tag{6.4}$$

where the index j must be understood in the obvious periodic way and $U_j^{n+1/2}$ stands for the average $(U_j^{n+1} + U_j^n)/2$. For $j = 1, \ldots, J$ set

$$U_j^0 - u^0(x_j) = 0. \tag{6.5}$$

Denote by Z_h the space of grid functions $\mathbf{U} = [U_1, \dots, U_J]$, endowed with the standard discrete L_2-norm. The scheme (6.4)–(6.5) may be rewritten

$$\mathbf{U}^0 - \mathbf{u}^0 = \mathbf{0}, \tag{6.6}$$

$$\frac{\mathbf{U}^{n+1} - \mathbf{U}^n}{k} + \mathbf{Q}(\mathbf{U}^{n+1/2}) = \mathbf{0}, \quad n = 0, \dots, N-1, \tag{6.7}$$

where $\mathbf{Q}$ is a nonlinear operator in Z_h. Next set again u_h equal to the grid restriction of u and $X_h = D_h = Y_h$ equal to the product of $N+1$ copies $Z_h \times \dots \times Z_h$. The norms in X_h and Y_h are derived from the L_2-norm in Z_h via (2.10) and (2.11). The scheme is easily seen to be consistent of the second order.

The scheme (6.6)–(6.7) is not Keller stable. In fact consider the case $u^0 \equiv 0$, $T = 1$. Introduce the vectors $V_h = 0 \in X_h$, $W_h = [\mathbf{W}^0, \dots, \mathbf{W}^N] \in X_h$, where $\mathbf{W}^n = \mathbf{0}$, $n = 0, 1, \dots, N-1$ and $\mathbf{W}^N$ has components $W_j^N = 0$, $j = 3, 4, \dots, J$, $W_1^N = 8/r$, $W_2^N = -8/r$. It is trivial to check that $\Phi_h(V_h) = \Phi_h(W_h) = 0 \in Y_h$. If the scheme were K-stable with threshold R then, by the local uniqueness of the discrete solution, $\|W_h - u_h\| \geq R$ (h sufficiently small). On the other hand,

$$\|W_h - u_h\| = \|W_h\| = \|\mathbf{W}^N\| = 8\sqrt{2h}/r,$$

and we have reached a contradiction.

This example shows that local uniqueness does not take place in an open ball around u_h of radius larger than $8\sqrt{2h}/r$. In this way we are led to the idea of h-dependent stability thresholds. In his thesis, J. C. López-Marcos[18] introduced the following definition.

Definition 6.1. *Suppose that, for each h in H, R_h is a value $0 < R_h \leq \infty$. The discretization (2.1) is said to be stable restricted to the thresholds R_h if there exist positive constants h_0 and S such that for h in H, $h \leq h_0$, the open ball $B(u_h, R_h)$ is contained in the domain D_h and for any V_h and W_h in $B(u_h, R_h)$ the bound (4.1) holds.*

For linear problems, this notion reduces to the standard (3.3). In general, this definition is weaker than that of Keller, so that it considers as stable schemes that are not K-stable. However the definition is strong enough to prove in some cases that consistency and stability lead to convergence. In fact the following theorem is a direct consequence of Stetter's lemma.

Theorem 6.2. *Assume that (2.1) is consistent and stable with thresholds R_h. If Φ_h is continuous in $B(u_h, R_h)$ and $\|\tau_h\| = o(R_h)$ as $h \to 0$, then:*

(i) For h sufficiently small the equations (2.1) possess a unique U_h solution in $B(u_h, R_h)$.

(ii) The global errors in the solutions considered in (i) have a bound $\|u_h - U_h\| \leq S\|\tau_h\|$, where S is the stability constant. In particular the discretization is convergent with an order not smaller than the order of consistency.

Example B (revisited). Let us apply this theorem to the discretization (6.6)–(6.7). We first show stability restricted to thresholds $\mu h^{3/2}$, where μ is any positive constant. If $V_h = [\mathbf{V}^0, \ldots, \mathbf{V}^N]$, $W_h = [\mathbf{W}^0, \ldots, \mathbf{W}^N]$, $\Phi_h(V_h) = [\boldsymbol{\rho}^0, \ldots, \boldsymbol{\rho}^N]$, $\Phi_h(W_h) = [\boldsymbol{\sigma}^0, \ldots, \boldsymbol{\sigma}^N]$, then

$$\mathbf{V}^0 - \mathbf{u}^0 = \boldsymbol{\rho}^0, \tag{6.8}$$

$$\frac{\mathbf{V}^{n+1} - \mathbf{V}^n}{k} + \mathbf{Q}(\mathbf{V}^{n+1/2}) = \boldsymbol{\rho}^{n+1}, \quad n = 0, \ldots, N-1, \tag{6.9}$$

$$\mathbf{W}^0 - \mathbf{u}^0 = \boldsymbol{\sigma}^0, \tag{6.10}$$

$$\frac{\mathbf{W}^{n+1} - \mathbf{W}^n}{k} + \mathbf{Q}(\mathbf{W}^{n+1/2}) = \boldsymbol{\sigma}^{n+1}, \quad n = 0, \ldots, N-1. \tag{6.11}$$

We need the following estimate, valid for any $\mathbf{V}, \mathbf{W}$ in Z_h:

$$| < \mathbf{Q}(\mathbf{V}) - \mathbf{Q}(\mathbf{W}), \mathbf{V} - \mathbf{W} > | \leq M\|\mathbf{V} - \mathbf{W}\|^2, \tag{6.12}$$

where

$$M = M(\mathbf{V}, \mathbf{W}) = \frac{3}{4} \max_j \left(\frac{|V_{j+1} - V_j|}{h} + \frac{|W_{j+1} - W_j|}{h} \right). \tag{6.13}$$

In (6.12), angular brackets denote the standard L_2-inner product. The proof of (6.12), not given here, follows standard finite-difference energy method manipulations, see e.g., López-Marcos[19]. Subtract (6.11) from (6.9) and take the inner product with $\mathbf{V}^{n+1/2} - \mathbf{W}^{n+1/2}$, to obtain

$$\frac{(e^{n+1})^2 - (e^n)^2}{2k} \leq M(\mathbf{V}^{n+1/2}, \mathbf{W}^{n+1/2})\|\mathbf{V}^{n+1/2} - \mathbf{W}^{n+1/2}\|^2 + \|\boldsymbol{\rho}^{n+1} - \boldsymbol{\sigma}^{n+1}\|\|\mathbf{V}^{n+1/2} - \mathbf{W}^{n+1/2}\|,$$

where we have used the abbreviation $e_n = \|\mathbf{V}_n - \mathbf{W}_n\|$. Next

$$\|\mathbf{V}^{n+1/2} - \mathbf{W}^{n+1/2}\| \leq \frac{e^{n+1} + e^n}{2},$$

so that

$$\frac{e^{n+1} - e^n}{k} \leq M(\mathbf{V}^{n+1/2}, \mathbf{W}^{n+1/2})\frac{e^{n+1} + e^n}{2} + \|\boldsymbol{\rho}^{n+1} - \boldsymbol{\sigma}^{n+1}\|. \tag{6.14}$$

The key observation is now that if V_h and W_h satisfy the threshold condition $V_h, W_h \in B(u_h, R_h)$, then $M(\mathbf{V}^{n+1/2}, \mathbf{W}^{n+1/2})$ can be bounded uniformly. This is because bounds $\|\mathbf{V} - \mathbf{u}^n\| = O(h^{3/2})$, $\|\mathbf{W} - \mathbf{u}^n\| = O(h^{3/2})$ imply that the components of $\mathbf{V}$, $\mathbf{W}$ are $O(h)$ away from the components of $\mathbf{u}^n$ and hence $M(\mathbf{V}, \mathbf{W})$ is $O(1)$ away from $M(\mathbf{u}^n, \mathbf{u}^n) = O(1)$. This is just an example of the use of inverse inequalities. The stability of the scheme restricted to the thresholds $R_h = \mu h^{3/2}$ may now be proved by a standard recursion in (6.14). Convergence of the second order (including the existence of solutions of the implicit equations for h small) follows from the theorem above. We emphasize that the study of the solvability of the equations (6.7) has required no separate proof.

To end the section we show that the restriction to thresholds of the form $\mu h^{3/2}$ is tight. Again we focus our attention on the case $u^0 \equiv 0$, $T = 1$. Fix a number η, $0 < \eta < 2$, and for each h in H, consider the recursion

$$\frac{\beta_{n+1} - \beta_n}{k} = \frac{(\beta_{n+1} + \beta_n)^2}{8}, \qquad k = rh,\ n = 0, \ldots, N-1,\ N = [1/k],$$

with initial condition $\beta_0 = \eta$. This is the implicit mid-point discretization of the problem $d\beta/dt = \beta^2/2$, $\beta(0) = \eta$, with solution $2\eta/(2 - \eta t)$, and therefore

$$\beta_N \to 2\eta/(2 - \eta) \quad \text{as } h \to 0. \tag{6.15}$$

Now consider vectors $V_h = 0 \in X_h$, $W_h = [\mathbf{W}^0, \ldots, \mathbf{W}^N] \in X_h$, where $\mathbf{W}^n = h\beta_n \mathbf{E}$, with $\mathbf{E} = [E_1, \ldots, E_J]$, $E_1 = 1$, $E_2 = -1$, $E_j = 0$, $j = 3, \ldots, J$. The vector $\mathbf{E}$ is an 'eigenfunction' of the quadratic opertor $\mathbf{Q}$ in (6.7) (see Vadillo and Sanz-Serna[46]). It is readily checked that W_h satisfy the equations (6.7). Thus $\|\Phi_h(W_h)\| = \|\mathbf{W}^0\| = \sqrt{2}h^{3/2}\eta$. On the other hand $\|W_h\| = \|\mathbf{W}^N\| = \sqrt{2}h^{3/2}\beta_N$, so that, on taking into account (6.15),

$$\frac{\|V_h - W_h\|}{\|\Phi_h(V_h) - \Phi_h(W_h)\|} = \frac{\|W_h\|}{\|\Phi_h(W_h)\|} \to \frac{2}{2-\eta}, \quad h \to 0.$$

By implication, if the discretization is stable for some thresholds $R_h > \|W_h - u_h\|$, then the corresponding stability constant S satisfies $S \geq 2/(2-\eta)$. Since S must be finite, η cannot be allowed to become arbitrarily close to 2. In other words, for η close to 2, and h small, W_h must violate the threshold condition, i.e.,

$$R_h \leq \|W_h - u_h\| = \|W_h\| = \sqrt{2}h^{3/2}\beta_N = O(h^{3/2}).$$

4.7 Linear investigation of nonlinear stability

In most practical situations, the mapping Φ_h in (2.1) is smooth, so that the (Fréchet) derivative (i.e., roughly the Jacobian matrix) $\Phi'_h(u_h)$ of Φ_h at u_h exists. Furthermore, if the discretization (2.1) is successful, a solution U_h of (2.1), close to u_h, exists. Thus

$$0 = \Phi_h(U_h) \approx \Phi_h(u_h) + \Phi'_h(u_h)(U_h - u_h)$$

and one is led to consider the linearized discretization

$$\Xi_h(U_h) \equiv \Phi_h(u_h) - \Phi'_h(u_h)u_h + \Phi'_h(u_h)U_h = 0. \qquad (7.1)$$

Our aim is to study the relation between the stability of (2.1) and that of its linearization (7.1). The main motivation for this sort of research is of course that the stability or otherwise of linear discretizations is more easily investigated than that of their nonlinear counterparts. Our presentation in this section follows closely that in López-Marcos and Sanz-Serna[21]. The following result shows that, under suitable technical assumptions, the stability of (7.1) and (2.1) are equivalent.

Theorem 7.1. *(i) Assume that for h in H, h sufficiently small, the mapping Φ_h in (2.1) is (Fréchet) differentiable at u_h. If (2.1) is stable restricted to some thresholds R_h, then (7.1) is stable.*
(ii) Assume that, for each h in H, h sufficiently small, the mapping Φ_h in (2.1) is (Fréchet) differentiable at each point v_h in an open ball $B(u_h, R_h)$. Suppose that (7.1) is stable with stability constant L and that there exists a constant Q, with $0 \leq Q < 1$, such that, for h in H, sufficiently small, and for each v_h in $B(u_h, R_h)$, we have

$$\|\Phi'_h(v_h) - \Phi'_h(u_h)\| \leq Q/L. \qquad (7.2)$$

Then (2.1) is stable with thresholds R_h and stability constant $L(1-Q)$.

Let us look closer at part (ii) above in the common case where Φ_h in (2.1) is (Fréchet) differentiable at each point v_h in an open ball $B(u_h, R_h)$ and

$$\|\Phi'_h(v_h) - \Phi'_h(u_h)\| \leq K_h\|v_h - u_h\|, \quad \text{for } v_h \in B(u_h, R_h). \qquad (7.3)$$

Choose a number S larger than the stability constant L of (7.1). If h is sufficiently small and $\|v_h - u_h\| \leq \min\{R_h, (L^{-1} - S^{-1})K_h^{-1}\}$, then

$$\|\Phi'_h(v_h) - \Phi'_h(u_h)\| \leq K_h(L^{-1} - S^{-1})K_h^{-1} = [(S-L)/S]/L,$$

so that (7.2) holds with $Q = (S - L)/S$. We conclude that, for discretizations (2.1) satisfying (7.3), linearized stability with constant L implies stability with constant $S > L$ and thresholds

$$\min\{R_h, (L^{-1} - S^{-1})K_h^{-1}\}. \qquad (7.4)$$

In particular (7.3) holds if the Φ_h are continuously differentiable. Thus for smooth discretizations linearized stability is equivalent to stability with some suitable (h-dependent) thresholds. According to (7.4), the size of the thresholds decreases when K_h in (7.3) increases, i.e., when the problem becomes more nonlinear.

It is perhaps useful to reconsider the stability of (2.8)–(2.9) with f smooth. If $V_h = [\Phi_h'(v_h) - \Phi_h'(u_h)]W_h$, where W_h is any vector in X_h; V_h has components

$$\begin{aligned} \mathbf{V}^0 &= \mathbf{0} \\ \mathbf{V}^n &= -(\mathrm{diag}[\mathbf{f}'(\mathbf{v}^n)] - \mathrm{diag}[\mathbf{f}'(\mathbf{u}^n)])\mathbf{W}^n, \quad n = 1, \ldots, N. \end{aligned} \qquad (7.5)$$

Note that only the nonlinear term in (2.9) has contributed here; the contributions of the linear terms in (2.9) to the Fréchet derivative cancel when subtracting $\Phi_h'(u_h)$ from $\Phi_h'(v_h)$. It is an easy task to see that (7.5) implies (7.3), where R_h can be chosen to be any (h-independent) positive number R and K_h can also be chosen to be independent of h (K_h is essentially the Lipschitz constant of f' in an R-neighbourhood of $\{u(x,t) : 0 \le x \le 1,\ 0 \le t \le T\}$).

When working similarly in the case of Example B, the components of $[\Phi_h'(v_h) - \Phi_h'(u_h)]W_h$ have negative powers of h, a fact that results, via (7.4) in thresholds that decrease with h.

4.8 Discussion: some open problems

The theory outlined in Sections 4.6 and 4.7 has been proved to be useful in the analysis of convergence of nonlinear discretizations, see the list of references before Example A in Section 4.2. The theory provides a systematic approach to the study of stability and convergence and replaces a number of problem-dependent tricks. Furthermore, with the methodology presented here, the existence of solutions is a consequence of stability and consistency and many *a priori* bounds can be done away with (Frutos and Sanz-Serna[10]).

The issue of whether stability in the senses of Keller or López-Marcos implies stability has not been discussed here. For consistent discretizations, equivalence theorems between convergence and stability can be proved. The reader is referred to the discussions in Sanz-Serna[29] and Palencia and Sanz-Serna[25]).

The theorem in Section 4.6, requires the hypothesis $\|\tau_h\| = o(R_h)$ as $h \to 0$. In practice, the thresholds are often of the form $R_h = \mu h^s$, and this hypothesis means that the order p of consistency should satisfy $p > s$. It is still possible, via an idea of Strang's[43], to prove convergence in some cases where $s \geq p$. A systematic treatment of Strang's idea can be seen in Spijker[41]; see also Sanz-Serna[28].

Let us examine some open questions. Consider again the vectors $W_h = W_h(\eta)$ introduced at the end of Section 4.6. We saw that any threshold condition that renders (6.7) stable ($u^0 \equiv 0$, $T = 1$) must exclude these vectors when η is large. On recalling that $\mathbf{W}^0 = h\eta\mathbf{E}$, we see that any smooth function v^0 whose restriction to the x_j nodes coincides with $\mathbf{W}^0$ satisfies $\max(dv^0/dx) \geq 2\eta$. Therefore, in the solution of (6.2)–(6.3) with initial condition v^0 the characteristics cross (Whitham[47]) before $t = 1/(2\eta)$, so that we should not expect the vectors $\mathbf{W}^n$ to behave smoothly if $\eta \geq 1/2$: they cannot be conceived as approximations of a smooth solution. Even though, as measured in the L_2-norm (in which convergence is proved), $\mathbf{W}^0$ is a small $O(h^{3/2})$ perturbation of $\mathbf{u}^0 = 0$, $\mathbf{W}^0$ is an $O(1)$ perturbation in the seminorm $\max|(dv^0/dx)|$. It is the latter seminorm which is relevant in deciding the fate of an initial profile. This suggests that the $O(h^{3/2})$ thresholds in the L_2-norm could be replaced by more meaningful h-independent thresholds in a Sobolev norm including the maximum of the first derivative (see also (6.13)). A theory could be considered where a norm in X_h is used to measure global errors and write the stability bound and a different norm is used to impose the threshold condition. Stability in this alternative setting could then be related to the well-posedness of the continuous problem whose solution u is being numerically approximated. Such a relation is not possible with the definition in Section 4.6, where it is difficult to find a continuous analogue to the h-dependency of the thresholds. It would also be interesting to investigate continuous versions of the linearization results in Section 4.7.

4.9 Stability of equilibria via Moser's twist theorem

We now leave the idea of stability in connection with convergence/error bounds and turn our attention to the idea of stability in connection with the long-time qualitative behaviour of discretizations of evolutionary problems.

We consider the familiar systems of ordinary differential equations that describe the motion of a pendulum

$$dp/dt = -\sin q, \quad dq/dt = p. \tag{9.1}$$

The system (9.1) is Hamiltonian with the Hamiltonian function (energy) $H = p^2/2 + (1 - \cos q)$. Let ϕ_t denote the flow of (9.1), i.e., the mapping that associates with each point (p^0, q^0) the value at time t, $(p(t), q(t))$, of the solution of (9.1) that satisfies $p(0) = p^0$, $q(0) = q^0$. Then, it is well known (see e. g. Arnold[3], Section 16) that ϕ_t is an *area-preserving* mapping, so that for each bounded open set Ω in the phase (p, q)-plane, the sets Ω and $\phi_t(\Omega)$ have the same area. Recall also that (9.1) has periodic solutions, whose trajectories in the phase plane are closed curves which fill the region $0 < H(p, q) < 2$. Hence H is a Liapunov function in the neighbourhood of the origin and the equilibrium $p = q = 0$ is stable.

The system (9.1) is discretized by the formulae

$$p^{n+1} - p^n = -k \sin q^n, \quad q^{n+1} - q^n = kp^{n+1}. \tag{9.2}$$

Our choice of numerical method has been determined by the fact that the transformation in phase space

$$p^+ = p - k \sin q, \quad q^+ = q + kp - k^2 \sin q \tag{9.3}$$

that maps (p^n, q^n) into (p^{n+1}, q^{n+1}) is area-preserving (its Jacobian determinant is 1).

We are interested in ascertaining whether the origin is a stable equilibrium of (9.2), i.e., whether numerical solutions remain in the neighbourhood of the origin provided that $|p^0|$, $|q^0|$ are small enough.

The linearization

$$p^+ = p - kq, \quad q^+ = q + kp - k^2 q \tag{9.4}$$

of (9.3) near the origin has eigenvalues

$$(1 - k^2/2) \pm \sqrt{-k^2 + k^4/4}.$$

For $k > 2$, the eigenvalues are real and one of them is greater than 1. The origin is therefore unstable, both for (9.4) and for the original (9.3). For $0 < k < 2$, we find unimodular complex conjugate eigenvalues λ and λ^*, with

$$\lambda = (1 - k^2/2) + ik\sqrt{1 - k^2/4}. \tag{9.5}$$

In this range of values of k, the linearization is neutrally stable and no conclusion on the stability of (9.3) can be obtained from the linear analysis. We are going to prove that, for $0 < k < 2$, and $k \neq \sqrt{2}, \sqrt{3}$, the origin is in fact a stable equilibrium of (9.2)–(9.3). The excluded values $\sqrt{2}, \sqrt{3}$ are those for which λ in (9.5) takes respectively the values i and $(-1 + i\sqrt{3})/2$ (roots of unity). Our first step is to find the *normal form* of (9.3) (Guckenheimer and Holmes[12], Sections 3.3 and 3.5; Arnold[2], Chapter 5, Arnold [3], Appendix 7). Essentially, this means changing variables in (9.3) so as to rewrite the problem in a form better suited for the analysis. The theory of normal forms shows that, because the eigenvalue λ satisfies

$\lambda^3 \neq 1$, $\lambda^4 \neq 1$, there exists an origin-preserving, invertible, cubic change of variables $P = P(p,q)$, $Q = Q(p,q)$, so that in the new variables P, Q the mapping (9.3) is given by

$$(P^+ + iQ^+) = \lambda \exp[i\gamma(P^2 + Q^2)](P + iQ) + O(4), \qquad (9.6)$$

where $O(4)$ denotes terms of order four and higher in the variables P and Q, and

$$\gamma = -(k/256)\sqrt{1 - k^2/4}\,\frac{16 + 4k^2 + 2k^4 - k^4(1 - k^2/2)}{(1 - k^2/4)}. \qquad (9.7)$$

In (9.6), we have used a complex form for convenience. A real form may clearly be obtained by separating real and imaginary parts. As k increases from 0 to 2, γ in (9.7) decreases monotonically from 0 to $-\infty$. The variables P, Q have been normalized so that, for $k = 0$, $P = p$, $Q = q$.

Let us know discard the $O(4)$ terms in (9.6) and introduce polar coordinates (R, Θ) with $P = R\cos\Theta$, $Q = R\sin\Theta$. We obtain

$$R^+ = R, \quad \Theta^+ = \Theta + \arg(\lambda) + \gamma R^2. \qquad (9.8)$$

This mapping clearly leaves invariant all circles $R =$ constant in the (P, Q)-plane. On each circle, the mapping acts as a rotation by an angle $\omega = \arg(\lambda) + \gamma R^2$ that, because $\gamma \neq 0$, varies with the radius of the circle. Such mappings are called *twists*. Clearly the origin is a stable equilibrium of (9.8), surrounded by invariant circles. If we change back to the original (p,q) variables, we obtain invariant closed curves of equation $P(p,q)^2 + Q(p,q)^2 =$ constant that surround the origin and hence stability can be concluded.

Note however that the preceding argument applies to (9.8), i.e., to the mapping (9.6) *after removal of the $O(4)$ terms.* The question arises as to whether the stability of the full (9.6) can be concluded from the stability of the truncated (9.8). At first glance, one may guess that the answer should be negative: since (9.8) is only neutrally stable, arbitrarily small perturbations may render the origin either unstable or asymptotically stable. Nevertheless the answer is positive: (9.6) is an *area-preserving* perturbation of a twist mapping, and, by restricting the attention to a sufficiently small neighbourhood of the origin, the size of the perturbation can be made arbitrarily small. Now according to Moser's twist theorem (see e.g. Siegel and Moser[40], Sections 31-34), if an area-preserving mapping is a sufficiently small perturbation of a twist, then it has an invariant curve surrounding the origin. Thus, in the announced range of values of k, the origin is a stable equilibrium of the numerical method.

The above technique is very general. Assume that we have a numerical method described by an *area-preserving* mapping in $\mathbb{R}^2$ for which the origin is an elliptic equilibrium (i.e., the eigenvalues are complex conjugate λ and λ^* with unit modulus). Then, except in the resonant cases where λ is a cubic or fourth root of unity, the mapping can be brought into the form (9.6) for a suitable real γ. The origin is therefore a stable equilibrium, except, perhaps, in the 'degenerate' case $\gamma = 0$. (Note that degeneracy means, that, except for $O(4)$ terms, the mapping acts as a *linear* rotation, i.e., a rotation where the angular velocity and hence the period are independent of the amplitude.)

4.10 Behaviour of numerical methods near elliptic equilibria

Let us return to the example (9.2). Now that we know that the origin is a stable equilibrium, we may ask ourselves about the qualitative behaviour of the computed points (p^n, q^n) near the origin. Judging by the situation in the truncation (9.8), one may guess that there exists a neighbourhood of the origin that is made up of invariant curves of (9.3). If this guess were true, near the origin, the qualitative behaviour of (9.2) would be the same as that of the problem (9.1) being discretized (cf. Beyn[4], Sanz-Serna[32]). However, it turns out that the dynamics of (9.2) is far more complicated that the dynamics of the flow of (9.1).

Let us reconsider the twist T in (9.8). Assume that, for a given radius R, $\omega = \arg(\lambda) + \gamma R^2$ is rational with respect to 2π, i.e., of the form $2\pi p/q$ with p, q integers, $q > 1$. Then, for each point in the circle of radius R, q applications of the twist send the point back to its initial position after having completed p revolutions around the origin. Hence T^q restricted to such a circle is the identity, a structurally unstable mapping whose dynamics is highly sensitive to perturbations. Accordingly, for such values of R, it can be shown that the invariant circle of the twist T disappears under the addition of the $O(4)$ perturbation leading to the full mapping (9.6). On the other hand, the circles of the twist for which ω is very irrational with respect to 2π are not destroyed under the perturbation, and give rise to invariant curves of (9.6). Here very $\omega/2\pi$ irrational means (Arnold[2], Sections 12-13) that for some suitable positive numbers K and s,

$$\left| \frac{\omega}{2\pi} - \frac{p}{q} \right| \geq \frac{K}{q^{2+s}}, \tag{10.1}$$

for all integers p and q, $q > 0$. (Recall that for any irrational number ν there exist infinitely many rational approximations whose error is less than the reciprocal value of the square of the denominator $|\nu - p/q| < 1/q^2$.) The set of invariant curves of (9.6) obtained in this way form a subset of the (P, Q)-plane with positive measure. In actual fact the complement of this subset in the circle $P^2 + Q^2 < \rho$ has an area $o(\pi\rho^2)$ so that the majority of points near the equilibrium belong to the set of invariant curves. In

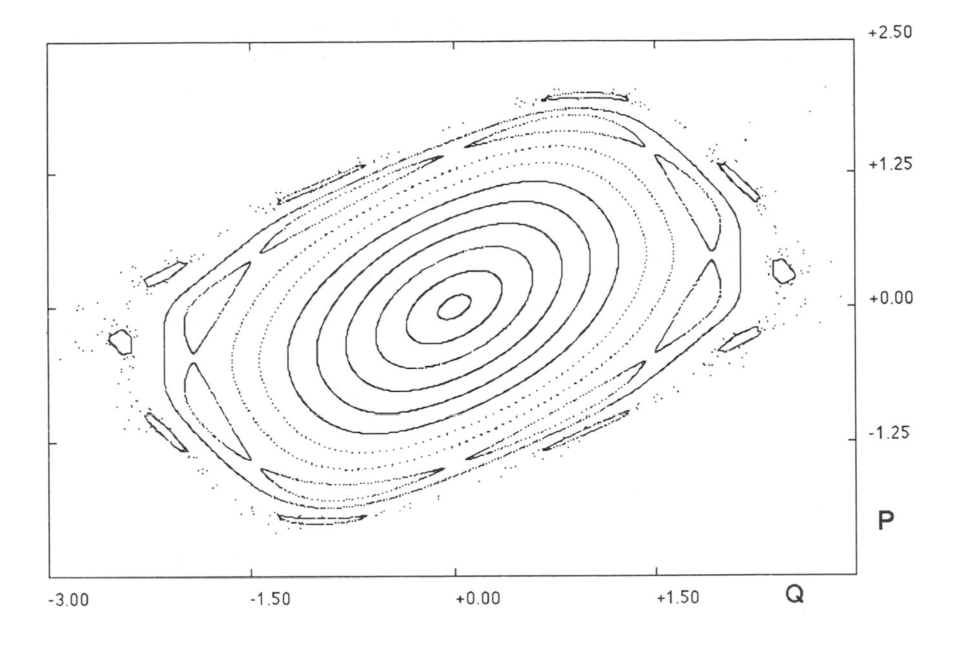

Fig. 4.1. Solutions of 9.2 with varying initial conditions

the gaps between the invariant curves the dynamics of (9.6) is exceedingly complicated (see e.g., Guckenheimer and Holmes[12], Section 4.8).

An illustration is provided in Figure 4.1, where the solutions of (9.2) corresponding to 12 different initial conditions have been presented. We have used $k = 1$ and for each solution 2000 points have been displayed. For the initial conditions (p^0, q^0) given by $(0, 0.1)$, $(0, 0.3)$, $(0, 0.5), \ldots, (0, 1.5)$ we observe that the solutions correspond to invariant curves. The initial condition $(0, 1.7)$ gives rise to an orbit consisting of eight suborbits (islands), so that after eight time-steps the point returns to the original suborbit. Thus, if only every eigth iterate were plotted or in other words the 8-th power M^8 of the mapping M in (9.3) were considered, then only one suborbit would be seen. In fact this suborbit is a twist theorem invariant curve of M^8 around an elliptic equilibria (i.e. around an elliptic 8-periodic equilibrium of (9.3)). For the initial condition $(0, 2.1)$ we see again an invariant curve and for $(.26, 2.55)$ a structure of 10 islands is found. Finally for $(.33, 2.55)$ the computed points behave erratically and eventually leave the plotting window.

Note that the dynamics depicted in Figure 4.1 cannot be the dynamics of the k-flow of a differential system: in the latter, all trajectories would lie on curves. This shows that, in general, numerical trajectories cannot be viewed as exact trajectories of a system of ordinary differential equations close to that being integrated.

4.11 An alternative application of Moser's twist theorem

It is useful to emphasize that the analysis above has only employed the properties of the numerical method (9.2), without taking into account any feature of the system (9.1) being approximated. Furthermore the analysis is valid for any fixed k in the announced range and (p, q) *sufficiently small.* There is an alternative way in which Moser's twist theorem could have been used (Sanz-Serna and Vadillo[36,37]). The starting point for the alternative approach is that in the region $H(p, q) < 2$, it is possible to change the dependent variables (p, q) of the continuous problem (9.1) into the so-called *action/angle* variables (I, α) (Arnold[3], Chapter 10). (The abstract angle α should not be confused with the physical angle q by which the pendulum deviates from the vertical axis.) Among the properties of (I, ϕ) we need the following three: (i) $p = p(I, \alpha)$, $q = q(I, \alpha)$ are 2π-periodic in α, i.e., α behaves as a genuine angle. (ii) I takes the value 0 at the origin and increases away from it. (iii) In the new variables (9.1) takes the simple form

$$dI/dt = 0, \quad d\alpha/dt = J(I), \tag{11.1}$$

where J is a known function of the action I. It is possible to give closed form expressions for $I(p, q)$, $\alpha(p, q)$, $J(I)$ in terms of elliptic integrals, but such expressions are not actually needed. The main advantage of the new variables is that now (11.1) is readily integrated to yield

$$I(t) = I(0), \quad \alpha(t) = J[I(0)]t + \alpha(0). \tag{11.2}$$

As a consequence, the k-flow of the system is now simply given by $(I, \alpha) \rightarrow (I, \alpha + J[I(0)]k)$, i.e., by a twist. By consistency, (9.3), when written in action/angle variables, is an $O(k^2)$ perturbation of this $O(k)$ twist. For $0 < H(p, q) < 2$, *k small enough*, Moser's theorem can be invoked to prove the existence of invariant curves.

We make a final remark. It is well known that the function $J(I)$ in (11.1) is decreasing so that the period $2\pi/J(I)$ of the oscillations increases with I (i.e., the larger the amplitude q_{max} of the swing of the pendulum the larger the period). This matches the fact that γ in (9.7) is negative. Moreover the classical theory of the motion of the pendulum shows that the period is given by $2\pi(1 + (1/16)q_{\text{max}}^2) + o(q_{\text{max}}^2)$. This should be compared with the value $2\pi(1+(1/16)q_{max}^2)+o(k+q_{max}^2)$ in the numerical (9.7)–(9.8). This agreement is of course no coincidence and could have been anticipated by the convergence of (9.2).

4.12 Symplectic numerical integrators

The material in the previous sections has used the twist theorem, a result restricted to dynamical systems with two dependent variables. In numerical analysis we are of course interested in mutidimensional systems, including those resulting from the space discretization of evolutionary partial differential equations. The KAM theory (see e.g. Arnold[3], Appendix 8, Guckenheimer and Holmes[12], Section 4.8) provides the multidimensional extension of the twist theorem. The role played above by area-preserving mappings is now played by *symplectic* mappings. A mapping $T : \mathbf{p}^+ = \mathbf{p}^+(\mathbf{p},\mathbf{q})$, $\mathbf{q}^+ = \mathbf{q}^+(\mathbf{p},\mathbf{q})$, $\mathbf{p} \in \mathbb{R}^g$, $\mathbf{q} \in \mathbb{R}^g$ is called symplectic or canonical (Arnold[3] Chapters 8-9) if it preserves the differential form

$$d\mathbf{p} \wedge d\mathbf{q} = dp^{(1)} \wedge dq^{(1)} + \ldots + dp^{(g)} \wedge dq^{(g)}, \qquad (12.1)$$

(superscripts denote components). In plain terms, this means that if we choose an open bounded set Ω in phase space $\mathbb{R}^g \times \mathbb{R}^g$, project it onto the g two-dimensional planes of the variables $(p^{(i)}, q^{(i)})$, $1 \le i \le g$ and sum the two-dimensional areas of the resulting projections, we obtain the same result for Ω and for the image $T(\Omega)$.

The flow ϕ of a differential system is canonical if and only if the system is Hamiltonian, i.e. there exists a real function $H(\mathbf{p},\mathbf{q})$ in phase space so that

$$dp^{(i)}/dt = -\partial H/\partial q^{(i)}, \qquad dq^{(i)}/dt = -\partial H/\partial p^{(i)}, \quad 1 \le i \le g.$$

Consequently the numerical integration of Hamiltonian systems is the natural setting in which symplectic mappings can occur in numerical analysis. A one-step method for the numerical integration of Hamiltonian system is said to be symplectic or canonical if when applied to any Hamiltonian system with any step-length it gives rise to a symplectic transformation in phase-space (Ruth[27], Feng[7], Sanz-Serna[31,33], Lasagni[17], Suris[45]).

For symplectic schemes, the KAM theory can be applied to obtain results like those in Sections 4.9–4.11. Furthermore, since the symplectic character of the flow characterizes Hamiltonian systems, the qualitative features of Hamiltonian dynamics derive from the conservation of (12.1). Hence the dynamics of symplectic schemes should be expected to mimic the qualitative features of the Hamiltonian flow. This point has been discussed in Sanz-Serna[33].

Let us present some examples of symplectic methods.

Example (A). An s-stage Runge-Kutta method of the form

$$\begin{aligned} \mathbf{Y}_i &= \mathbf{y}^n + k \sum_{1 \le j \le s} a_{ij} \mathbf{F}(\mathbf{Y}_j), \quad 1 \le i \le s, \\ \mathbf{y}^{n+1} &= \mathbf{y}^n + k \sum_{1 \le i \le s} b_i \mathbf{F}(\mathbf{Y}_i), \end{aligned}$$

is canonical provided that for each $i, j = 1, \ldots, s$,

$$b_i a_{ij} + b_j a_{ji} - b_i b_j = 0, \tag{12.2}$$

(see Sanz-Serna[31], Lasagni[17], Suris[45]). The Gauss-Legendre methods with s stages and order $2s$ are canonical. This includes the standard mid-point rule. It is interesting to note that, as shown by Sanz-Serna and Abia[34]), when condition (12.2) holds, the conditions for the RK method to be of order p are notably simplified. Rather than a condition per rooted tree with p or fewer nodes (Butcher[6]), we have a condition per so-called non-superfluous tree. For instance, for order $p = 5$, 17 order conditions are required in general and only 6 when (12.2) holds.

Example (B). The explicit mid-point rule is often used to advance in time the solution of systems of ordinary differential equations arising from the space-discretization of time-dependent partial differential equations (leap-frog schemes). Since this rule is not a one-step method, it does not give rise directly to a mapping in phase space and the definition of canonicity is not applicable. Sanz-Serna[30] noticed that it is possible to rewrite the rule as one-step convergent discretization. The technique is as follows. Write two consecutive steps of the explicit mid-point rule

$$\mathbf{y}^{2n+2} = \mathbf{y}^{2n} + 2k\mathbf{F}(\mathbf{y}^{2n+1}), \quad \mathbf{y}^{2n+3} = \mathbf{y}^{2n+1} + 2k\mathbf{F}(\mathbf{y}^{2n+2}), \tag{12.3}$$

as applied to a general system of ordinary differential equations

$$d\mathbf{y}/dt = \mathbf{F}(\mathbf{y}), \quad (\mathbf{y} \in \mathbb{R}^d). \tag{12.4}$$

On denoting $\mathbf{y}^{2n} = \mathbf{u}^n$, $\mathbf{y}^{2n+1} = \mathbf{v}^n$, n integer, (12.3) becomes

$$\mathbf{u}^{n+1} = \mathbf{u}^n + 2k\mathbf{F}(\mathbf{v}^n), \quad \mathbf{v}^{n+1} = \mathbf{v}^n + 2k\mathbf{F}(\mathbf{u}^{n+1}). \tag{12.5}$$

Now (12.5) is a convergent one-step method for the integration of the $2d$-dimensional system

$$d\mathbf{u}/dt = \mathbf{F}(\mathbf{v}), \quad d\mathbf{v}/dt = \mathbf{F}(\mathbf{u}). \tag{12.6}$$

This is called the augmented system corresponding to the original system (12.4). While (12.3) and (12.5) differ only in notation, the interpretation associated with (12.5) should be preferred in the study of the dynamics of the solution. In fact (12.3) is a linear two-step method violating the strong root condition and hence gives rise to a dynamics widely different from that of the original (12.4). On the other hand, (12.5), being of a one-step nature, inherits many features of the dynamics of the system (12.6) it approximates (see Beyn[4,5], Sanz-Serna[32] and the references therein). Sanz-Serna and Vadillo[37] show that, if the original system is Hamiltonian, then the augmented system is also Hamiltonian and that, in such a case, (12.5) is a symplectic scheme for the approximation of the augmented system.

For other symplectic methods see Sanz-Serna[33] and references therein. There is much work to be done in constructing, implementing and testing symplectic formulae.

Acknowledgement

The author has been partly supported by project CICYT PB-86-0313. He is most thankful to Dr. L. Abia for his help in the preparation of Section 4.9.

References

[1] L. Abia, and J. M. Sanz-Serna, *The spectral accuracy of a fully-discrete scheme for a nonlinear third order equation.* Computing, 44, 1990, 187-196.

[2] V. I. Arnold, *Geometrical methods in the theory of ordinary differential equations.* Springer-Verlag, New York, 1988.

[3] V. I. Arnold, *Mathematical methods of classical mechanics* (2nd ed.). Springer-Verlag, Berlin, 1989.

[4] W.-J. Beyn, *On the numerical approximation of phase portraits near stationary points.* SIAM J. Numer. Anal., 24, 1987, 1095-1113.

[5] W.-J. Beyn, *On invariant closed curves for one-step methods.* Numer. Math., 51, 1987, 103-122.

[6] J. C. Butcher, *The numerical analysis of ordinary differential equations.* John Wiley, Chichester, 1987.

[7] K. Feng, *Difference schemes for Hamiltonian formalism and symplectic geometry.* J. Comput. Mat., 4, 1986, 279-289.

[8] B. Fornberg, *On the instability of leap-frog and Crank-Nicolson approximations of a nonlinear partial differential equation.* Math. Comput., 27, 1973, 45-57.

[9] J. de Frutos and J. M. Sanz-Serna, *Split-step spectral schemes for nonlinear Dirac systems.* J. Comput. Phys., 83, 1989, 407-423.

[10] J. de Frutos and J. M. Sanz-Serna *h-dependent stability thresholds avoid the need for a priori bounds in nonlinear convergence proofs.* In *Computational Mathematics III, Proceedings of the Third International Conference on Numerical Mathematics and its Applications,* January 1988, Benin City, Nigeria, (in the press).

[11] J. de Frutos, T. Ortega and J. M. Sanz-Serna, *A Hamiltonian, explicit algorithm with spectral accuracy for the 'good' Boussinesq system.* Comput. Methods Appl. Mech. Engrg., 80, 1990, 417-423.

[12] J. Guckenheimer and P. Holmes, *Nonlinear oscillations, dynamical systems, and bifurcations of vector fields.* Springer-Verlag, New York, 1983.

[13] E. Hairer, A. Iserles and J. M. Sanz-Serna, *Equilibria of Runge-Kutta methods.* University of Cambridge, Numerical Analysis Report DAMTP/1989NA 4, 1989.

[14] E. Hairer, S. P. Nørsett and G. Wanner, *Solving ordinary differential equations I. Nonstiff problems.* Springer-Verlag, Berlin, 1987.

[15] A. Iserles, *Stability and dynamics of numerical methods for nonlinear ordinary differential equations.* IMA J. Numer. Anal., 10, 1990, 1-30.

[16] H. B. Keller, *Approximation methods for nonlinear problems with application to two-point boundary value problems.* Math. Comput., 29, 1975, 464-474.

[17] F. Lasagni, *Canonical Runge-Kutta methods.* ZAMP, 39, 1988, 952-953.

[18] J. C. López-Marcos, *Estabilidad de discretizaciones no lineales.* Ph. D. Thesis, Universidad de Valladolid, Valladolid, 1985.

[19] J. C. López-Marcos, *A difference scheme for a nonlinear partial integrodifferential equation.* SIAM J. Numer. Anal., 27, 1990, 20-31.

[20] J. C. López-Marcos and J. M. Sanz-Serna, *A definition of stability for nonlinear problems.* In *Numerical treatment of differential equations* (Strehmel, K. ed.), pp. 216-226. Teubner, Leipzig, 1988.

[21] J. C. López-Marcos and J. M. Sanz-Serna, *Stability and convergence in Numerical Analysis III: Linear investigation of nonlinear stability.* IMA J. Numer. Anal., 8, 1988, 71-84.

[22] T. Murdoch and C. Budd, *Convergent and spurious solutions of nonlinear elliptic equations.* Preprint, 1989.

[23] T. Ortega and J. M. Sanz-Serna, *Nonlinear stability and convergence of finite-difference methods for the 'good' Boussinesq equation.* Numer. Math., (in the press).

[24] C. Palencia and J. M. Sanz-Serna, *An extension of the Lax-Richtmyer theory.* Numer. Math., 44, 1984, 279-283.

[25] C. Palencia and J. M. Sanz-Serna *Equivalence theorems for incomplete spaces: an appraisal.* IMA J. Numer. Anal., 4, 1984, 109-115.

[26] R. D. Richtmyer and K. W. Morton, *Difference methods for initial-value problems.* John Wiley, New York, 1967.

[27] R. Ruth, *A canonical integration technique.* IEEE Trans. Nucl. Sci., 30, 1983, 2669-2671.

[28] J. M. Sanz-Serna, *Convergence of the Lambert-McLeod trajectory solver and of the CELF method.* Numer. Math., 45, 1984, 173-182.

[29] J. M. Sanz-Serna, *Stability and convergence in numerical analysis I: Linear problems, a simple comprehensive account.* In *Nonlinear differential equations* (Hale, J. K. and Martínez Amores, P. eds.), pp. 64-113. Pitman, Boston, 1985.

[30] J. M. Sanz-Serna, *Studies in numerical nonlinear instability I. Why do leapfrog schemes go unstable?* SIAM J. Sci. Stat. Comput., 6, 1985, 923-938.

[31] J. M. Sanz-Serna, *Runge-Kutta schemes for Hamiltonian systems.* BIT, 28, 1988, 877-883.

[32] J. M. Sanz-Serna, *Numerical ordinary differential equations vs. dynamical systems.* Universidad de Valladolid, Applied Mathematics and Computation Report 1990/3, 1990.

[33] J. M. Sanz-Serna, *The numerical integration of Hamiltonian systems.* To appear.

[34] J. M. Sanz-Serna and L. Abia, *Order conditions for canonical Runge-Kutta schemes.* Universidad de Valladolid, Applied Mathematics and Computation Report 1990/1, 1990.

[35] J. M. Sanz-Serna and C. Palencia, *A general equivalence theorem in the theory of discretization methods.* Math. Comput., 45, 1985, 143-152.

[36] J. M. Sanz-Serna and F. Vadillo, *Nonlinear instability, the dynamic approach.* In *Numerical Analysis* (Griffiths, D. F. and Watson, G. A. eds.), pp. 187-199. Longman, London, 1986.

[37] J. M. Sanz-Serna and F. Vadillo, *Studies in nonlinear instability III: augmented Hamiltonian systems.* SIAM J. Appl. Math., 47, 1987, 92-108.

[38] J. M. Sanz-Serna and J. G. Verwer, *A study of the recursion* $y_{n+1} = y_n + \tau y_n^m$. J. Math. Anal. Appl., 116, 1986, 456-464.

[39] L. F. Shampine and M. K. Gordon, *Computer solution of ordinary differential equations.* W. H. Freeman and Co., San Francisco, 1975.

[40] C. L. Siegel and J. K. Moser, *Lectures on Celestial Mechanics.* Springer, Berlin-Heidelberg-New York, 1971.

[41] M. N. Spijker, *Equivalence theorems for nonlinear finite difference methods.* In *Numerische Behandlung nichtlinearer Integrodifferential und Differential Gleichgungen* (Ansorge, R. and Törnig, W., eds.), pp. 109-122. Springer, Berlin, 1974.

[42] H. J. Stetter, *Analysis of discretization methods for ordinary differential equations.* Springer Verlag, Berlin-Heidelberg-New York, 1973.

[43] G. Strang, *Accurate partial difference methods II, nonlinear problems.* Numer. Math., 6, 1964, 37-46.

[44] E. E. Süli, *Convergence and nonlinear stability of the Lagrange Galerkin method for the Navier-Stokes equations.* Numer. Math., 53, 1988, 459-483.

[45] Y. B. Suris, *Canonical transformations generated by methods of Runge-Kutta type for the numerical integration of the system* $x'' = -\partial U/\partial x$. Zh. Vychisl. Mat. i Mat. Fiz., 29, 1989, 202-211 (in Russian).

[46] F. Vadillo and J. M. Sanz-Serna, *Studies in nonlinear instability II. A new look at* $u_t + uu_x = 0$. J. Comput. Phys., 66, 1986, 225-238.

[47] G. B. Whitham, *Linear and nonlinear waves.* Wiley-Interscience, New York, 1974.

Professor J. M. Sanz-Serna
Dpto. Matematica Aplicada y Computacion
Facultad de Ciencias
Universidad de Valladolid
Valladolid
Spain

5

Numerical Methods for Dynamical Systems

Wolf-Jürgen Beyn

These lectures are intended to give a survey of numerical methods for analyzing dynamical systems. The growing interest in these systems, in particular in their chaotic behaviour, has stimulated an immense amount of theoretical as well as numerical investigation. Therefore we restrict ourselves to a few typical phenomena which are only first steps to chaos, but whose numerical properties are fairly well understood. Among the topics covered are the numerical computation of invariant sets such as stationary points, periodic orbits and tori, the transitions between these objects in parametrized systems and the analysis of the longtime behaviour of numerical trajectories which are generated with sufficiently small step-size.In particular, we discuss in some detail the numerical computation of singular points and of homoclinic orbits via defining equations.

5.1 Basic phenomena and numerical problems

In this chapter we discuss some basic notions and results from dynamical systems theory. These will help us in motivating the numerical questions relevant in this context. Since there is a vast literature on the subject, we do not attempt to give complete references, but rather follow our personal view. In many cases the references cited may be taken as a starting point for further study.

As some general references for the numerical part we quote here the monographs by Kubiček and Marek[63], Rheinboldt[80], Seydel[87] and the special volumes Küpper, Mittelmann and Weber[65], Küpper, Seydel and Troger[66] and J. of Comp. Appl. Math. 26 (1989).

5.1.1 Dynamical systems

We consider the time dependence of a system which can be described by an N-dimensional state vector

$$u(t) = (u_1(t), \ldots, u_N(t)) \in \mathbb{R}^N, \ t \in \mathbb{R}.$$

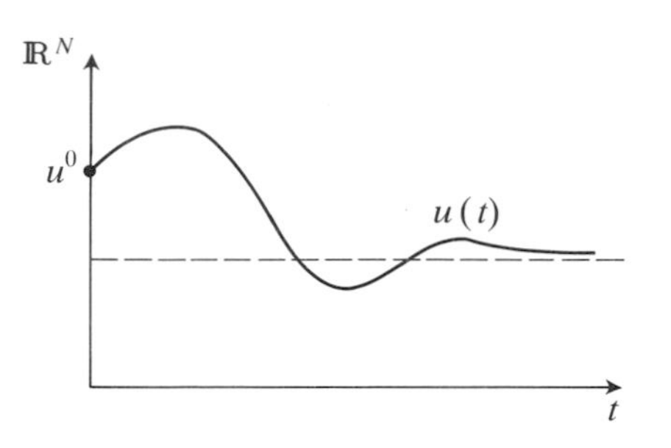

Fig. 5.1. Time diagram

We will also assume that the function $u(t)$ is determined by a dynamical system, i.e., a first order autonomous ordinary differential equation

$$\dot{u}(t) = f(u(t)). \tag{1.1}$$

Here the function $f : \mathbb{R}^N \to \mathbb{R}^N$ describes the mechanism of the underlying system and we assume it to be sufficiently smooth. The unique solution of the initial value problem

$$\dot{u} = f(u), \ u(0) = u^0 \in \mathbb{R}^N \tag{1.2}$$

exists in some maximal open interval $J(u^0) \subset \mathbb{R}$. It will be denoted by $u(t)$ or $\Phi(t, u^0)$ or $\Phi(t, u^0, f)$, if the dependence on u^0 or f is of importance. For fixed t, the map

$$u^0 \to \Phi(t, u^0)$$

is called the *t-flow* of the system (1.1). This notion is made clear in fluid dynamics (cf. Kreiss and Lorenz[62], Chapter 1.2). There $f(u)$ denotes the velocity field at position u, so that a particle starting at position u^0 at time $t = 0$ will be at position $\Phi(t, u^0)$ at time t. Correspondingly, the curve

$$\gamma(u^0) = \{\Phi(t, u^0) : t \in J(u^0)\} \tag{1.3}$$

is called the *orbit* or the *trajectory of* u^0.

There are essentially two ways of visualizing the solutions of (1.2), either in a *time diagram*, where $u(t)$ or some functional of it is plotted versus time, or in a *phase diagram*, where the orbits are drawn (see Figures 5.1 and 5.2). The basic problem in dynamical systems is to describe the asymptotic behaviour

$$\Phi(t, u^0) \to ? \text{ as } t \to \infty$$

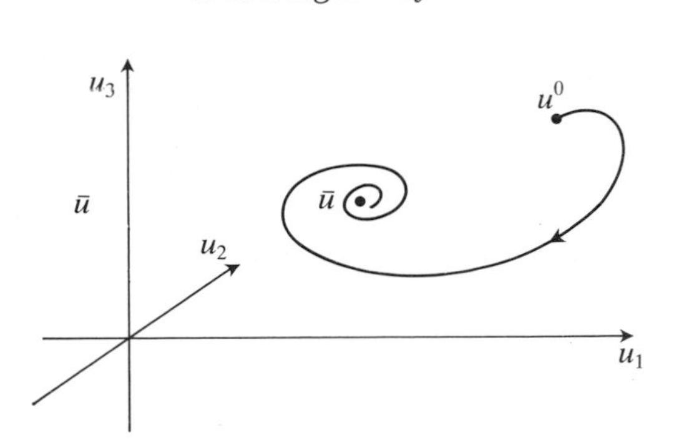

Fig. 5.2. Phase diagram

for as many initial values u^0 as possible. In some sense this asymptotic behaviour is captured by the *ω-limit set*

$$\omega(u^0) = \{v \in \mathbb{R}^N : \Phi(t_k, u^0) \to v \text{ as } k \to \infty \text{ for some sequence } t_k \to \infty\}.$$

The simplest case occurs if the solution becomes stationary, i.e.

$$\Phi(t, u^0) \to \overline{u} \quad \text{as } t \to \infty.$$

Then obviously $\omega(u^0) = \{\overline{u}\}$ and $f(\overline{u}) = 0$. This case is shown in Figures 5.1 and 5.2. Any vector $\overline{u} \in \mathbb{R}^N$ which satisfies $f(\overline{u}) = 0$ is called a *stationary point* or a *steady state*.

5.1.2 Two numerical approaches

The numerical analysis of the longtime behaviour of the flow Φ usually follows two complementary approaches:

Methods of type I (sometimes called *direct methods*)

Set up and solve numerically so-called *defining equations* for possible ω-limit sets of (1.1).
For example, set up the stationary system $f(v) = 0$ and use Newton's method

$$v^{n+1} = v^n - [f'(v^n)]^{-1} f(v^n). \tag{1.4}$$

It is well-known, that this method converges locally; i.e., the sequence will converge to some stationary point $\overline{u}$, if v^0 is sufficiently close to $\overline{u}$ and if

$$f'(\overline{u}) \text{ is nonsingular .} \tag{1.5}$$

Once a stationary point has been found, we may also determine its stability characteristics. If, in addition, the system has parameters we may continue this point into a branch of stationary points and detect singular points at which the stability characteristics change (see section 5.1.5 below).

Methods of type II (sometimes called *indirect methods*)

Solve the initial value problem (1.2) by some numerical integration method. In the simplest case this may be a one-step method with constant step size h

$$u^{n+1} = \varphi(h, u^n), \; n = 0, 1, 2, \ldots, \; u^0 \in \mathbb{R}^N. \tag{1.6}$$

For example $\varphi(h, u) = u + hf(u)$ in the case of Euler's method. The mapping $\varphi(h, \cdot)$ is the *discrete h-flow*, which is taken as an approximation of the continuous h-flow $\Phi(h, \cdot)$. A method of order p is obtained, if

$$\varphi(h, v) = \Phi(h, v) + O(h^{p+1}) \tag{1.7}$$

holds uniformly in some bounded set $\Omega \subset \mathbb{R}^N$ which contains the forward orbit $\{\Phi(t, u^0) : t \geq 0\}$. If, in addition, $\varphi(h, \cdot)$ has a uniform Lipschitz constant L in Ω, then classical estimates of the global discretization error are of the form (see e.g. Isaacson and Keller[52])

$$||u(nh) - u^n|| \leq C \; h^p e^{Lnh}. \tag{1.8}$$

Clearly, these estimates becomes useless if $L > 0$ and $nh \gg 1$. One might then ask, if there are systems for which (1.8) can be shown to hold with $L = 0$ or even $L < 0$. This is in fact the case (see e.g. Stetter[93, Chapter 3.5]), however, the assumptions on the system usually require that all trajectories converge to one and the same stable stationary point. If we are also interested in unstable phenomena, then there will be some exponential divergence of trajectories (at least locally) and (1.8) cannot hold with $L \leq 0$. Therefore, we think, one should rather ask completely different questions, such as the following:

- Define the *discrete ω-limit set* of the numerical sequence u^n from (1.6) as

  $$\omega_h(u^0) = \{v \in \mathbb{R}^N : u^{n_k} \to v \text{ as } k \to \infty \text{ for some sequence } n_k \to \infty\}.$$

  Can anything be said about the distance between $\omega(u^0)$ and $\omega_h(u^0)$?
- Can we obtain the estimate (1.8) with $L \leq 0$ if we allow different initial values for the discrete and the continuous trajectories?

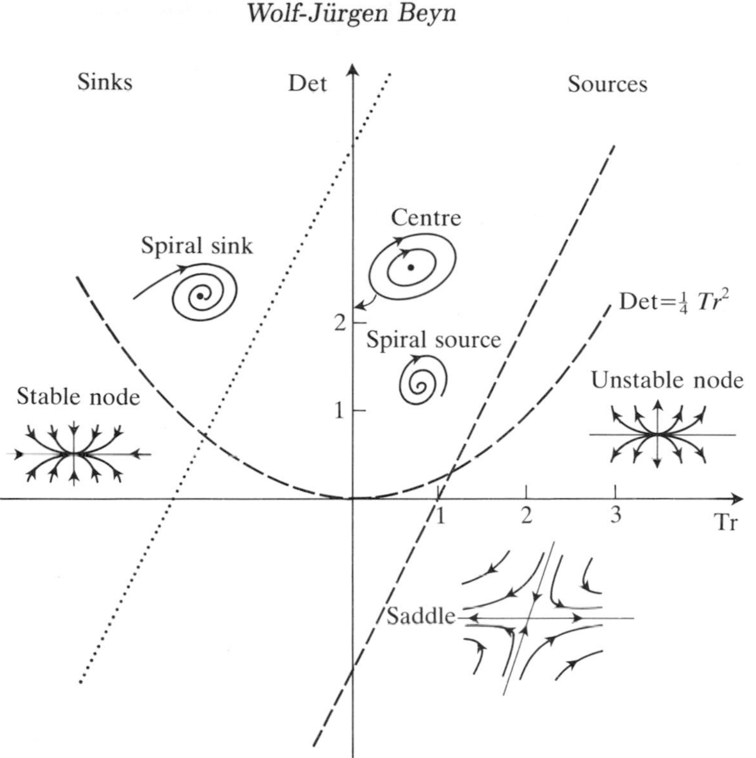

Fig. 5.3. Phase diagrams of 2-dimensional linear systems

This topic will be taken up in the last chapter. There we will concentrate on some positive answers for sufficiently small h. We notice, however, that there is also a growing literature which discusses the failures in the asymptotic behaviour of the numerical sequence u^n if h increases (see e.g., Brezzi, Fujii and Ushiki[17], Sanz-Serna[85], Stuart [96], Iserles, Peplow and Stuart[53]). Indirect methods usually provide some information on the *global* behaviour of trajectories for a few initial values. In contrast to this, direct methods only give some *local* information, which, however, is valid for a solid neighbourhood of initial values. In the following we will be concerned with these direct methods.

5.1.3 Some examples

The following three examples have been selected in order to demonstrate certain dynamic features and the typical structure of large dynamical systems ($N \gg 1$). A great variety of further examples can be found in the books cited at the beginning of this chapter.

Example 1

$$\dot{x} = y, \quad \dot{y} = 1 + y - x^2 + xy. \tag{1.9}$$

This is a seemingly simple example in 2 dimensions with $u = (x, y)$. There are two stationary points $(-1, 0)$ and $(1, 0)$.

Example 2 The linear two-dimensional system

$$\dot{u} = Au, \; A = \begin{pmatrix} a & b \\ c & d \end{pmatrix} \tag{1.10}$$

The phase portrait depends on the eigenvalues of A

$$\mu_{\pm} = \frac{1}{2}(Tr \pm \sqrt{Tr^2 - 4\; Det}), \; Tr = a + d, \; Det = ad - bc$$

and the various cases are best shown in a Tr-Det diagram as in Figure 5.3 (see Hirsch see Smale[48], Chapter 5). The two lines shown will be used in later examples.

Example 3 Diffusion-reaction systems

$$\begin{array}{lll} v_t = Dv_{xx} + g(x, v), & 0 \leq x \leq 1, & t \geq 0 \\ v(x, 0) = v^0(x), & 0 \leq x \leq 1 & \\ v(0, t) = \gamma_0, & v(1, t) = \gamma_1, & t \geq 0. \end{array}$$

Here $v(x, t) \in \mathbb{R}^n$ is a vector describing the concentrations of n reactants at time t and location $x \in [0, 1]$. The reaction term g couples the various concentrations (in the case of bimolecular reactions it is a quadratic term in v), while D is a diagonal matrix containing the diffusion coefficients. The standard method of lines approach introduces a spatial grid $x_i = i\Delta x$, $\Delta x = 1/(m + 1)$ and approximates the parabolic system above by

$$\begin{aligned} v_t(x_i, t) = {} & (\Delta x)^{-2} D(v(x_{i-1}, t) - \\ & 2v(x_i, t) + v(x_{i+1}, t)) + g(x_i, v(x_i, t)), \; i = 1, \ldots m. \end{aligned} \tag{1.11}$$

Here $v(x_0, t)$ and $v(x_{m+1}, t)$ are replaced by the given boundary values. Introducing the vector

$$u(t) = (v(x_1, t), \ldots, v(x_m, t)) \in \mathbb{R}^{mn}$$

yields a dynamical system of the form (1.1) where the Jacobian $f'(u)$ is a large matrix with tridiagonal block structure

Fig. 5.4. A periodic orbit and an invariant torus

$f'(u) =$ 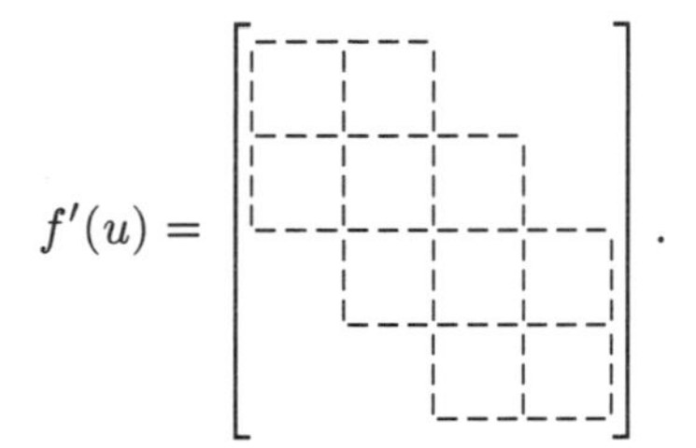 .

This example opens up the road to partial differential equations and it suggests that direct numerical methods should take advantage of the sparsity of $f'(u)$.

5.1.4 Fundamental notations and results

At the beginning of this chapter we referred to a few books on numerical methods for dynamical systems. Similarly, we mention here some monographs on the theory of dynamical systems Hale[44], Arnold[5], Hirsch and Smale[48], Irwin[51], Chow and Hale[20], Guckenheimer and Holmes[43], Amann[3]. These will be freely used without giving the particular reference at any instant.

Given a dynamical system (1.1), an arbitrary set $M \subset \mathbb{R}^N$ is called *invariant* if

$$\Phi(t, u^0) \in M \quad \text{for all} \quad t \in \mathbb{R} \quad \text{whenever} \quad u^0 \in M.$$

If this holds only for $t \geq 0$, then M is called *positively invariant.* It is easily seen that any ω-limit set $\omega(u^0)$ is invariant and closed. Hence it is also compact, if the positive trajectory stays bounded.

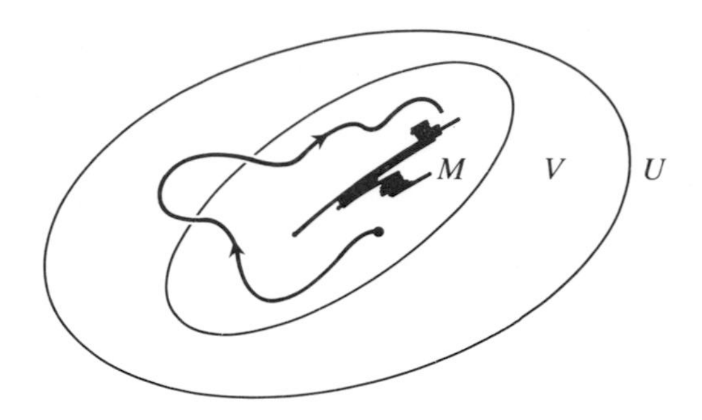

Fig. 5.5. Stability of a compact invariant set

The most frequent compact invariant sets next to stationary points are *periodic orbits* and *invariant tori* (see Figure 5.4). Suppose that $\Phi(T, u^0) = u^0$ for some $T > 0$ and that T is the smallest number with this property. Then

$$\gamma(u^0) = \{\Phi(t, u^0) : 0 \le t \le T\}$$

is a periodic orbit of period T and we have

$$\Phi(t + T, v) = \Phi(t, v) \quad \text{for all} \quad t \in \mathbb{R}, \ v \in \gamma(u^0).$$

An invariant torus is of the form

$$M = \{P(\Theta_1, \Theta_2) : 0 \le \Theta_1 \le 2\pi, 0 \le \Theta_2 \le 2\pi\}$$

where $P : \mathbb{R}^2 \to \mathbb{R}^N$ is 2π-periodic in Θ_1 and Θ_2 and where

$$\Phi(t, P(\Theta)) = P(\Phi_M(t, \Theta)), \ t \in \mathbb{R}, \Theta = (\Theta_1, \Theta_2)$$

for some mapping Φ_M. Notice that the invariant set M can no longer be parametrized by time. Some nontrivial dynamics on M remains, given by the reduced flow $\Phi_M(t, \cdot)$.

Let $M \subset \mathbb{R}^N$ be some compact invariant set. This set can only be "observed" in a real system, i.e., in mathematical terms it appears as an ω-limit set for sufficiently many initial values, if it attracts nearby trajectories or at least if it keeps them close. This motivates the following definition.

The set M is called *stable* (compare Figure 5.5), if for any neighbourhood U of M there exists another neighbourhood V of M such that

$$u^0 \in V \Rightarrow \Phi(t, u^0) \in U \quad \text{for all} \quad t \ge 0.$$

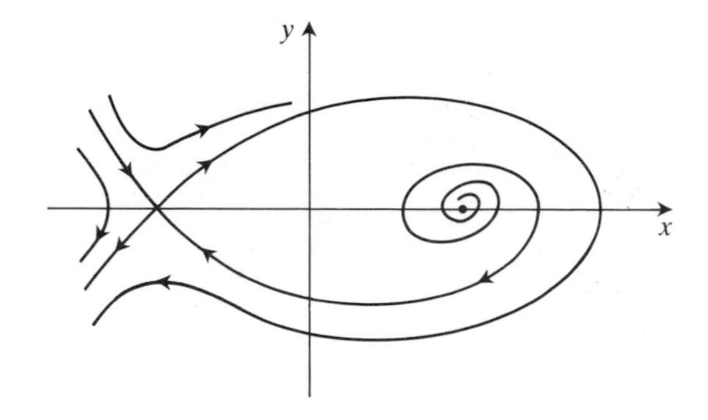

Fig. 5.6. Global phase diagram for example 1

M is called *unstable*, if it is not stable. Finally, M is called *asymptotically stable* if it is stable and if there exists a neighbourhood U of M such that

$$u^0 \in U \Rightarrow \operatorname{dist}(\Phi(t, u^0), M) \to 0 \quad \text{as} \quad t \to \infty.$$

The basic stability result for stationary points is

Theorem 1.1. *Let $\overline{u} \in \mathbb{R}^N$ be a stationary point of $\dot{u} = f(u)$. Then $\overline{u}$ is asymptotically stable if Re $\mu < 0$ for all eigenvalues $\mu \in C$ of $f'(\overline{u})$, and it is unstable, if Re $\mu > 0$ for at least one eigenvalue of $f'(\overline{u})$.*

As a consequence, we find for our example 2(see Figure 5.3) that the origin is asymptotically stable, if (Tr, Det) is in the open upper left quadrant, and it is unstable if Tr > 0 or Det < 0. The center occuring on the semi-axis Tr $= 0$, Det > 0 is stable but not asymptotically stable.

The stationary point $\overline{u} \in \mathbb{R}^N$ of (1.1) is called *hyperbolic*, if $f'(\overline{u})$ has no eigenvalue on the imaginary axis. In this case the phase diagram of the nonlinear system resembles at least locally that of the linearized system according to the following theorem.

Theorem 1.2. *(Hartman, Grobman). Let $\overline{u} \in R^N$ be a hyperbolic stationary point. The flows of $\dot{u} = f(u)$ and $\dot{u} = f'(\overline{u})u$ are locally flow equivalent, more precisely there exists a homeomorphism h from some neighbourhood of 0 onto some neighbourhood of $\overline{u}$ such that*

$$\Phi(t, h(u^0), f) = h(\Phi(t, u^0, f'(\overline{u}))) = h(e^{tf'(\overline{u})}u^0).$$

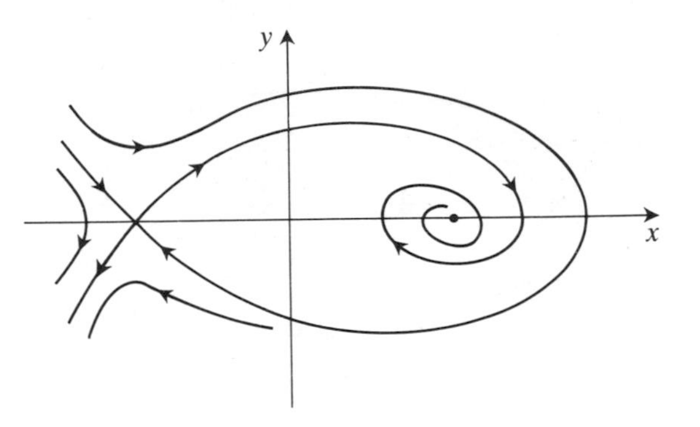

Fig. 5.7. Global phase diagram for example 1(-)

Let us apply these results to example 1 in 5.1.3. We have

$$f'(x,y) = \begin{pmatrix} 0 & 1 \\ -2x+y & 1+x \end{pmatrix}, \text{ Tr} = 1+x, \text{ Det} = 2x-y.$$

Hence, Tr $= 0$, Det $= -2$ at $(x,y) = (-1,0)$ and Tr $= 2$, Det $= 2$ at $(x,y) = (1,0)$.

Therefore, $(-1,0)$ is a saddle and $(1,0)$ is a spiral source. The global phase diagram is shown in Figure 5.6. We have included in Figure 5.7 the slightly modified example

Example 1(-)

$$\dot{x} = y, \; \dot{y} = 1 - 2y - x^2 + xy. \tag{1.12}$$

Here, $(-1,0)$ remains a saddle, but $(1,0)$ has turned into a spiral sink.

5.1.5 Parameters and bifurcations

In this section we consider dynamical systems with one parameter λ,

$$\dot{u} = f(u,\lambda). \tag{1.13}$$

Changing the parameter may drive the system from one asymptotic behaviour to another. At certain values of λ one type of invariant set (e.g., a stationary point) may loose its asymptotical stability and a new type of invariant set (e.g., a periodic orbit) may be created which takes over the

stability. In a loose sense these are the so-called *bifurcations*. There need not be an exchange of stability at a bifurcation. However, in applications this is the most important effect.
We briefly review here the various bifurcation phenomena in one parameter systems which connect stationary points, periodic orbits and tori. In Table 5.1 some typical stability assignments are shown using the following conventions (cf. Doedel and Kernevez[28]).

———————	stable stationary points
- - - - - - - - - - - - - - -	unstable stationary points
• • • • • • • • • •	stable periodic orbits
○ ○ ○ ○ ○ ○ ○ ○ ○ ○	unstable periodic orbits
⊙ ⊙ ⊙ ⊙	stable tori

For a numerical analysis of these bifurcations we are confronted with the following tasks:

- detect a bifurcation point while following a branch of invariant objects
- accurately locate the bifurcation point by a *defining equation*
- create a good initial approximation for starting a branch of the new invariant objects (*branch switching*).

It has been pointed out by Seydel[87], Chapter 5.3, that in many practical problems the accurate location of bifuration points is not really necessary (some interpolation on the branch will be sufficient). However, it becomes important if we introduce a second parameter and try to follow a branch of bifurcation points. On such a branch we may well encounter a new bifurcation – a so-called *codimension 2 singularity* – and all the above questions arise again. We will treat in this paper only one such codimension 2 singularity – the so-called *Takens-Bogdanov singularity* or B-point (see e.g. Fiedler[34]). Even higher singularities have been detected for example by Khibnik, Bykov and Yablonskii[58], De Dier, Roose and Van Rompay[23] and Khibnik[59].

Let us finally add an example of a torus bifurcation.

Example 4

$$\dot{r} = r(\lambda - r^2),\ \dot{\Theta}_1 = a > 0,\ \dot{\Theta}_2 = b. \tag{1.14}$$

Here Θ_1, Θ_2 are assumed to be 2π-periodic, i.e., $\Theta_1, \Theta_2 \in S_{2\pi} := \mathbb{R}/2\pi\mathbb{Z}$. For the r-equation an asymptotically stable stationary point $\sqrt{\lambda}$ bifurcates from the origin at $\lambda = 0$. If we transform (1.14) into Cartesian coordinates via

$$(x, y, z) = ((1 - r\cos\Theta_2)\cos\Theta_1,\ (1 + r\cos\Theta_2)\sin\Theta_1,\ r\sin\Theta_2)$$

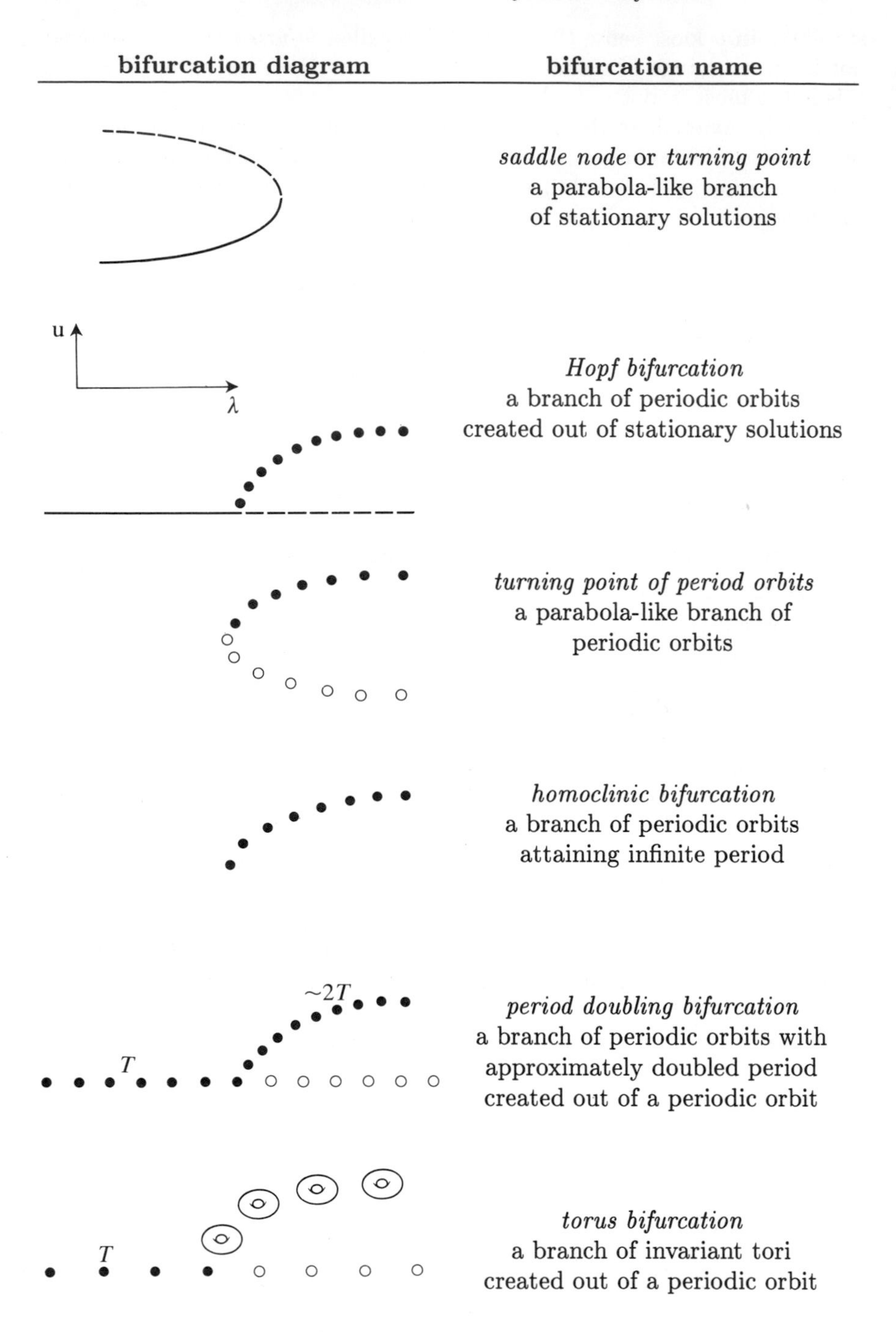

bifurcation diagram	bifurcation name
	saddle node or *turning point* a parabola-like branch of stationary solutions
u, λ	*Hopf bifurcation* a branch of periodic orbits created out of stationary solutions
	turning point of period orbits a parabola-like branch of periodic orbits
	homoclinic bifurcation a branch of periodic orbits attaining infinite period
T, $\sim 2T$	*period doubling bifurcation* a branch of periodic orbits with approximately doubled period created out of a periodic orbit
T	*torus bifurcation* a branch of invariant tori created out of a periodic orbit

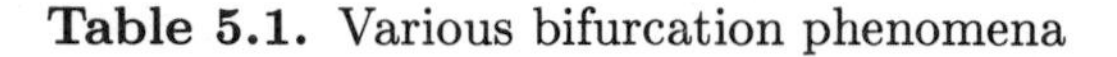

Table 5.1. Various bifurcation phenomena

we find that the periodic orbit

$$\{(\cos(at), \sin(at), 0) : 0 \le t \le 2\pi/a\}$$

bifurcates at $\lambda = 0$ into an asymptotically stable torus

$$\Big\{(1 - \sqrt{\lambda}\ \cos\Theta_2)\ \cos\Theta_1, \\ (1 + \sqrt{\lambda}\ \cos\Theta_2)\ \sin\Theta_1,\ \sqrt{\lambda}\ \sin\Theta_2) :\ 0 \le \Theta_1,\ \Theta_2 \le 2\pi\Big\}.$$

5.2 The direct computation of stationary points, periodic orbits and more general invariant manifolds

5.2.1 Stationary points

The basic features of Newton's method for solving the stationary equation $f(u) = 0$ have already been mentioned in 1.2. We notice here that the nonsingularity assumption on $f'(\overline{u})$ (see (1.5)) is slightly weaker then the hyperbolicity assumption which guarantees the persistence of the local phase diagram (Theorem 1.2). Further details on Newton's method, update methods and methods for following branches can be found in Ortega and Rheinboldt[74], Stoer and Bulirsch[95], Rheinboldt[80] Seydel[87] and Allgower and Georg[1] The stability of $\overline{u}$ can be analysed by calculating the eigenvalues of $f'(\overline{u})$ (see Theorem 1.1).

5.2.2 Periodic orbits

In order to compute a periodic orbit we have to find a period $T > 0$ and a solution $u(t) \subset \mathbb{R}^N$ of the boundary value problem

$$\dot{u} = f(u),\ t \in [0, T],\ u(0) = u(T). \tag{2.1}$$

Since T is one of the unknowns we introduce the scaled function $v(t) = u(tT)$, $t \in [0, 1]$, for which we have the boundary value problem

$$\dot{v} = T\ f(v),\ t \in [0, 1] \tag{2.2}$$

$$v(0) - v(1) = 0. \tag{2.3}$$

Here we have $N+1$ unknowns $v(t), T$, but only N boundary conditions. In fact, if $v(t)$ is a solution of (2.2, 2.3), then we can extend it 1-periodically to $t \in \mathbb{R}$ and find that any phase shifted function $v(t+q)$, $q \in \mathbb{R}$ also solves

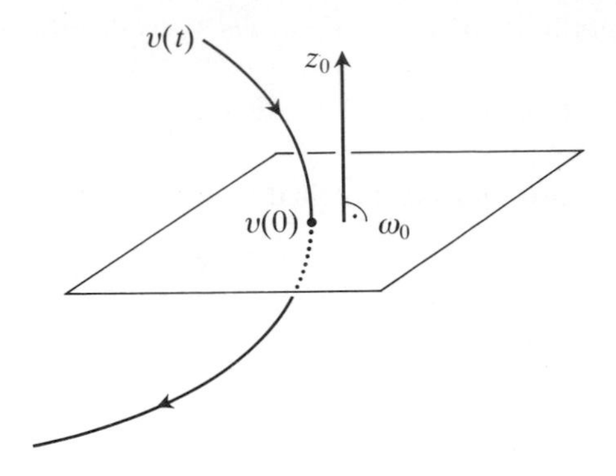

Fig. 5.8. Classical phase condition

(2.2, 2.3). This arbitrariness can be eliminated by imposing an $(N+1)$-st boundary condition, the so-called *phase condition.* We take it in the general form

$$\Psi(v) = 0, \tag{2.4}$$

where $\Psi : C_1^1[0,1] \to \mathbb{R}$ is any functional and $C_1^1[0,1]$ denotes the space of 1-periodic C^1 functions from $[0,1]$ to $\mathbb{R}^N$. The *classical phase condition* is obtained if we fix $v(0)$ in a hyperplane through some approximate vector w_0 and orthogonal to some direction z_0 (see Figure 5.8), i.e.,

$$\Psi(v) = z_0^T(v(0) - w_0). \tag{2.5}$$

A good choice usually is $z_0 = f(w_0)$. Finally, adding the equation $\dot{T} = 0$, we may now apply any of the available boundary value solvers to (2.2, 2.3, 2.4) (see e.g., Ascher, Mattheij and Russell[6], Seydel[87]). These codes usually do not allow to take advantage of some sparsity of f' (cf. 5.1.3, Example 3), so that specialized programs have been developed (Holodniok, Knedlik and Kubiček[49]).

An integral phase condition has been set up and implemented by Doedel[27] (see also Doedel and Kernevez[28]). In continuation problems this condition usually allows for larger step sizes and is more robust than (2.5). It has the form

$$\Psi(v) = \int_0^1 z_0^T(t)(v(t) - v_0(t))\, dt, \tag{2.6}$$

where v_0 is an approximation from the predictor and $z_0 = \dot{v}_0 \approx f(v_0)$. The motivation for (2.6) comes from the following:

Lemma 2.1. *Suppose that $v_0 \in C_1^1[0,1]$ is nonconstant. Then there is a neighbourhood U of v_0 in $C_1^1[0,1]$, such that for any $v \in U$ the L_2-distance*

$$\int_0^1 ||v(t+q) - v_0(t)||_2^2 \, dt, \; ||\cdot||_2 = \textit{Euclidean norm} \tag{2.7}$$

has a unique local minimum at some $q = q(v)$ close to zero. At the minimum we have

$$\int_0^1 \dot{v}_0^T(t)(v(t+q) - v_0(t)\, dt = 0.$$

Proof. Let $F(q,v)$ denote the L_2-norm from (2.7). Then we find

$$\frac{\partial F}{\partial q}(q,v) = 2\int_0^1 \dot{v}(t+q)^T(v(t+q) - v_0(t))\, dt.$$

Using the periodicity of v and partial integration we obtain

$$\frac{\partial F}{\partial q}(q,v) = 2\int_0^1 \dot{v}_0^T(t)v(t+q)\, dt.$$

Therefore, $\frac{\partial F}{\partial q}(0,v_0) = 0$ and

$$\frac{\partial^2 F}{\partial q^2}(0,v_0) = 2\int_0^1 ||\dot{v}_0(t)||_2^2\, dt > 0.$$

We may now apply the implicit function theorem to the equation $\frac{\partial F}{\partial q}(q,v) = 0$ and find all our assertions. ∎

Determining the stability of a periodic orbit can be a considerable numerical task. One has to calculate the so-called *monodromy matrix* $Y(T) \in \mathbb{R}^{N,N}$ which is the T-value of the fundamental matrix $Y(t)$, $0 \le t \le T$ of the system

$$\dot{Y}(t) = f'(u(t))Y(t), \; 0 \le t \le T, \; Y(0) = I. \tag{2.8}$$

The eigenvalues of $Y(T)$ are called *Floquet multipliers.* By differentiating (2.1) we easily find

$$Y(t)\dot{u}(0) = \dot{u}(t) \text{ and } Y(T)\dot{u}(0) = \dot{u}(T) = \dot{u}(0).$$

Hence 1 is always a Floquet multiplier and stability is determined by the remaining ones (cf. Hirsch and Smale[48], Amann[3]).

Theorem 2.2. *Let $\gamma = \{u(t) : 0 \le t \le T\}$ be a T-periodic orbit of $\dot{u} = f(u)$ with monodromy matrix $Y(T)$. Then the orbit is asymptotically stable if 1 is a simple eigenvalue of $Y(T)$ and if $|\mu| < 1$ for all other eigenvalues μ. The orbit is unstable if $|\mu| > 1$ for at least one Floquet multiplier μ.*

The monodromy matrix can often be determined as a by-product of the numerical solution of the boundary value problem (2.2, 2.3, 2.4) (see Doedel and Kernevez[28], Seydel[87], Chapter 7). This is possible because the discretization of (2.2, 2.3, 2.4) is linearized during some Newton step and the matrix obtained is close to a discretization of the linearized system (2.8). Doing some kind of 'forward integration' with this system provides an approximation for $Y(T)$. However, it is also well-known that this shooting type approach completely fails if there are Floquet multipliers which are very small or very large in modulus (cf. Doedel and Kernevez[28], Stiefenhofer[94]). This *stiff periodic case* typically occurs with relaxation oscillations.

We finally mention a result of Keller and Jepson[57]. It characterizes the admissible phase conditions (2.4) which lead to a regular boundary value problem (2.2, 2.3, 2.4). By *regular* we mean here that if we write (2.2, 2.3, 2.4) as an operator equation

$$F(v,T) = (\dot{v} - T\ f(v), v(0) - v(1), \Psi(v)) = 0$$

where $F : C^1[0,1] \times \mathbb{R} \to C^0[0,1] \times \mathbb{R}^{N+1}$, then we require that the Frechét derivative of F (in some suitable norms) at the solution is a homeomorphism.

Theorem 2.3. *Let $u(t)$ be a solution of (2.1) and let $v(t) = u(tT)$. Then (2.2, 2.3, 2.4) is a regular boundary value problem for (v,T) if and only if the following two conditions hold*

(i) 1 is a simple Floquet multiplier
(ii) $\Psi'(v)\dot{v} \neq 0$.

Condition (ii) is a rather mild requirement; for the classical phase condition (2.5) it means $z_0^T \dot{v}(0) \neq 0$, while (2.6) requires

$$\int_0^1 z_0^T(t)\dot{v}(t)\, dt \neq 0.$$

Further, as in the stationary case condition (i) is weaker than the *hyperbolicity* of the orbit, which requires that 1 is not only a simple Floquet multiplier, but also the only one on the unit circle. Again hyperbolicity guarantees the persistence of the dynamic behaviour under perturbations.

5.2.3 More general invariant manifolds

Smooth invariant manifolds are an important tool in the analysis of dynamical systems. On the one hand, these are applied for lowering the dimensionality of the system without loosing the ω-limit sets (or attractors) under investigation. Examples are:

- *center manifolds* (see e.g. Carr[18], Guckenheimer and Holmes[43])
- *inertial manifolds* (e.g. Temam[98]).

On the other hand, invariant manifolds also serve to understand the dynamics in low-dimensional systems by either bounding domains of attraction or appearing as limit sets themselves such as

- *separatrices* consisting of *stable and unstable manifolds*
- *invariant tori.*

The numerical approximation of these invariant manifolds has been undertaken just recently (see Kevrekidis, Aris, Schmidt and Pelikan[60], van Veldhuizen[99], Foias, Jolly, Kevrekidis, Sell and Titi[35], Doedel and Friedman[29,30,31], Dieci, Lorenz and Russell[25], Beyn[11]). Some special separatrices – the homoclinic orbits – will be discussed in the next chapter. In this section we outline the method of Dieci, Lorenz and Russell[25] (see also Lorenz and Van de Velde[70]) for calculating invariant manifolds in the special case of a two-dimensional torus.

They assume, that the given dynamical system has been subject to a coordinate transformation $u \to (\Theta, R)$, after which it takes the form

$$\dot{\Theta} = f(\Theta, R), \quad \dot{R} = g(\Theta, R). \tag{2.9}$$

Here $R \in \mathbb{R}^{N-2}$ and $\Theta \in T^2 := (\mathbb{R}/2\pi\mathbb{Z})^2$ is already a toroidal coordinate. Our example 4 (section 5.1.5) is of this type with f being constant and g depending only on R. We look for an invariant manifold of the form

$$M = \{(\Theta, r(\Theta)) : \Theta \in T^2\}. \tag{2.10}$$

Decomposing the flow $\Phi = (\Phi_\Theta, \Phi_R)$ according to (2.9) we may write the invariance condition as

$$\Phi_R(t, \Theta, R) = r(\Phi_\Theta(t, \Theta, R)) \quad \text{for } (\Theta, R) \in M.$$

Differentiating this identity with respect to t and setting $t = 0$ yields

$$g(\Theta, R) = r'(\Theta) f(\Theta, R) = f_1(\Theta, R) \frac{\partial r}{\partial \Theta_1} (\Theta) + f_2(\Theta, R) \frac{\partial r}{\partial \Theta_2} (\Theta).$$

Inserting $R = r(\Theta)$ gives the following first order system of nonlinear partial differential equations for the function $r : T^2 \to \mathbb{R}^{N-2}$

$$g(\Theta, r) = f_1(\Theta, r)\frac{\partial r}{\partial \Theta_1} + f_2(\Theta, r)\frac{\partial r}{\partial \Theta_2}. \tag{2.11}$$

The requirement that r is a smooth function of $\Theta \in T^2$ can be written as

$$\Theta \in [0, 2\pi] \times [0, 2\pi], r(0, \Theta_2) = r(2\pi, \Theta_2),\ r(\Theta_1, 0) = r(\Theta_1, 2\pi). \tag{2.12}$$

The boundary value problem (2.11), (2.12) is discretized with *central differences* (*leap frog*)

$$g(\Theta, r(\Theta)) = f_1(\Theta_1 r(\Theta))D_1 r(\Theta) + f_2(\Theta, r(\Theta))D_2 r(\Theta),\ \Theta \in \Omega_h \tag{2.13}$$

where

$$\Omega_h = \{(n_1 h_1, n_2 h_2) : n_j \in Z(\text{mod } N_j),\ j = 1, 2\},\ h_j = \frac{2\pi}{N_j}$$

and

$$\begin{aligned} D_1 r(\Theta) &= \frac{1}{2h_1}\left(r(\Theta_1 + h_1, \Theta_2) - r(\Theta_1 - h_1, \Theta_2)\right) \\ D_2 r(\Theta) &= \frac{1}{2h_2}\left(r(\Theta_1, \Theta_2 + h_2) - r(\Theta_1, \Theta_2 - h_2)\right). \end{aligned}$$

The periodicity condition is built into the definition of Ω_h. Now Newton's method can be applied to (2.13), and the linear systems arising in each step are of block tridiagonal form (cf. 5.1.3) with an additional block in the upper right and lower left corner.

If these linear systems are solved in a shooting type manner (called *compactification*), then ill-conditioned matrices arise, and a modification of the discretization avoiding these failures is proposed in Dieci, Lorenz Russell[25]. The compactification of these linear systems also elucidates the relations to another approach for invariant tori (see van Veldhuizen[99]). In that method the torus is computed via its intersection with a given plane, which is an invariant curve for the Poincaré map. This increases the geometric flexibility of the method if compared to the rather restrictive pre-transformation (2.9). On the other hand, the invariant circle methods rely to some extent on the asymptotic stability of the torus and the dynamics on it, while the PDE approach (2.11), (2.13) seems to be independent of it.

Of course, all these methods run into difficulties if the torus looses smoothness and breaks up. This is probably one of the possible routes to chaos (cf. Newhouse, Ruelle and Takens[73]).

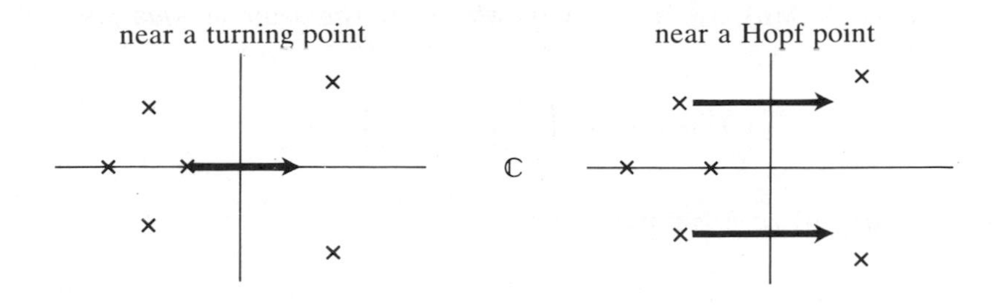

Fig. 5.9. Eigenvalues of f_u

5.3 Singular points in one-parameter systems

5.3.1 The loss of hyperbolicity on a branch

In this chapter we take up the questions of detecting and locating singular points in one-parameter systems (cf. section 5.1.5)

$$\dot{u} = f(u, \lambda), \ \lambda \in \mathbb{R}, \ u(t) \in \mathbb{R}^N. \tag{3.1}$$

Any stationary point of (3.1), which is not hyperbolic, will be called *singular*. Let $(u(s), \lambda(s))$ be a smooth branch of stationary points parametrized by $s \in (-s_0, s_0) \subset \mathbb{R}$, $s_0 > 0$. Looking at the spectrum of $f_u(u(s), \lambda(s))$, we can imagine basically two ways in which the stationary points loose hyperbolicity. Either a real eigenvalue crosses zero or a pair of complex conjugate eigenvalues crosses the imaginary axis (see Figure 5.9). In the first case we have a *turning point* (or a *saddle node* in the language of dynamical systems), and in the second case we have a *Hopf point.*

5.3.2 Turning points (saddle nodes)

The situation near a turning point is described in the following Theorem:

Theorem 3.1. *Let $(u(s), \lambda(s))$, $s \in (-s_0, s_0)$ be a smooth stationary branch of (3.1) with $(u'(s), \lambda'(s)) \neq 0$ for all s. Further assume that $f_u^0 := f_u(u(0), \lambda(0))$ has a simple eigenvalue 0 with eigenvector ϕ_0 normalized by $\phi_0^T \phi_0 = 1$ and that $(f_u^0 \ f_\lambda^0)$ has rank N. Then $\lambda'(0) = 0$ holds and the following conditions are equivalent*

- *(i) $\lambda''(0) \neq 0$*
- *(ii) $\mu'(0) \neq 0$, where $\mu(s)$ denotes the smooth continuation of the eigenvalue 0 for $f_u^s := f_u(u(s), \lambda(s))$*

(iii) $(u(0), \lambda(0), \phi_0)$ *is a regular solution of the defining equation*

$$T(u, \lambda, \phi) = \begin{pmatrix} f(u, \lambda) \\ f_u(u, \lambda)\phi \\ \phi^T \phi - 1 \end{pmatrix} = 0, \tag{3.2}$$

i.e., $T'(u(0), \lambda(0), \phi_0)$ *is nonsingular.*

Remark: If any of these three conditions is satisfied, then $(u(0), \lambda(0))$ is called a *quadratic* (or *simple*) *turning point.* The obvious reason is the geometric condition (i), according to which the branch is locally a parabola either turning to the right or to the left (see 5.1.5). Condition (i) is the characterization in terms of eigenvalues and condition (iii) is the first example of a defining equation given by Seydel[86] and further analyzed by Moore and Spence[72]. We will discuss a special aspect of defining systems in the next section, but for a broad overview we refer to Seydel[87]. For later reference we indicate here the proof of (i) $\Leftrightarrow$ (ii).

Proof. (i) $\Leftrightarrow$ (ii) Differentiating

$$f(u(s), \lambda(s)) = 0$$

yields

$$f_u^s u'(s) + f_\lambda^s \lambda'(s) = 0.$$

Taking $s = 0$ we obtain from our assumptions that $\lambda'(0) = 0$ and $u'(0) = c\,\phi_0$ for some $c \in \mathbb{R}$. We differentiate again with respect to s and find at $s = 0$,

$$c^2 f_{uu}^0 \phi_0^2 + f_u^0 u''(0) + f_\lambda^0 \lambda''(0) = 0.$$

Consequently, condition (i) is equivalent to

$$f_{uu}^0 \phi_0^2 \notin R(f_u^0) \tag{3.3}$$

where $R(f_u^0)$ denotes the range of f_u^0.
Similarly, the equivalence of condition (ii) and (3.3) can be shown by differentiating the following eigenvalue equation at $s = 0$

$$f_u(u(s), \lambda(s))\phi(s) = \mu(s)\phi(s),$$

where $\mu(0) = 0,\ \phi(0) = \phi_0.$ ■

Let us illustrate the phase diagrams near a turning point by introducing a parameter λ into example 1 (see (1.9))

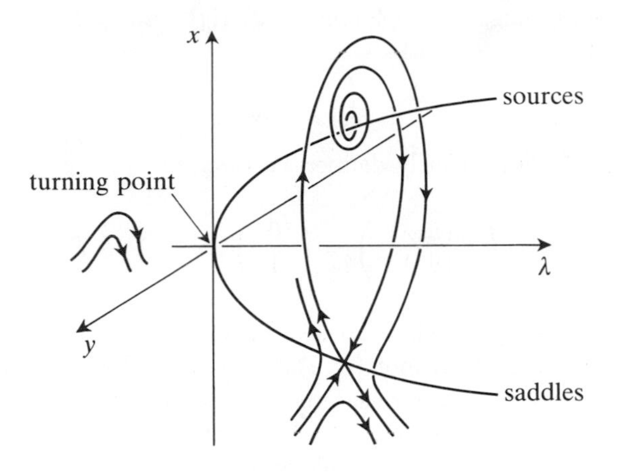

Fig. 5.10. Phase diagrams for Example $1(\lambda)$

Example 1(λ)

$$\dot{x} = y, \ \dot{y} = \lambda + y - x^2 + xy. \tag{3.4}$$

Here the stationary branch is

$$(x(s), y(s), \lambda(s)) = (s, 0, s^2)$$

and for the linearization f_u^s we find

$$\mathrm{Tr}(s) = 1 + s, \quad \mathrm{Det}(s) = 2s.$$

This is shown as a dashed linc in Figurc 5.3. The change of phase diagram with λ is illustrated in Figure 5.10. Notice that, according to Figure 5.3, the spiral source at $(\sqrt{\lambda}, 0)$ becomes an unstable node before coalescing with the saddle at $\lambda = 0$.

5.3.3 Defining equations

One Newton step for the defining equation (3.2) involves the solution of a linear system of dimension $2N + 1$, but, as Moore and Spence[72] showed, this can be reduced to 4 linear systems of dimension $N + 1$ with the same matrix. This matrix is nonsingular at the turning point and has the form

$$A(u, \lambda) = \begin{pmatrix} f_u(u, \lambda) & b_0 \\ c_0^T & 0 \end{pmatrix}, \ b_0, c_0 \in \mathbb{R}^N. \tag{3.5}$$

This result clearly suggests replacing (3.2) right away by a system of dimension $N+1$:

$$T(u,\lambda) = \begin{pmatrix} f(u,\lambda) \\ g(u,\lambda) \end{pmatrix} = 0, \tag{3.6}$$

where $g(u,\lambda) \in \mathbb{R}$ is implicitly defined through

$$A(u,\lambda) \begin{pmatrix} v(u,\lambda) \\ g(u,\lambda) \end{pmatrix} = \begin{pmatrix} 0 \\ 1 \end{pmatrix}, \quad v(u,\lambda) \in \mathbb{R}^N. \tag{3.7}$$

In fact, this is the approach proposed by Griewank and Reddien[41,42] Similar systems are given Abbott[2], Pönisch and Schwetlick[78], and even further reduced equations are considered in Jepson and Spence[55], Beyn[7]. Though many of these methods lead to comparable numerical effort, we think that the approach of Griewank and Reddien has several appealing features, conceptually as well as computationally:

(a) The function g can be used for detection as well as accurate location of a singular point. This generalizes to other singularities (see below).
(b) The system is written in such a way that the linear algebra of a Newton step suggests itself. In particular it can be seen that several linear systems with the same matrix $A(u,\lambda)$ or its adjoint have to be solved (see Lemma 3.2).
(c) If one wants to retain the original equation $f=0$, then (3.6) is the minimal extension possible.

Let us explain the argument b) in more detail. First, we have to evaluate the derivative

$$g_z(z) = (g_u(u,\lambda), g_\lambda(u,\lambda)), \;\; z=(u,\lambda)$$

which appears in a Newton step for (3.6). Here the following Lemma is useful (cf. Griewank and Reddien[41,42]). For later purposes we will formulate it for the case of $p \geq 1$ parameters.

Lemma 3.2. *Let $\Omega \subset \mathbb{R}^{N+p}$ be open and $f : \Omega \to \mathbb{R}^N$ be smooth such that*

$$A(z) := \begin{pmatrix} f_u(z) & B_0 \\ C_0^T & 0 \end{pmatrix} \in \mathbb{R}^{N+p,N+p}, \;\; B_0, C_0 \in \mathbb{R}^{N,p} \text{ fixed} \tag{3.8}$$

is nonsingular for all $z=(u,\lambda) \in \Omega$. Define the functions

$$U, V : \Omega \to \mathbb{R}^{N,p} \;\; \text{and} \;\; G, \widetilde{G} : \Omega \to \mathbb{R}^{p,p}$$

by

$$A(z)\begin{pmatrix} V(z) \\ G(z) \end{pmatrix} = \begin{pmatrix} 0 \\ I_p \end{pmatrix}, \tag{3.9}$$

$$(U^T(z)\ \widetilde{G}(z))A(z) = (0\ I_p). \tag{3.10}$$

Then the following relations hold in Ω

$$G = \widetilde{G} = -U^T f_u V \tag{3.11}$$

$$G_z = -U^T f_{uz} V. \tag{3.12}$$

Proof. Let us write down (3.9), (3.10) explicitly

$$\begin{aligned} f_u V + B_0 G = 0, \quad & C_0^T V = I_p \\ U^T f_u + \widetilde{G} C_0^T = 0, \quad & U^T B_0 = I_p. \end{aligned}$$

Multiplying the first equation by U^T from the left and the third by V from the right immediately yields (3.11). The same operation on the derived terms gives

$$U^T f_{uz} V + U^T f_u V_z + G_z = U_z^T f_u V + U^T f_{uz} V + G_z = 0.$$

We then arrive at (3.12) by combining these formulas with the formula obtained by direct differentiation of (3.11). ∎

We notice that G_z can be easily evaluated from f_z by numerical differentiation, e.g., if $p = 1$, we may use

$$g_z(z) \sim -\frac{1}{h} U^T(f_z(u + hV, \lambda) - f_z(u, \lambda)). \tag{3.13}$$

In the Newton step for (3.6) we have to solve a linear system with

$$T'(u, \lambda) = \begin{pmatrix} f_u & f_\lambda \\ g_u & g_\lambda \end{pmatrix}(u, \lambda).$$

We now insist that this is done with the help of the matrix (3.5). One reason for this is that the user may provide a black box routine for solving with the bordered matrix (3.5) (or (3.8) in the general case). The following Lemma shows how to reduce the solution of one bordered linear system to the solution of another one. It is in some sense implicitly contained in Griewank and Reddien[42].

Lemma 3.3. *Let* $E_0 = \begin{pmatrix} A_0 & B_0 \\ C_0^T & 0 \end{pmatrix} \in \mathbb{R}^{N+p,N+p}$ *be nonsingular and let*

$E = \begin{pmatrix} A_0 & B \\ C^T & D \end{pmatrix} \in R^{N+k,N+k}$ *be another bordering of* A_0. *Solve the following* $p+k$ *linear systems with* E_0

$$E_0 \begin{pmatrix} X_0 \\ Y_0 \end{pmatrix} = \begin{pmatrix} 0 \\ I_p \end{pmatrix}, \quad E_0 \begin{pmatrix} X \\ Y \end{pmatrix} = \begin{pmatrix} B \\ 0 \end{pmatrix}. \tag{3.14}$$

Then E *is nonsingular iff* Δ *is nonsingular, where*

$$\Delta = \begin{bmatrix} Y & Y_0 \\ C^T X - D & C^T X_0 \end{bmatrix} \in \mathbb{R}^{p+k,p+k}. \tag{3.15}$$

The solution of a system $E \begin{pmatrix} x \\ y \end{pmatrix} = \begin{pmatrix} f \\ g \end{pmatrix}$ *can be written as*

$$x = x_0 - Xy - X_0\, d_0 \tag{3.16}$$

where $E_0 \begin{pmatrix} x_0 \\ y_0 \end{pmatrix} = \begin{pmatrix} f \\ 0 \end{pmatrix}$ *and* $\Delta \begin{pmatrix} y \\ d_0 \end{pmatrix} = \begin{pmatrix} y \\ C^T x_0 - g \end{pmatrix}$.

The proof is rather easily obtained by inserting the formula (3.16) into the given linear system, and we omit the details. If $k = p$ and $B = B_0$, then we can take $X = 0$, $Y = B_0$. We also notice, that in the extreme case $p = 0$, Lemma 3.3 reduces to the *block elimination method* (e.g. Keller[56]). In the turning point case one might use this Lemma with $B = B_0 = f_\lambda$ and, noticing the coincidence of (3.7) and (3.14), one ends up, for one Newton step, with 2 linear systems in A and one in A^T as well as 2 evaluations of (f_u, f_λ) (see (3.13)). We finally notice that several algorithms have been proposed for solving systems with bordered almost singular matrices as in (3.5). The emphasis here of course is on methods which exploit sparsity of f_u or even work with a black box solver for f_u (see e.g. Rheinboldt[79], Chan[19] and the remarkably simple, recent approach in Govaerts[39]).

5.3.4 Hopf points

We return to the second mechanism of loosing hyperbolicity, i.e., through two complex eigenvalues crossing the imaginary axis. This situation is described by the classical Hopf bifurcation theorem (see e.g. Hassard, Kazarinoff and Wan[46], Amann[3]).

Theorem 3.4. *Let* $(u(\lambda), \lambda)$, *be a smooth stationary branch of (3.1). Assume that at some* $\lambda = \lambda_0$ *the matrix* $f_u^0 = f_u(u(\lambda_0), \lambda_0)$ *has a simple* *eigenvalue $i\omega_0, \omega_0 \neq 0$ with eigenvector $x_0 + iy_0$ and no eigenvalue of the type $ik\omega_0,\ k = 0, 2, 3, \ldots$ Finally, assume*

$$Re\ \mu'(\lambda_0) \neq 0, \tag{3.17}$$

where $\mu(\lambda)$ is the continuation of the simple eigenvalue $i\omega_0$ for $f_u(u(\lambda), \lambda)$. Then there exists an $a_0 > 0$ and a smooth branch of $T(a)$-periodic solutions $(u(t,a)(0 \le t \le T(a)),\ \lambda(a)), |a| < a_0$ for $\dot{u} = f(u, \lambda)$ with the following properties

$$\begin{aligned} u(t,a) &= u(\lambda(a)) + a(\cos(\omega_0 t)x_0 - \sin(\omega_0 t)y_0) + \mathcal{O}(a^2) \\ \lambda(a) &= \lambda_0 + \mathcal{O}(a^2), T(a) = \frac{2\pi}{\omega_0} + \mathcal{O}(a^2). \end{aligned} \tag{3.18}$$

Compared to Theorem 3.1, this theorem only describes the geometric setting, i.e., the periodic orbits created out of $(u(\lambda_0), \lambda_0)$ and parametrized by the amplitude a.

For an illustration we now insert a parameter λ into example 1(–) (see 5.1.4).

Example 1(-λ)

$$\dot{x} = y,\ \dot{y} = \lambda - 2y - x^2 + xy. \tag{3.19}$$

As in example 1(λ) the stationary branch is

$$(x(s), y(s), \lambda(s)) = (s, 0, s^2).$$

For the linearization f_u^s at these points we obtain

$$\mathrm{Tr}(s) = -2 + s, \quad \mathrm{Det}(s) = 2s.$$

This is the dotted line shown in Figure 5.3. We have a saddle for $s < 0$, a sink for $0 < s < 2$ and a source for $s > 2$. Consequently, we still find a turning point at $s = 0$, but in addition to this, there is a Hopf bifurcation at $s = 2$. It turns out that the periodic orbits created at $s = 2$ are asymptotically stable and that they exist in a certain interval for $\lambda > 4$. We have sketched the changes in the phase diagram up to this region in Figure 5.11.

In our next step we relate the eigenvalue condition (3.17) of the Hopf Theorem to the regularity of a defining equation. This is the analogy to the equivalence (ii) $\Leftrightarrow$ (iii) in Theorem 3.1.

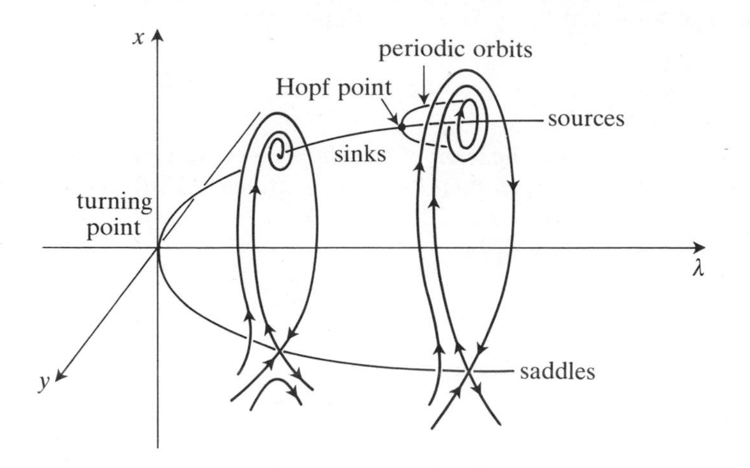

Fig. 5.11. Phase diagrams for Example $1(-\lambda)$

Theorem 3.5. *Let the assumptions of Theorem 3.4 hold with the exception of (3.17) and let $c \in \mathbb{R}^N$ be given such that*

$$c^T x_0 = 0, \; c^T y_0 = 1.$$

Then the eigenvalue condition (3.17) holds if and only if

$$(u(\lambda_0), \lambda_0, x_0, y_0, \omega_0) \in \mathbb{R}^{3N+2}$$

is a regular solution of the defining equation

$$T(u, \lambda, x, y, \omega) = \begin{pmatrix} f(u, \lambda) \\ f_u(u, \lambda)x + \omega y \\ f_u(u, \lambda)y - \omega x \\ c^T x \\ c^T y - 1 \end{pmatrix} = 0. \tag{3.20}$$

The system (3.20) was set up and analyzed by Jepson[54]. Further investigations are due to Griewank and Reddien[40], who showed that the linearized system can be reduced to solving several systems with a bordering of $f_u^2(u, \lambda) + \omega^2 I$.

In the spirit of the previous section it seems therefore reasonable to replace (3.20) by the $(N+2)$-dimensional system

$$T(u, \lambda, \omega) = \begin{pmatrix} f(u, \lambda) \\ g(u, \lambda, \omega) \end{pmatrix} = 0, \tag{3.21}$$

where $g(u, \lambda, \omega) \in \mathbb{R}^2$ is defined by

$$\begin{pmatrix} f_u^2(u,\lambda) + \omega^2 I & B_0 \\ C_0^T & 0 \end{pmatrix} \begin{pmatrix} v(u,\lambda,\omega) \\ g(u,\lambda,\omega) \end{pmatrix} = \begin{pmatrix} 0 \\ 0 \\ 1 \end{pmatrix} \tag{3.22}$$

Since, at the Hopf point $f_u^2 + \omega_0^2 I$ has a two-dimensional null space, spanned by x_0, y_0, it is clear that $B_0, C_0 \in \mathbb{R}^{N,2}$ can be chosen in such a way that (3.22) is nonsingular close to the Hopf-point. Similar strategies, which employ the matrix $f_u^2 + \omega^2 I$ or its characteristic polynomial were proposed by Kubićek and Holodniok[64] and Roose and Hlavacek[83]. Let us consider here one Newton step for (3.20).

As in Lemma 3.2 we solve the adjoint system

$$(W^T G) \begin{pmatrix} f_u^2 + \omega^2 I & B_0 \\ C_0^T & 0 \end{pmatrix} = (0 \; I_2), \; W \in \mathbb{R}^{N,2}, \; G \in \mathbb{R}^{2,2}$$

and find

$$g = -W^T(f_u^2 + \omega^2 I)v = \text{ second column of } G \tag{3.23}$$

$$g_z\zeta = -W^T[f_{uz}(f_u v, \zeta) + f_u \; f_{uz}(v, \zeta)], \; z = (u, \lambda), \zeta \in \mathbb{R}^{N+1}, \tag{3.24}$$

$$g_\omega = 2\omega \; W^T v. \tag{3.25}$$

Again, g_z can be approximated by difference quotients, and with this information we can set up the bordered matrix

$$T' = \begin{pmatrix} f_u & f_\lambda & 0 \\ g_u & g_\lambda & g_\omega \end{pmatrix}$$

At a Hopf point f_u is regular, so we can solve with T' by a block elimination method. However, as Griewank and Reddien[40] pointed out, turning points are also solutions of (3.20) (and hence of (3.21)), namely with $\omega = 0$. Therefore, one of the algorithms mentioned at the end of the last section is also recommended for this system. Of course, (3.22) requires the computation of f_u^2, but for large banded matrices the computing effort still grows only linearly with N. We should add here, that up to now we have no numerical experience with the system (3.21).

We close this section with some comments on the problem of detecting Hopf points (compare Jepson[54] or Seydel[87] for a discussion). Since the function g in (3.21) is two-dimensional, this problem amounts to watching for a zero of

$$g(u(s), \lambda(s), \omega) \text{ at some } \omega \in \mathbb{R}.$$

Clearly, this is a difficult task. One can simply compute all eigenvalues of $f_u(u(s), \lambda(s))$. But then the computational costs usually dominate

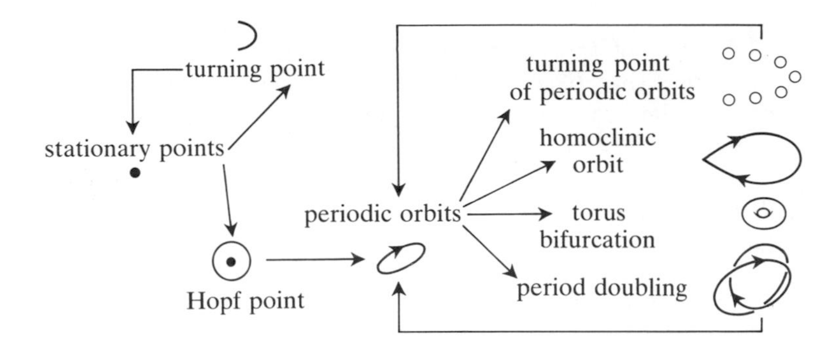

Fig. 5.12.

those for the branch itself, which is particularly awkward for N large and f_u sparse. For a recent attack to this difficult problem by using inserve subspace iteration see Garratt, Moore and Spence[37]. A further possibility for line systems (see 5.1.3, Example 3) is recommended by Seydel[87]. He proposes to use a coarser spatial grid for the Hopf detection.

5.3.5 Singular periodic orbits

Once a branch of periodic orbits has been created at a Hopf point one would like to continue these periodic orbits (cf. 5.2.2) and detect bifurcations on the branch. In Figure 5.12 we have sketched the possible further scenarios by using the list of bifurcations from 5.1.5. We may either find a turning point of periodic orbits, a period doubling, a homoclinic orbit or a torus bifurcation. The last three phenomena already represent final steps on the route to chaotic behaviour, such as a period doubling sequence, periodic forcing of systems with homoclinics and torus breakdown (see Guckenheimer and Holmes[43]).

In what follows, we will focus on only one of the possibilities in Fig. 10, viz. the homoclinic bifurcation.

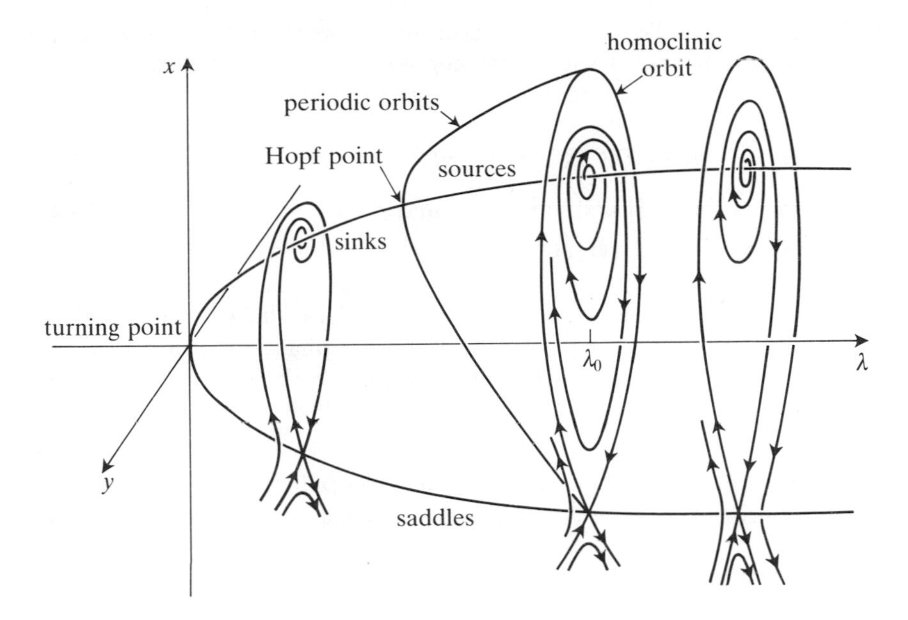

Fig. 5.13. Phase diagram for Example $1(-\lambda)$

5.3.6 The homoclinic bifurcation

Let us continue example $1(-\lambda)$ from section 5.3.4 and observe the fate of the periodic orbit created at the Hopf point. With increasing λ the periods also increase, while the periodic orbits in phase space approach on one side the saddle point. Finally, at some $\lambda = \lambda_0$ the period becomes infinite and the perodic orbit turns into a homoclinic orbit, connecting the unstable saddle with itself in infinite time. Increasing λ beyond λ_0 just leaves us with the two unstable stationary points, and all trajectories not starting at these two points or on their stable manifolds eventually escape to infinity. The phase diagrams are shown schematically in Figure 5.13. Because of the global change of phase diagram at the homoclinic orbit, this transition is often called a *global bifurcation.* Before we can state some basic results we have to introduce further notions. Let $\overline{u}(t), t \in \mathbb{R}$ be a solution of $\dot{u} = f(u, \lambda)$ at some $\lambda = \overline{\lambda}$ such that there exists a $\overline{v} \in \mathbb{R}^N$ with

$$\overline{u}(t) \to \overline{v} \quad \text{as} \quad t \to \infty \quad \text{and as} \quad t \to -\infty. \tag{3.26}$$

Then $\{\overline{u}(t) : t \in \mathbb{R}\}$ is called a *homoclinic orbit* with *base point* $\overline{v}$, and the pair $(\overline{u}, \overline{\lambda})$ is called a *homoclinic orbit pair* (HOP). Obviously, $\overline{v}$ has to be an unstable stationary point of $\dot{u} = f(u, \overline{\lambda})$.

Similar to periodic orbits (see Theorem 2.3) we want to recognize $(\overline{u}, \overline{\lambda})$ as a regular solution of some operator equation. For this purpose, the following Banach spaces are useful.

$$\begin{aligned} B_0 &= \{u \in C(\mathbb{R}, \mathbb{R}^N) : \lim_{t\to\infty} u(t) \text{ and } \lim_{t\to-\infty} u(t) \text{ exist }\} && (3.27) \\ ||u||_0 &= \sup \ \{||u(t)|| : t \in \mathbb{R}\} \text{ and} && (3.28) \\ B_1 &= \{u \in C^1(\mathbb{R}, \mathbb{R}^N) : u, \dot{u} \in B_0\}, \ ||u||_1 = ||u||_0 + ||\dot{u}||_0. && (3.29) \end{aligned}$$

Next we notice that homoclinic orbits may be phase shifted just as periodic ones (cf. 5.2.2), so for uniqueness we impose a phase condition $\Psi(u) = 0$ where $\Psi : B_0 \to \mathbb{R}$ is assumed to be smooth. We then have the following characterization (see Beyn[11]).

Theorem 3.6. *Let $(\overline{u}, \overline{\lambda})$ be a homoclinic orbit pair with a hyperbolic base point. Then $(\overline{u}, \overline{\lambda}) \in B_1 \times \mathbb{R}$ is a regular solution of the operator equation*

$$F(u, \lambda) = \begin{pmatrix} \dot{u} - f(u, \lambda) \\ \Psi(u) \end{pmatrix} = 0, \ F : B_1 \times \mathbb{R} \to B_0 \times \mathbb{R} \qquad (3.30)$$

if and only if the following two conditions hold

(i) the only solutions $(v, \mu) \in B_1 \times \mathbb{R}$ of the variational equation

$$\dot{v} = f_u(\overline{u}, \overline{\lambda})v + f_\lambda(\overline{u}, \overline{\lambda})\mu \qquad (3.31)$$

are $v = c\dot{\overline{u}}$ $(c \in \mathbb{R})$, $\mu = 0$.
(ii) $\Psi(\overline{u}) = 0$, $\Psi'(\overline{u})\dot{\overline{u}} \neq 0$.

Remark: Condition (i) is plausible from the fact that $v = c\dot{\overline{u}}$, $\mu = 0$ always solves (3.31), as can be seen by differentiating $\dot{\overline{u}} = f(\overline{u}, \overline{\lambda})$. Moreover, this condition can be characterized by a transversal intersection of certain stable and unstable manifolds (Beyn[12]).

We will not prove Theorem 3.6. However, we would like to draw the reader's attention to an important technical tool which is employed in this as well as the following theorems. The linearization (3.31) suggests that we first have to study the behaviour of linear differential operators

$$Lu = \dot{u} - A(t)u, \quad A(t) \in \mathbb{R}^{N,N} \text{ continuous in } t \in \mathbb{R}.$$

For these, the notion of an *exponential dichotomy* (Coppel[21], Palmer[76]) is of utmost importance. Let $Y(t), t \in \mathbb{R}$ be a fundamental matrix of L normalized by $Y(0) = I$. Then L is said to have an exponential dichotomy on some interval $J \subset \mathbb{R}$, if there exist $K, \alpha > 0$ and a projection P in $\mathbb{R}^N$ such that for all $t, s \in J$

$$||Y(t)PY(s)^{-1}|| \leq Ke^{-\alpha(t-s)}, \quad s \leq t \tag{3.32}$$
$$||Y(t)(I-P)Y(s)^{-1}|| \leq Ke^{-\alpha(s-t)}, \quad t \leq s. \tag{3.33}$$

This means that we can decompose the solution operator associated with L into a part which decays exponentially in forward time and another part which decays exponentially in backward time. This property holds for the linear differential operator L obtained by setting

$$A(t) = f_u(\overline{u}(t), \overline{\lambda}),$$

where $(\overline{u}, \overline{\lambda})$ is the HOP. Using the hyperbolicity of the base point, one can show that L has an exponential dichotomy on both $J = [0, \infty)$ and $J = (-\infty, 0]$ (but not on $J = \mathbb{R}$!). These dichotomies may then be used to set up a linear Fredholm theory for the operator $L : B_1 \to B_0$. In fact, in our homoclinic case L turns out to have Fredholm index zero.

Theorem 3.6 gives rise to the following definition. A HOP $(\overline{u}, \overline{\lambda})$ is called *nondegenerate*, if the base point is hyperbolic and if condition (i) of Theorem 3.6 holds (see Beyn[11] for the general case of connecting orbits). It is remarkable that this basic assumption suffices to guarantee a branch of periodic orbits created out of the homoclinic orbit according to the following *homoclinic bifurcation theorem.*

Theorem 3.7. *Let $(\overline{u}, \overline{\lambda})$ be a nondegenerate homoclinic orbit pair of*

$$\dot{u} = f(u, \lambda). \tag{3.34}$$

Then there exists a $T_0 > 0$ and a branch of $2T$-periodic solutions

$$(u(t, T)(|t| \leq T), \lambda_T), \quad T \geq T_0$$

of the system (3.34) with the following estimates

$$\begin{aligned} ||\overline{u}(t) - u(t, T)|| &\leq C\, e^{-\alpha T} \quad \text{for all } t \in [-T, T] \\ |\overline{\lambda} - \lambda_T| &\leq C\, e^{-2\alpha T} \end{aligned} \tag{3.35}$$

where $\alpha < |Re\ \mu|$ for all eigenvalues μ of the linearization at the base point.

Surprisingly, this general theorem seems not to have been noticed until recently (Lin[69], Beyn[12]), although there is a long history of homoclinic bifurcation results (cf. Andronov, Leontovich, Gordon and Maier[4], Chow and Hale[20], Guckenheimer and Holmes[43], Wiggins[100]). One reason may be that the cited references try to discuss simultaneously the dynamics

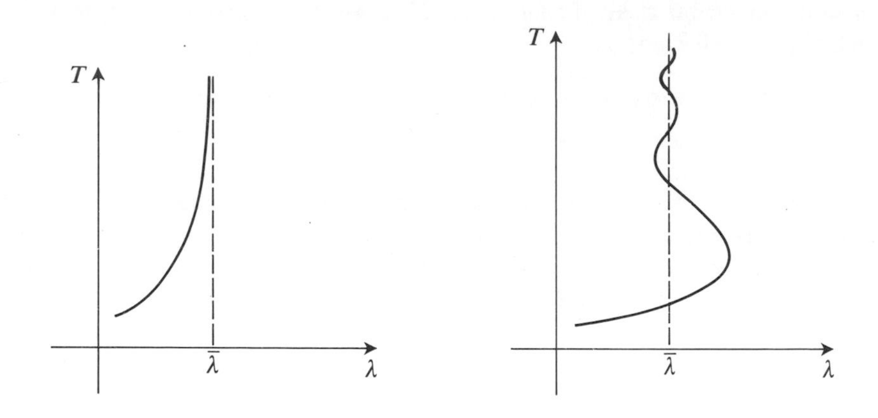

Fig. 5.14. Typical diagrams at a homoclinic bifurcation

of (3.34) for λ close to $\overline{\lambda}$. However, these can be very complicated in dimensions $N \geq 3$. In Figure 5.14 we sketch two basic behaviours of the periodic branch in a (λ, T) diagram. Let us denote those eigenvalues at the hyperbolic base point as *critical* which have smallest positive or largest negative real part. Then case 1 in Figure 5.14 usually occurs when the two critical eigenvalues are real, while in case 2 there is a real and a complex conjugate pair of critical eigenvalues. This last case is called the Shil'nikov bifurcation, see Shil'nikov[88], Glendinning,[38], Wiggins[100], Lin[69]. One of its remarkable features is that there exist infinitely many periodic orbits at $\lambda = \overline{\lambda}$.

Theorem 3.7 also has an impact on numerical calculations. It shows that HOP's may well be approximated by periodic orbits of large but fixed periods, and this approach has been successfully used by Doedel and Kernevez[28].

However, it is possible to replace the periodic boundary conditions by more efficient ones, the so-called *projection boundary conditions*, which lead to even better convergence than (3.35) and hence allow for smaller time intervals. We will briefly discuss here the method of Beyn[11,12], and we mention that a slightly different method was developed by Doedel and Friedman[29,30] (see also Kuznetsov[67], Rodrigues-Luis, Freire and Ponce[81]). Both approaches allow for the computation of more general *connecting orbits*, i.e., orbits which connect two possibly different stationary points.

We know from Theorem 3.6 that we have to solve a defining equation, which is a parametrized boundary value problem on the real line

$$\dot{u} = f(u, \lambda), \quad u \in B_1, \lambda \in \mathbb{R}, \tag{3.36}$$

$$\Psi(u) = 0. \tag{3.37}$$

For numerical purposes we have to truncate this boundary value problem to a finite interval $J = [T_-, T_+]$, and we have to set up boundary conditions at T_-, T_+ which catch the asymptotic behaviour of the solution (notice that in (3.36) the boundary conditions are hidden in the space B_1). For the case of semi-infinite intervals such boundary conditions have been set up by de Hoog and Weiss[22] and Lentini and Keller[68]. Here we generalize the first approach to our problem (3.36).

We require that a branch of stationary hyperbolic points $v(\lambda)$, is known (or can at least be computed numerically) which contains the candidates for the base points (compare Figure 5.13). Consider the *stable subspace* $Y_s(\lambda)$ of $f_u(v(\lambda), \lambda)$, which contains all generalized eigenvectors belonging to the eigenvalues with negative real part, and similarly let $Y_u(\lambda)$ be the *unstable subspace*. Then we impose the projection boundary conditions

$$u(T_+) - v(\lambda) \in Y_s(\lambda), \quad u(T_-) - v(\lambda) \in Y_u(\lambda). \tag{3.38}$$

These conditions are very natural since the homoclinic orbit must leave $v(\lambda)$ via its unstable manifold, which is tangent to $Y_u(\lambda)$, and must approach $v(\lambda)$ again via its stable manifold, which is tangent to $Y_s(\lambda)$ (see Figure 5.15 and, for example, Irwin[51]). For a numerical implementation we construct full rank matrices

$$P_s(\lambda) \in \mathbb{R}^{N_u, N}, \; P_u(\lambda) \in \mathbb{R}^{N_s, N}, \; N_u = N - N_s$$

such that

$$Y_s(\lambda) = \{z : P_s(\lambda)z = 0\}, \; Y_u(\lambda) = \{z : P_u(\lambda)z = 0\}$$

and rewrite (3.38) as

$$P_s(\lambda)(u(T_+) - v(\lambda)) = 0, \; P_u(\lambda)(u(T_-) - v(\lambda)) = 0. \tag{3.39}$$

These projection type matrices can be computed numerically in such a way that the smooth dependence on λ is guaranteed (see Beyn[11]).

Finally, bearing in mind the discussion of phase conditions in 5.2.2, we replace (3.37) by an integral condition

$$\Psi_I(u) = \int_{T_-}^{T_+} z_0^T(t)(u(t) - u_0(t))\, dt = 0, \tag{3.40}$$

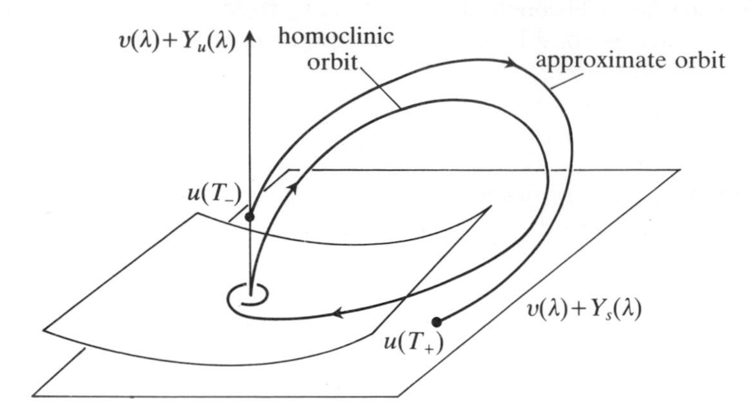

Fig. 5.15. Illustration of projection boundary conditions

where $u_0(t)$ is some initial approximation and $z_0(t) \approx \dot{u}_0(t)$. Summing up, we have to solve the $(N+1)$-dimensional system

$$\dot{u} = f(u, \lambda), \;\; \dot{\lambda} = 0 \;\; \text{for} \;\; t \in [T_-, T_+] \tag{3.41}$$

subject to the $N+1$ boundary conditions (3.39), (3.40). Notice that (3.39) is linear in u but nonlinear in λ.

As in Theorem 3.7 the approximation error, due to the truncation to the finite interval, can be estimated for $[T_-, T_+]$ sufficiently large (see Beyn[11, 12]).

Theorem 3.8. *Let $(\overline{u}, \overline{\lambda})$ be a nondegenerate homoclinic orbit pair and assume that in the phase condition (3.40) we use functions $z_0, u_0 \in B_0$ such that*

$$\int_{-\infty}^{+\infty} z_0^T(t)(\overline{u}(t) - u_0(t))\, dt = 0, \;\; \int_{-\infty}^{\infty} z_0^T(t)\dot{\overline{u}}(t)\, dt \neq 0. \tag{3.42}$$

Then there exists a $T_0 > 0$, such that for $T_+, -T_- \geq T_0$ the boundary value problem (3.39)–(3.41) has a unique solution $(u_{[T_-,T_+]}, \lambda_{[T_-,T_+]})$ close to $(\overline{u}|_{[T_-,T_+]}, \overline{\lambda})$ and the following estimates hold for a suitable phase shift $\tau = \tau(T_-, T_+)$

$$\begin{aligned} ||\overline{u}(t+\tau) - u_{[T_-,T_+]}(t)|| &\leq C\, e^{-2\alpha \, \mathrm{Min}(|T_-|, T_+)}, \;\; t \in [T_-, T_+] \\ |\overline{\lambda} - \lambda_{[T_-,T_+]}| &\leq C\, e^{-3\alpha \, \mathrm{Min}(|T_-|, T_+)} \end{aligned} \tag{3.43}$$

where α is a given constant with $\alpha < |Re\; \mu|$ for all eigenvalues μ of the linearization at the base point.

The requirements for the phase fixing of the continuous homoclinic orbit are not really restrictive, since we are only interested in approximating it up to a phase shift. We notice that (3.43) has better exponents that (3.35) and this difference can be clearly seen in practical computations. Also, for both types of boundary conditions we have a superconvergence phenomenon in the parameter. This also shows up numerically, and in some cases the λ-error is even smaller than the prediction from (3.43).

We consider as a final example the Lorenz equations (cf. Sparrow[90]).

Example 5

$$\dot{x} = \sigma(y - x), \ \dot{y} = \lambda\, x - y - xz, \ \dot{z} = -\mu z + xy.$$

First, at the Lorenz values $\sigma = 10$, $\mu = \frac{8}{3}$ the homoclinic orbit connecting the origin with itself and the λ-value ($= 13.926557$) were computed. Due to the symmetry in the Lorenz equation this homoclinic orbit has a symmetric companion orbit, and both together create a strange invariant set (see Sparrow[90], Glendinning[38]) which at higher λ-values stabilizes to an attractor.

We then continued this HOP into a branch of HOP's by freeing the parameter μ. In addition, the automatic adaptation strategy for $[T_-, T_+]$, as developed in Beyn[11], was used. Some of the homoclinic orbits from this branch are shown in an xy-projection in Figure 5.16.

5.4 Two–parameter problems

In this chapter we discuss some aspects of two parameter systems

$$\dot{u} = f(u, \lambda), \ \lambda = (\lambda_1, \lambda_2) \in \mathbb{R}^2, \ u(t) \in \mathbb{R}^N. \tag{4.1}$$

It is evident that in such systems singular points with higher degeneracies are possible, and one might well ask why one should analyze or even numerically compute these points. The answer is that these singular points serve as *organizing centers* for dynamic features, which are obtained under parametric perturbations. And these dynamic features may well be generic, i.e., they appear in 0-parameter problems (e.g., invariant tori), and so we may start a branch of these objects at the singularity. In addition, from two-parameter problems onwards, chaotic behaviour is possible near singularities, so that there is some chance for an analytical treatment. This is one of the important discoveries of dynamical systems theory.

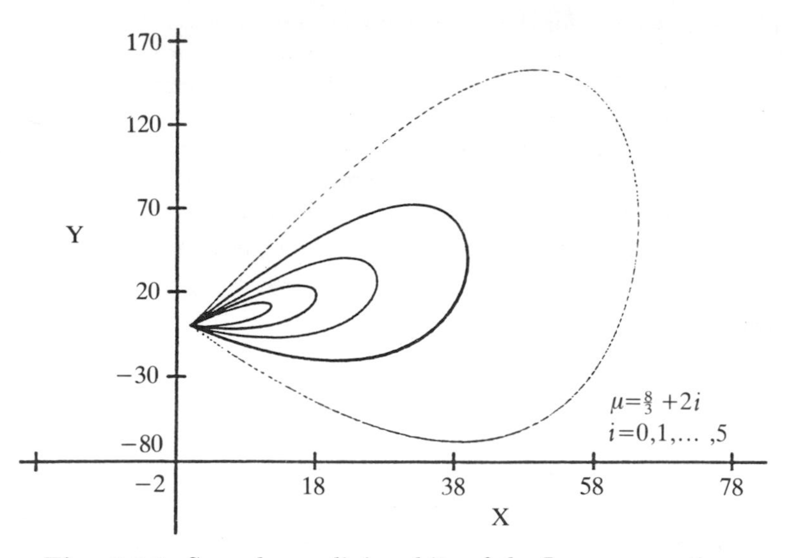

Fig. 5.16. Some homoclinic orbits of the Lorenz equations

5.4.1 The possibilities

Suppose that we continue a branch of stationary points $(u(s), \lambda(s))$, $s \in \mathbb{R}$ of (4.1) which are singular (u, λ_1)–points, i.e., either a branch of Hopf points or turning points. Thus the Jordan normal form J^s of $f_u^s = f_u(u(s), \lambda(s))$ is of the type

$$J^s = \begin{pmatrix} 0 & 0 \\ 0 & H(s) \end{pmatrix} \text{ or } J^s = \begin{bmatrix} 0 & \omega(s) & 0 \\ -\omega(s) & 0 & 0 \\ 0 & 0 & H(s) \end{bmatrix}$$

For most of the points we expect the submatrix $H(s)$ to be hyperbolic, but at some values of s it may be nonhyperbolic. However, the reader is cautioned that the Jordan structure (along with some nondegeneracy conditions for quadratic terms) is no longer sufficient for classifying the singularities in two-parameter problems. For example, on a branch of turning points we may find a degeneracy in the quadratic terms, viz.(compare (3.3)),

$$f_{uu}^s \, \phi_s^2 \in R(f_u^s),$$

where ϕ_s spans the null space of f_u^s. This is the well-known *cusp point* (see Guckenheimer and Holmes[43], Chapter 7.1 and for numerical methods Spence and Werner[91], Roose and Piessens[84], Pönisch[77], Griewank and Reddien[42]). For the remaining three possibilities we have a singular block in the Jordan form of one of the following types

$$\begin{pmatrix} 0 & 1 \\ 0 & 0 \end{pmatrix}, \begin{pmatrix} 0 & \omega & 0 \\ -\omega & 0 & 0 \\ 0 & 0 & 0 \end{pmatrix}, \begin{pmatrix} 0 & \omega_1 & 0 & 0 \\ -\omega_1 & 0 & 0 & 0 \\ 0 & 0 & 0 & \omega_2 \\ 0 & 0 & -\omega_2 & 0 \end{pmatrix}. \tag{4.2}$$

The analysis of the first case is generally attributed to Bogdanov[15] and Takens[97]. Consequently, we call these points *Takens-Bogdanov singularities* or *TB-points.* A numerical method for computing TB-points was given by Roose[82]; see also Khibnik[59] and the further references therein. Here we will pursue the corresponding defining equation from the viewpoint taken in section 5.3, i.e. we will follow Griewank and Reddien[42]. It will become clear from the following presentation and from the treatment of Hopf points in 5.3.4, how to construct defining equations for the remaining cases in (4.2). We notice that these bifurcations can only occur in systems of dimensions $N \geq 3$ and $N \geq 4$ and a complete analysis of these cases is still not available, mainly due to the local occurrence of chaotic behaviour (Guckenheimer and Holmes[43]).

5.4.2 The Takens-Bogdanov singularity

Let us first consider a two-dimensional example which contains all the variations of example 1 (see 5.1.3, 5.1.4, 5.3.2, 5.3.4, 5.3.6).

Example 1(λ_1, λ_2)

$$\dot{x} = y, \ \dot{y} = \lambda_1 + \lambda_2 y - x^2 + xy. \tag{4.3}$$

In 5.3.2, 5.3.4, 5.3.6 we studied the cases $\lambda_2 = 1$ and $\lambda_2 = -2$ with varying λ_1. In the general case of (4.3) the stationary points lie on the folded surface

$$S = \{(x, y, \lambda_1, \lambda_2) = (s, 0, s^2, \lambda_2) : s, \lambda_2 \in \mathbb{R}\}$$

and the trace and determinant of f_u at an arbitrary point on S are given by

$$\mathrm{Tr}(s, \lambda_2) = \lambda_2 + s, \quad \mathrm{Det}(s, \lambda_2) = 2s.$$

Varying s and λ_2, these values now cover the whole plane in Figure 5.3. Hopf bifurcation occurs on the half-parabola

$$\lambda_2 = -s, \lambda_1 = s^2, \ s > 0. \tag{4.4}$$

The periodic orbits created at these Hopf points vanish through a homoclinic orbit on a curve which in a neighbourhood of $\lambda_1 = \lambda_2 = 0$ is approximately given by

$$\lambda_1 = \left(\frac{7}{5}\,\lambda_2\right)^2, \ \lambda_2 < 0.$$

System (4.3) is in fact a so-called *unfolding* of the system

$$\dot{x} = y, \ \dot{y} = -x^2 + xy,$$

which has 0 as a TB-point (see Guckenheimer and Holmes[43], Chapter 7 and notice that we slightly transformed the system). The qualitative information on the various phase diagrams is contained in Figure 5.17.

For the numerical computation of a TB-point let us start as in (4.1) with a branch of λ_1-turning points $(u(s), \lambda(s))$, obtained by solving

$$T(u, \lambda) = \begin{pmatrix} f(u, \lambda) \\ g(u, \lambda) \end{pmatrix} = 0,$$

where

$$A(u, \lambda) \begin{pmatrix} v(u, \lambda) \\ g(u, \lambda) \end{pmatrix} = \begin{pmatrix} 0 \\ 1 \end{pmatrix}, \ A(u, \lambda) = \begin{pmatrix} f_u(u, \lambda) & b_0 \\ c_0^T & 0 \end{pmatrix} \tag{4.5}$$

and

$$(\Psi^T(u, \lambda), g(u, \lambda)) \ A(u, \lambda) = (0, 1). \tag{4.6}$$

Compare with (3.6), but notice that now $\lambda \in \mathbb{R}^2$.

A test function for detecting a TB-point is then (cf. Roose[82], Spence, Cliffe and Jepson[92])

$$\tau(s) = \Psi^T(u(s), \lambda(s)) \ v(u(s), \lambda(s)), \tag{4.7}$$

because we expect the left eigenvector of f_u to be orthogonal to the right eigenvector at a TB-point.

For the accurate location of the TB-point we now set up the defining system

$$S(u, \lambda) = \begin{pmatrix} f(u, \lambda) \\ g(u, \lambda) \\ h(u, \lambda) \end{pmatrix} = 0 \tag{4.8}$$

where $h(u, \lambda)$ is defined by

$$A(u, \lambda) \begin{pmatrix} w(u, \lambda) \\ h(u, \lambda) \end{pmatrix} = \begin{pmatrix} v(u, \lambda) \\ 0 \end{pmatrix}. \tag{4.9}$$

It is also natural to solve the adjoint system

$$(\zeta^T(u, \lambda), \widetilde{h}(u, \lambda)) \ A(u, \lambda) = (\Psi^T(u, \lambda), 0). \tag{4.10}$$

By similar manipulations as in the proof of Lemma 3.2, we then obtain the following result (see Griewank and Reddien[42]).

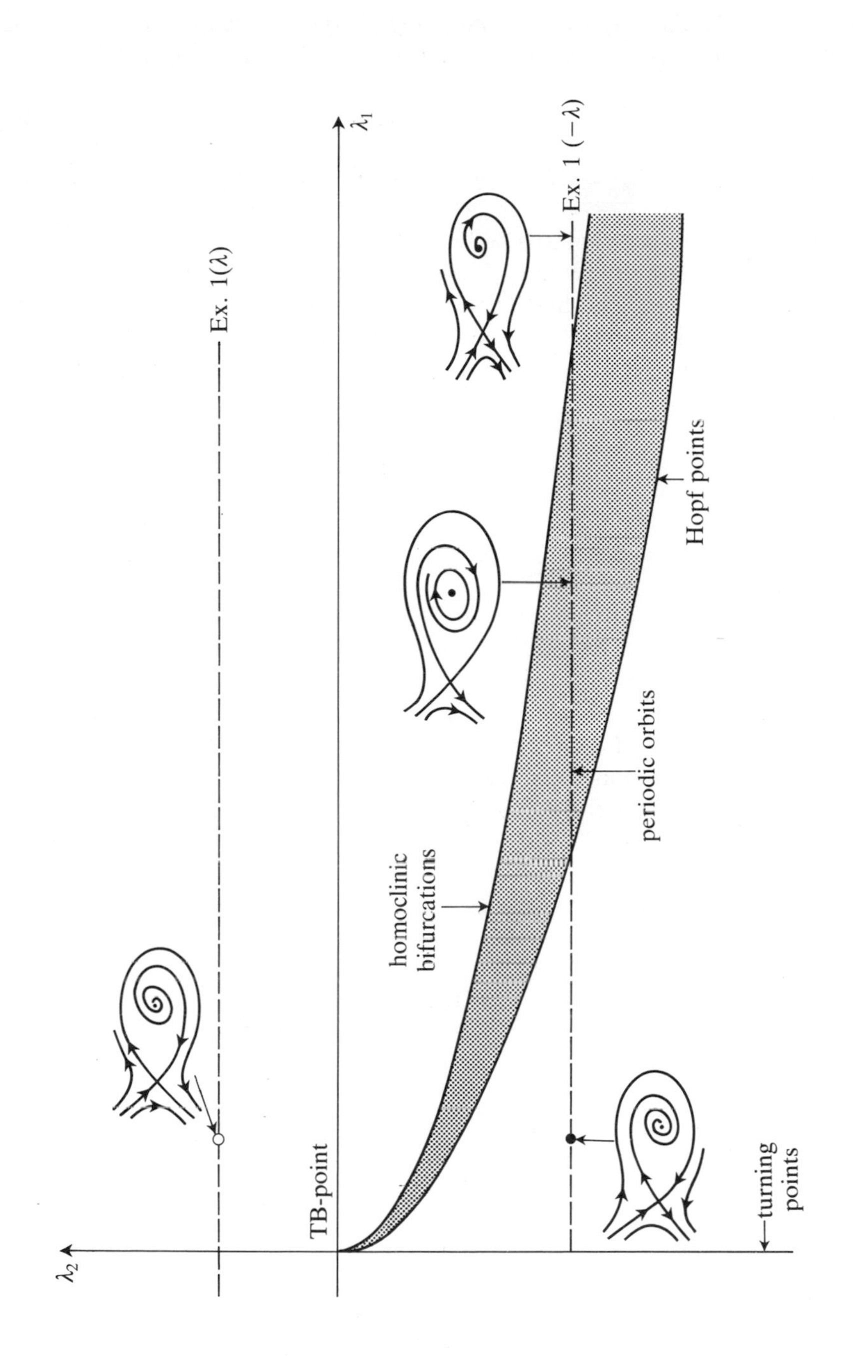

Fig. 5.17. Unfolding picture at a TB-point

Lemma 4.1. *For any* (u, λ), *where* $A(u, \lambda)$ *is nonsingular, the functions defined in (4.5), (4.6) and (4.9), (4.10) satisfy the following relations*

$$h = \widetilde{h} = \Psi^T v \tag{4.11}$$

$$h_z = -\Psi^T f_{uz} w - \zeta^T f_{uz} v, \; z = (u, \lambda). \tag{4.12}$$

From this we see that the test function in (4.7) is in fact obtained by evaluating h along the branch. Though this could have been done without solving (4.9), (4.10), we find from (4.12) that the derivatives of h can be easily expressed in terms of those solutions (using a difference formula for the second derivative of f, if necessary). Thus we can evaluate the Jacobian

$$S' = \begin{pmatrix} f_u & f_\lambda \\ g_u & g_\lambda \\ h_u & h_\lambda \end{pmatrix}$$

and we may now solve for the Newton step by invoking Lemma 3.3 with $p = 1$ and $k = 2$.

A careful count of the numerical work for one Newton step (notice the coincidence of (4.5) with the first system in (3.14)) yields 5 linear systems in A, two in A^T and 4 evaluations of (f_u, f_λ).

Similar to Roose[82] the regularity of (4.8) at some solution can be related to a nonvanishing derivative of the test function (4.7) under further assumptions. We don't discuss any details here. Instead, let us write down the various functions above for the simple example 1 (λ_1, λ_2). With $u = (x, y)$, $b_0 = \begin{pmatrix} 0 \\ 1 \end{pmatrix}$, $c_0^T = (1, 0)$ we find

$$v(u,\lambda) = \begin{pmatrix} 1 \\ 0 \end{pmatrix}, \qquad g(u,\lambda) = 2x - y,$$
$$\Psi^T(u,\lambda) = (-\lambda_2 - x, 1), \quad h(u,\lambda) = -\lambda_2 - x$$

and thus

$$S'(u,\lambda) = \begin{pmatrix} 0 & 1 & 0 & 0 \\ -2x + y & \lambda_2 + x & 1 & y \\ 2 & -1 & 0 & 0 \\ -1 & 0 & 0 & -1 \end{pmatrix},$$

and

$$S'(0,0) = \begin{pmatrix} 0 & 1 & 0 & 0 \\ 0 & 0 & 1 & 0 \\ 2 & -1 & 0 & 0 \\ -1 & 0 & 0 & -1 \end{pmatrix}.$$

Clearly, $S'(0,0)$ is nonsingular.

5.4.3 Starting 1-parameter singularities at a TB-point

In the last section we discussed the passage from a branch of 1-parameter singularities to a 2-parameter singularity. If we think of the singularities being ordered in a hierarchy, then we may call this step *'descending the hierarchy'*, a term coined by Jepson and Spence[55] for stationary problems. See also the hierarchy described by Khibnik[59]. Of equal importance is the reverse step of *ascending the hierarchy*, i.e., starting a branch of 1-parameter singularities from a 2-parameter singularity. In fact, the calculation of TB-points was taken up by Roose[82] in order to start branches of Hopf-points (see also Spence, Cliffe and Jepson[92]).

If we look at the unfolding picture Figure 5.17, then this suggests to try starting a branch of homoclinic orbits at a TB-point. This was first carried out for planar systems with the help of Melnikov's method by Freire, Ponce and Rodriguez-Luis[36], see also Rodriguez-Luis, Freire and Ponce[81]. We will briefly outline here the algorithm for the general case (more details are contained in Beyn[13]). Though this algorithm involves several steps, it is not very difficult to implement. Its derivation, however, needs all the theoretical machinery developed for proving the existence of a branch of homoclinic orbits, such as *center manifolds*, *normal forms* and *Melnikov's method.*

We assume, that we have computed a TB-point $(u_0, \lambda_0) \in \mathbb{R}^{N+2}$ as in the last section, and we will use the matrix $A_0 = A(u_0, \lambda_0)$ as well as the (generalized) right and left eigenvectors v, w and Ψ, ζ of f_u^0 (see (4.5), (4.6), (4.9), (4.10)). We also assume that these vectors are normalized such that

$$\zeta^T v = \Psi^T w = 1, \;\; \Psi^T v = \zeta^T w = 0. \tag{4.13}$$

This can always be achieved by replacing Ψ, ζ by $\alpha\Psi$, $\alpha\zeta + \beta\Psi$ for some suitable α, β. Finally, we require $\Psi^T f_\lambda^0 \neq 0$ and without loss of generality we can assume

$$\delta = \Psi^T f_{\lambda_1}^0 \neq 0. \tag{4.14}$$

Otherwise, we exchange the roles of λ_1 and λ_2. We then proceed as follows:

Step 1 Linear normal form:

Introduce new coordinates $z \in \mathbb{R}^N$, $\mu \in \mathbb{R}^2$ via a linear transformation

$$\begin{pmatrix} u \\ \lambda \end{pmatrix} = \begin{pmatrix} u_0 \\ \lambda_0 \end{pmatrix} + \begin{pmatrix} R & v & w & D_1 & D_2 \\ 0 & 0 & 0 & B_1 & B_2 \end{pmatrix} \begin{pmatrix} z \\ \mu \end{pmatrix}, \tag{4.15}$$

with $R \in \mathbb{R}^{N,N-2}$, $D_1, D_2 \in \mathbb{R}^N$, $B_1, B_2 \in \mathbb{R}^2$, such that the system $\dot{u} = f(u, \lambda)$ takes the form

$$\dot{z} = g(z, \mu) \tag{4.16}$$

with

$$g_z(0,0) = \begin{bmatrix} H & 0 \\ 0 & J \end{bmatrix}, \; g_\mu(0,0) = \begin{bmatrix} 0 \\ J^T \end{bmatrix},$$

and

$$H \in \mathbb{R}^{N-2,N-2} \text{ hyperbolic }, \; J = \begin{bmatrix} 0 & 1 \\ 0 & 0 \end{bmatrix}.$$

This form is obtained by setting

$$B_1 = \begin{pmatrix} B_{11} \\ 0 \end{pmatrix} = \begin{pmatrix} \delta^{-1} \\ 0 \end{pmatrix}, \; B_2 = \begin{pmatrix} B_{12} \\ 1 \end{pmatrix} = \begin{pmatrix} -\delta^{-1}\Psi^T f^0_{\lambda_2} \\ 1 \end{pmatrix}$$

and by solving the two linear systems

$$A_0 \begin{pmatrix} D_1 \\ \alpha_1 \end{pmatrix} = \begin{pmatrix} \zeta - B_{11} f^0_{\lambda_1} \\ 0 \end{pmatrix}, \; A_0 \begin{pmatrix} D_2 \\ \alpha_2 \end{pmatrix} = \begin{pmatrix} -B_{12} f^0_{\lambda_1} - f^0_{\lambda_2} \\ 0 \end{pmatrix}. \tag{4.17}$$

Step 2 Center manifold reduction:

Write $z = (\eta, \xi) \in \mathbb{R}^{N-2} \times \mathbb{R}^2$, then there are locally invariant two-dimensional center manifolds which are graphs of the type $\eta = F(\xi, \mu)$. Inside the manifolds the system (4.16) reduces to

$$\dot{\xi} = \begin{pmatrix} g_{N-1} \\ g_N \end{pmatrix} (F(\xi, \mu), \xi, \mu) =: h(\xi, \mu). \tag{4.18}$$

Furthermore, h has a Taylor expansion

$$h(\xi, \mu) = \begin{pmatrix} \xi_2 \\ \mu_1 \end{pmatrix} + Q(\xi, \mu) + \mathcal{O}((||\xi|| + ||\mu||)^3),$$

where Q contains the quadratic terms. Only the following ones are needed for our further calculation

$$Q(\xi, \mu) = \begin{pmatrix} p_{11}\xi_1^2 + p_{14}\xi_1\mu_2 + \dots \\ q_{11}\xi_1^2 + q_{12}\xi_1\xi_2 + q_{14}\xi_1\mu_2 + q_{24}\xi_2\mu_2 + q_{44}\mu_2^2 + \dots \end{pmatrix},$$

and these are given by

$$\begin{aligned} p_{11} &= \zeta^T f^0_{uu} v^2, \\ p_{14} &= \zeta^T (f^0_{uu} v D_2 + f^0_{u\lambda} v B_2) \\ q_{11} &= \Psi^T f^0_{uu} v^2, \; q_{12} = \Psi^T f^0_{uu} v w \\ q_{14} &= \Psi^T (f^0_{uu} v D_2 + f^0_{u\lambda} v B_2) \\ q_{24} &= \Psi^T (f^0_{uu} w D_2 + f^0_{u\lambda} w B_2) \\ q_{44} &= \Psi^T (f^0_{uu} D_2^2 + 2 f^0_{u\lambda} D_2 B_2 + f^0_{\lambda\lambda} B_2^2). \end{aligned}$$

Usually, the next step consists in a normal form transformation of the quadratic terms. However, we omit this step, since many of the quadratic terms prove to be irrelevant after the following scaling transformation.

Step 3 Scaling transformation:

Again new coordinates are introduced

$$(\xi_1, \xi_2) \to (x, y), \ (\mu_1, \mu_2) \to (\epsilon, \tau)$$

such that the two dimensional system (4.17) assumes the form

$$\begin{aligned} \dot{x} &= y + \epsilon(a_1 x^2 + a_2 \tau x + a_3 \tau^2) + \mathcal{O}(\epsilon^2) \\ \dot{y} &= x^2 - 4 + \epsilon(b_1 xy + b_2 \tau y) + \mathcal{O}(\epsilon^2) \end{aligned} \tag{4.19}$$

The transformation is of the form

$$\begin{aligned} \mu_1 &= (\alpha_0 + \alpha_1 \tau^2)\epsilon^4, \ \mu_2 = \tau\epsilon \\ \xi_1(t) &= \alpha_2 \epsilon^2 (x(\alpha_3 \epsilon t) + \alpha_4 \tau) \\ \xi_2(t) &= \alpha_5 \epsilon^3 y(\alpha_3 \epsilon t). \end{aligned}$$

The constants α_i, $i = 0, \ldots, 5$ are determined in such a way that the special system (4.19) is obtained.

The coefficients a_i and b_j can be explicitly expressed in terms of the quadratic coefficients from the last step. But we don't write down these relations because a complete set of formulas for an approximate homoclinic orbit of (4.17) will be given below.

Step 4 Melnikov's method:

First notice that the unperturbed system (4.19) ($\epsilon = 0$) is Hamiltonian and has the homoclinic orbit

$$(\overline{x}(t), \overline{y}(t)) = 2(1 - 3\mathrm{sech}^2(t), \ 6\mathrm{sech}^2(t)\tanh(t)).$$

Melnikov's method (Melnikov[71], Guckenheimer and Holmes[43]) can now be used to show how this homoclinic orbit survives for $\epsilon \neq 0$. Instead of the above two references we strongly recommend Hale's reformulation of Melnikov's method as a problem of bifurcation from the trivial solution (Hale[45]). In fact, taking τ as the bifurcation parameter, we see that $(\overline{x}, \overline{y}, \epsilon = 0)$ is a trivial branch of HOP's for (4.19). After some analysis, which employs the techniques from 5.3.6, we then find that bifurcation from a simple eigenvalue occurs at

$$\tau_0 = \frac{10}{7} \frac{2a_1 + b_1}{a_2 + b_2}.$$

Taking the last two steps together, we find an approximate homoclinic orbit for (4.17) by choosing some small ϵ and setting

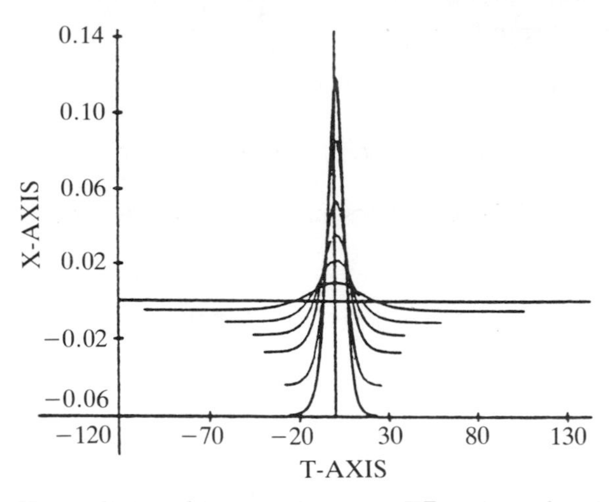

Fig. 5.18. Homoclinic orbits starting at a TB-point: the time diagram, $\lambda_1 = 0.25 \cdot 10^{-4}, \ldots, 0.35 \cdot 10^{-2}$.

$$\begin{aligned}
\mu_1 &= \sigma_0 \epsilon^4, \ \mu_2 = \tau_0 \epsilon^2, \\
\xi_1(t) &= \frac{\epsilon^2}{q_{11}} \left(1 - 3 \operatorname{sech}^2(\frac{\epsilon}{2}\, t) - q_{14}\tau_0\right) \\
\xi_2(t) &= \frac{3\epsilon^3}{q_{11}} \operatorname{sech}^2 \left(\frac{\epsilon}{2}\, t\right) \tanh \left(\frac{\epsilon}{2}\, t\right),
\end{aligned}$$

where

$$\tau_0 = \frac{5}{7} \, \frac{p_{11} + q_{12}}{q_{11}p_{14} - p_{11}q_{14} + q_{24}q_{11} - q_{14}q_{12}}, \ \sigma_0 = \frac{1}{2q_{11}} \left((q_{14}^2 - q_{11}q_{44})\tau_0^2 - 1\right) \tag{4.20}$$

Of course, we require the two denominators in (4.20) to be nonzero. These are the crucial conditions which in conjunction with the assumptions on f_u^0 define a *nondegenerate TB-point.* For our model example (4.3) they are satisfied because the coefficients in front of x^2 and xy do not vanish.

The above formulae can finally be used to set up an approximate homoclinic orbit for the original problem via

$$\begin{aligned}
\widetilde{u}(t) &= \xi_1(t)v + \xi_2(t)w + D_1\mu_1 + D_2\mu_2 + u_0 \\
\widetilde{\lambda} &= \mu_1 B_1 + \mu_2 B_2 + \lambda_0 \quad \text{(cf. step 1)}\ .
\end{aligned}$$

With this approximation we can employ the numerical method for homoclinic orbit pairs from (3.6). In the beginning we look for a HOP (u, λ_2) close to $(\widetilde{u}, \widetilde{\lambda}_2)$ with $\lambda_1 = \widetilde{\lambda}_1$ fixed. Then λ_1 is freed and used for continuation. This approach has been tried successfully on a series of examples.

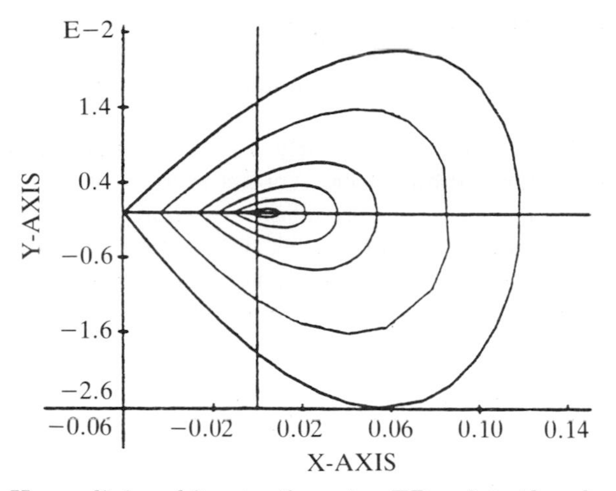

Fig. 5.19. Homoclinic orbits starting at a TB-point: the phase diagram, $\lambda_1 = 0.25 \cdot 10^{-4}, \ldots, 0.35 \cdot 10^{-2}$.

However, sometimes difficulties arise with the realization of the projection boundary conditions since we are very close to the fold of the stationary surface (cf. 5.4.2). We display in Figures 5.18 and 5.19 some results for our model example (5.4.3). The computation was started with $\epsilon = 0.1$ and then continuation with respect to λ_1 and the automatic truncation strategy was used (cf. Figure 5.16).

5.5 The longtime behaviour of integration methods

5.5.1 Comparison of discrete and continuous flows

In section 5.1.2 we already considered the basic problems, which occur when comparing the longtime dynamics of a one-step method with that of the dynamical system. Of course, in practice one would prefer to use a sophisticated code (variable step size, variable order) for the initial value problem

$$\dot{u} = f(u), \; u(0) = u^0 \in \mathbb{R}^N. \tag{5.1}$$

The user prescribed tolerances of these codes however, can only control the local discretization error, and in principle there is no escape from the exponential growth of the global error as signalled by the estimate (1.8). It may occur just at a later time than with a simple minded one-step method. Whether this blow-up of the error really occurs, clearly depends on the dynamics of the system (5.1) itself. In what follows we will focus on the behaviour of the sequence

$$u^{n+1} = \phi(h, u^n), \; n = 0, 1, 2, \ldots, \; u^0 \in \mathbb{R}^N, \tag{5.2}$$

if n tends to infinity, but h is sufficiently small (this simulates the accurate ordinary differential equation-solver). For line systems of partial differential equations, even this may be an unrealistic situation.

We start with a discussion of the relations between the discrete h-flow $\phi(h, \cdot)$ and the continuous h-flow $\Phi(h, \cdot)$. For an illustration we use *m-stage explicit Runge-Kutta methods*, which are of the form

$$\phi(h, u) = u + h \sum_{i=0}^{m} \beta_i k_i(h, u) \tag{5.3}$$

with $k_0(h, u) = f(u)$ and

$$k_i(h, u) = f(u + h \sum_{j=0}^{i-1} \beta_{ij} k_j(h, u)), \; i = 1, \ldots m.$$

As in (1.7), let us assume that the one step mapping $\phi : [0, h_0] \times \mathbb{R}^N \to \mathbb{R}^N$ is smooth and defines a method of order p, i.e., let

$$\phi(h, v) = \Phi(h, v) + \mathcal{O}(h^{p+1}) \tag{5.4}$$

holds uniformly in any bounded v-set. For Runge-Kutta methods this condition is used for determining the coefficients. Due to the smoothness of ϕ and Φ we may restate (5.4) as

$$\frac{\partial^i \phi}{\partial h^j} (0, v) = \frac{\partial^j \Phi}{\partial h^j} (0, v), \; j = 0, \ldots p, \; v \in \mathbb{R}^N. \tag{5.5}$$

Differentiating this relation with respect to v, we find by a Taylor-expansion at $h = 0$ that also

$$\phi_v(h, v) = \Phi_v(h, v) + \mathcal{O}(h^{p+1}) \tag{5.6}$$

holds uniformly in bounded v-sets. Therefore, the sensitivity of the discrete flow to perturbations of the initial value is close to the sensitivity of the continuous flow with the same order of accuracy. This argument can be continued for higher derivatives, but of course the constants in front of h^{p+1} may grow in general. Since discrete and continuous h-flows are so close, one might think, that it is possible to interpret a discrete h-flow as the continuous h-flow of a perturbed dynamical system. More explicitly, given the system (5.1) and the one-step method (5.2), do there exist smooth functions

$$F_h : \mathbb{R}^N \to \mathbb{R}^N$$

such that for sufficiently small h

$$\phi(h, u, f) = \Phi(h, u, F_h), \ u \in \mathbb{R}^N? \tag{5.7}$$

This would allow us to use the perturbation results on dynamical systems, in particular those on structural stability, for analyzing one-step methods. However, (5.7) is false in general. For example, take $N = 1$, $f(u) = u^2$ and Euler's method. Then $\phi(h, u, f) = u + hu^2$ is not a diffeomorphism, but $\Phi(h, \cdot, F_h)$ is a diffeomorphism, whatever F_h looks like. This example relies on the global behaviour of the discrete flow, but it is also very likely that (5.7) does not hold locally, i.e. for some u-neighbourhood. But we don't know of any rigorous proof.

In contrast to the negative statements above, there is an elementary class of functions f, where (5.7) does hold. This is the linear case $f(u) = Au$, $A \in \mathbb{R}^{N,N}$. For all common methods, e.g. the Runge Kutta methods, the discrete h-flow is of the form

$$\phi(h, u, A) = g(hA)u$$

where $g(z)$ is a complex function, which is holomorphic in a neighbourhood of zero. g is usually called the *growth function* of the one-step method. If the method is of order p, then we obtain from (5.6) the well-known relation

$$g(hA) = e^{hA} + \mathcal{O}(|h|^{p+1}). \tag{5.8}$$

Now we use the logarithm, holomorphic near 1, in order to define

$$A_h = \frac{1}{h} \ \ln[g(hA)], \ h > 0, \ A_0 = A.$$

From (5.8) we find

$$A_h = \frac{1}{h} \ (hA + \mathcal{O}(h^{p+1})) = A + \mathcal{O}(h^p)$$

and the relation (5.7)

$$\Phi(h, u, A_h) = e^{hA_h}u = g(hA)u = \phi(h, u, A).$$

In other words, one-step methods applied to linear systems, yield h-flows of a perturbed linear system which is close to the original one within the order of the method.

5.5.2 Persistence of compact invariant sets under discretization

In general we would like to know, if for any compact invariant set M of the flow Φ we can find a compact set $M_h \subset \mathbb{R}^N$ which is *invariant under* $\phi(h,\cdot)$ (i.e. $v \in M_h \Rightarrow \phi(h,v) \in M_h$ and $\phi(h,w) \in M_h \Rightarrow w \epsilon M_h$) and which satisfies

$$H(M, M_h) \to 0 \quad \text{as} \quad h \to 0. \tag{5.9}$$

Here $H(M_1, M_2)$ is the *Hausdorff distance* of two closed sets $M_1, M_2 \subset \mathbb{R}^N$ given by

$$\begin{aligned} H(M_1, M_2) &= \operatorname{Max}(\operatorname{dist}(M_1, M_2),\ \operatorname{dist}(M_2, M_1)) \\ \operatorname{dist}(M_1, M_2) &= \sup_{u \in M_1} \inf_{v \in M_2} ||u - v||. \end{aligned}$$

In addition, we would like M_h to inherit the stability properties of M. For stationary points such an asymptotic result is easily established (see Theorem 5.1 below). This contrasts with the variety of spurious solutions that may arise with growing h (cf. the references cited in 5.1.2). In spite of these spurious effects, let us notice that most one-step methods have all stationary points of $\dot{u} = f(u)$ as exact fixed points for all $h > 0$. For example, this is easily verified for the Runge-Kutta methods (5.3). Nevertheless, the following global result is instructive.

Theorem 5.1. *Let $\Omega \subset \mathbb{R}^N$ be compact and assume that (5.1) has finitely many stationary points v_i, $i = 1, \ldots K$ in the interior of Ω (which are unique in Ω) and assume these to be regular, i.e.,*

$$f'(v_i) \text{ is invertible for } i = 1, \ldots K.$$

Let ϕ be a smooth one-step method of order $p \geq 1$. Then there exists an $h_0 > 0$, such that the discrete h-flow $\phi(h,\cdot)$, $h \leq h_0$, has exactly K fixed points $v_i(h)$, $i = 1, \ldots K$ in Ω and these satisfy

$$v_i(h) = v_i + \mathcal{O}(h^p), \quad i = 1, \ldots, K. \tag{5.10}$$

Moreover, if Re $\mu > 0$ for some eigenvalue μ of $f'(v_i)$ then $v_i(h)$ is an unstable fixed point for $\phi(h,\cdot)$, and if Re $\mu < 0$ for all eigenvalues μ of $f'(v_i)$ then it is an asymptotically stable fixed point.

Proof. For the construction of the fixed points let us define the smooth function

$$g(h,v) = \frac{1}{h}\,(\phi(h,v) - v) = \int_0^1 \frac{\partial \phi}{\partial h}\,(sh, v)\,ds. \tag{5.11}$$

Using (5.5) we obtain

$$g(0,v) = \frac{\partial \phi}{\partial h}(0,v) = \frac{\partial \Phi}{\partial h}(0,v) = f(v) \tag{5.12}$$

and $g_v(0,v) = f'(v)$. Therefore we can apply the implicit function theorem to the equation $g(h,v) = 0$, with (h,v) in a neighbourhood of $(0,v_i)$. This gives us the existence and local uniqueness of the fixed points $v_i(h)$. Moreover, (5.10) follows from

$$g(h,v_i) = \frac{1}{h}\left(\Phi(h,v_i) + \mathcal{O}(h^{p+1}) - v_i\right) = \mathcal{O}(h^p).$$

Now we take any sequence v_h of fixed points for $\phi(h,\cdot)$ in Ω where $h \to 0$. Then we can assume $v_h \to \overline{v} \in \Omega$ for some subsequence $h \to 0$ and find from (5.12)

$$0 = g(h,v_h) \to g(0,\overline{v}) = f(\overline{v}) \quad \text{as } h \to 0.$$

Thus $\overline{v} = v_i$ for some i, and v_h must enter the uniqueness neighbourhood for $v_i(h)$ and hence coincide with $v_i(h)$. This establishes the global uniqueness of the fixed points $v_i(h)$ in Ω. Finally, by differentiating (5.11) we obtain

$$A_h := g_v(h,v_i(h)) = \frac{1}{h}\left(\phi_v(h,v_i(h)) - I\right) \to g_v(0,v_i) = f'(v_i)$$

as $h \to 0$. Therefore,

$$\phi_v(h,v_i(h) = I + hA_h$$

has an eigenvalue of modulus larger than 1, if $f'(v_i)$ has an eigenvalue with positive real part. Similarly, all eigenvalues of $\phi_v(h,v_i(h))$ lie inside the unit circle if those of $f'(v_i)$ are in the negative half plane. The standard analogoue of Theorem 1.1 for the stability of fixed points (see Irwin[51]) then yields the desired result. ■

A corresponding result for periodic orbits is considerably more involved. It was shown by Braun and Hershenov[16] that a one-step method has an invariant circle for sufficiently small h close to an asymptotically stable periodic orbit. This was generalized by Doan[26] to the hyperbolic case and further details, in particular on the estimates, were developed by Beyn[8], Eirola[32,33]. The general result is

Theorem 5.2. *Assume that $\dot{u} = f(u)$ has a hyperbolic periodic orbit (cf. 2.2, 2.3, 2.4)*

$$\gamma = \{\overline{u}(t) : 0 \le t \le T\}$$

and let ϕ be a smooth one-step method of order p. Then for h sufficiently small there exists an invariant curve γ_h for the discrete h-flow which is $\mathcal{O}(h^p)$ close to γ. More precisely, we have

$$\gamma_h = \{\overline{u}_h(t) : 0 \leq t \leq T\},$$

where $\overline{u}_h : \mathbb{R} \to \mathbb{R}^N$ is a T-periodic Lipschitz function, and

$$||\overline{u}(t) - \overline{u}_h(t)|| \leq C\ h^p \text{ for } 0 \leq t \leq T \tag{5.13}$$

$$\phi(h, \overline{u}_h(t)) = \overline{u}_h(\widetilde{\phi}(t,h)),\ \widetilde{\phi}(t,h) = t + h + \mathcal{O}(h^{p+1})(t \in \mathbb{R}). \tag{5.14}$$

Clearly, (5.13) gives us the desired result $H(\gamma, \gamma_h) = \mathcal{O}(h^p)$, but some more information is contained in (5.14). The mapping $t \to \widetilde{\phi}(t,h)$ may be regarded as the reduced discrete h-flow on the invariant curve and its *rotation number* is found to be $\frac{h}{T} + \mathcal{O}(h^{p+1})$ (see Beyn[8]). It is also true that, if the stability or instability criteria from Theorem 2.2 are satisfied for γ, then also γ_h is asymptotically stable respectively unstable. Consider a system with an asymptotically stable periodic orbit and apply a one-step method. If we plot the points of the iteration on a screen then we usually see the invariant curve gradually filled up by pixels (cf. Brezzi, Fujii and Ushiki[17], Beyn[8] for some illustrations). The reason for the fill up is that the rotation number of $\widetilde{\phi}$, obtained by a random choice of h, is 'sufficiently irrational'. For larger values of h, however, periodic orbits with a finite number of points are quite typical.

Let us mention that Eirola[33] used in his proof a general theorem of Hirsch, Pugh and Shub[47] on the persistence of so-called *normally hyperbolic invariant manifolds.* This suggests that Theorem 5.2 may be generalized to this type of invariant manifolds. But no detailed investigations seem to be available up to now.

For the specific case of center manifolds, however, there is a corresponding result by Beyn and Lorenz[14]. Let $v \in \mathbb{R}^N$ be a stationary but nonhyperbolic point of (5.1). Then the linearization $f'(v)$ induces a splitting $\mathbb{R}^N = X \oplus Y$ where X (respectively Y) are invariant subspaces spanned by the (generalized) eigenvectors which belong to the eigenvalues on (respectively off) the imaginary axis. Under these assumptions there exists a center manifold, i.e. a locally invariant manifold of the form

$$M = \{u = (x, y(x)) : x \in X,\ ||x - v|| < \epsilon\}$$

with $y'(v) = 0$. Under some technical assumptions, it is then shown that any p-th order one-step method also has a locally invariant manifold of the form

$$M_h = \{u = (x, y_h(x)) : x \in X,\ ||x - v|| < \epsilon\},$$

such that $y_h(x) = y(x) + \mathcal{O}(h^p)$. This again gives us $H(M, M_h) = \mathcal{O}(h^p)$. The result should be carefully interpreted, however, for the following two reasons. First, the center manifold M is usually not unique, though all center manifolds are tangent to each other to all orders. The manifolds M and M_h are therefore selected out of a possible continuum of manifolds. Second, M_h is not necessarily a center manifold for $\phi(h, \cdot)$, because eigenvalues μ of $f'(v)$ with $Re\ \mu = 0$ might lead to stable or unstable eigenvalues of the linearization $\phi_v(h, v)$ (see the proof of Theorem 5.1).

In view of our treatment of direct methods in chapters 2 to 4, it is natural to continue the previous discussion for parametrized systems

$$\dot{u} = f(u, \lambda), \ u(t) \in \mathbb{R}^N, \ \lambda \in \mathbb{R}. \tag{5.15}$$

There are still many open questions in this field and so we just briefly mention a few results and problems. As mentioned above, stationary points are usually reproduced exactly as fixed points for common one-step methods. This also applies to the stationary bifurcation diagram of (5.15), so that there are no problems.

The next step is the analysis of one-step methods

$$u^{n+1} = \phi(h, u^n, \lambda) \tag{5.16}$$

in the neighbourhood of some λ_0, where the system (5.15) undergoes a Hopf bifurcation (cf. 5.3.4). For this case a corresponding result is given (with a sketch of proof) for Euler's method in Brezzi, Fujii and Ushiki[17]. According to their result, a branch of invariant circles for the map $\phi(h, \cdot)$ bifurcates off at some $\lambda_h = \lambda_0 + \mathcal{O}(h)$.

Let us write down for an illustration Euler's method for the example 1(-λ) in 5.3.4

$$\phi(h, u) = (x + hy, y + h(\lambda - 2y - x^2 + xy)), \ u = (x, y).$$

At the stationary points $(\sqrt{\lambda}, 0)$ we find

$$\phi_u(h, \sqrt{\lambda}, 0) = \begin{pmatrix} 1 & h \\ -2h\sqrt{\lambda} & 1 + h(\sqrt{\lambda} - 2) \end{pmatrix}.$$

A short calculation reveals that this matrix has two complex conjugate eigenvalues for λ close to 4, and these two eigenvalues cross the unit circle at

$$\lambda_h = \left(\frac{2}{1 + 2h}\right)^2 = 4 + \mathcal{O}(h).$$

At this point the invariant curves are born according to the theorem of *Hopf bifurcation for maps* (see e.g., Iooss[50]) and this is the technique used by Brezzi, Fujii and Ushiki[17].

An attempt to understand (5.16) in the neighbourhood of a homoclinic bifurcation was made in Beyn[9]. From some numerical experiments it was conjectured that the homoclinic structure is preserved by the one-step mapping in the case of a smooth system (5.15) but destroyed otherwise. This conjecture seems to be false, but the case is still under investigation and details will appear elsewhere. For the remaining two bifurcations with periodic orbits, i.e. the period doubling and the torus bifurcation, we don't know of any results concerning the behaviour of the one-step mapping.

Let us conclude this section with a reference to the paper of Kloeden and Lorenz[61]. It is the only one which deals with a general attracting set of the dynamical system. They assume that the given system (5.1) has a compact invariant set $M \subset \mathbb{R}^N$ which is *uniformly asymptotically stable*, i.e., there exists a $\delta > 0$ and for each $\epsilon > 0$ a time $T(\epsilon)$ such that that

$$\operatorname{dist}(\Phi(t, u^0), M) \leq \epsilon \text{ if } t \geq T(\epsilon) \text{ and } \operatorname{dist}(u^0, M) \leq \delta.$$

Under this assumption they show that a one-step method of order p has a compact positively invariant set of the form

$$M_h = \{u \in \mathbb{R}^N : u \in U,\ V(u) \leq C\ h^p\}. \tag{5.17}$$

Here U is some suitable open neighbourhood of M and

$$V : U \to \mathbb{R}$$

is a Lipschitz continuous Liapunov function which decreases along trajectories, is identically zero on M and satisfies an estimate

$$\alpha(\operatorname{dist}(u, M)) \leq V(u) \leq \beta(\operatorname{dist}(u, M)) \tag{5.18}$$

where $\alpha, \beta : \mathbb{R}_+ \to \mathbb{R}_+$ are continuous, strictly increasing functions. The existence of such a Liapunov function follows from the stability assumption by a general theorem of Yoshizawa[101]. From (5.17) one obtains the convergence in the Hausdorff distance, more precisely

$$H(M, M_h) = \mathcal{O}(\alpha^{-1}(h^p)).$$

Of course, in general V, α, β are not known explicitly. Also the positively invariant sets are rather 'thick', because they consist of shrinking neighbourhoods of the continuous attractor. In fact, they absorb the discrete trajectories in a uniform time. Nevertheless, this kind of result seems to be the only one possible under such general assumptions.

5.5.3 Comparison of trajectories

Let us finally return to the problem of estimating the global discretization error on the positive real axis (compare (1.8)). If the system $\dot{u} = f(u)$ has some kind of sensitive dependence on initial conditions, then it is conceivable that also discrete and continuous trajectories will go apart after some time even if the step-size h is small and if the continuous trajectory stays bounded for all times (see also Fig. 5.20 below). The simplest situation of this type occurs in the neighbourhood of a hyperbolic unstable stationary point. There the longtime behaviour of a trajectory depends on the position of the initial value above, below or on the stable manifold. Let us consider for illustration the scalar example

$$\dot{u} = \lambda u, \quad u(0) = u^0, \tag{5.19}$$

where $\lambda \in C$ and $u(t) \in C^N$, and take the familiar one-step method ($\Theta \in [0, 1]$)

$$\frac{u^{n+1} - u^n}{h} = \Theta(\lambda u^{n+1}) + (1 - \Theta)(\lambda u^n), \; n = 0, 1, 2, \ldots. \tag{5.20}$$

For the solutions

$$\begin{aligned} \Phi(t, u^0) &= e^{\lambda t} u^0, \\ \Psi(nh, u^0) &:= u^n = [g(h\lambda)]^n u^0, \; g(z) = \frac{1 + (1 - \Theta)z}{1 - \Theta z} \end{aligned} \tag{5.21}$$

one readily verifies in the case $Re\ \lambda < 0, \; |u^0| \leq 1$

$$|\Phi(nh, u^0) - \Psi(nh, u^0)| \leq C_\lambda h^p \text{ for } n \geq 0 \text{ and } 0 < h \leq h_0. \tag{5.22}$$

Here C_λ is a constant independent of Θ and h, and $p = 2$ if $\Theta = \frac{1}{2}$ and $p = 1$ otherwise.

In the unstable case $Re\ \lambda > 0$ an estimate of type (5.22) is impossible, even if we restrict n, h such that $\Phi(nh, u^0)$ is bounded, say $|\Phi(nh, u^0)| \leq 1$. Take, for example, $\lambda \in \mathbb{R}$, $\lambda > 0$ and Euler's method. Let n be the integer part of $1 + 1/(h\lambda)^2$ and consider the initial value $u^0 = \exp(-nh)$. Then, by construction $\Phi(nh, u^0) = 1$, but

$$\begin{aligned} \Psi(nh, u^0) &= \exp(n(\ln(1 + h\lambda) - h\lambda)) \\ &= \exp(n(-\frac{1}{2}(h\lambda)^2 + \mathcal{O}(h\lambda)^3)) \\ &\leq \exp(-\frac{1}{2} + \mathcal{O}(h\lambda)). \end{aligned}$$

Therefore $\Phi(nh, u^0) - \Psi(nh, u^0)$ is of order 1, no matter how small h is chosen. However, it is possible in the unstable case $Re\ \lambda > 0$ to find a suitable initial value $v^0 \in \mathbb{R}$ such that

$$|\Phi(nh, u^0) - \Psi(nh, v^0)| \le C_\lambda h^p \tag{5.23}$$

holds for $|u^0| \le 1$ and for all n, h with $|\Phi(nh, u^0)| \le 1$. In the case $u^0 = 0$ we can take $v^0 = 0$, but in the case $0 < |u^0| \le 1$ we construct v^0 as follows. The solution of (5.19) reaches the unit circle at time $T = -(\ln |u^0|)/(Re\lambda)$; let n_0 be the integer part of T/h and define

$$v^0 = u^0 \left(\frac{\exp(h\lambda)}{g(h\lambda)} \right)^{n_0}. \tag{5.24}$$

Then we have identical final values $\Phi(n_0 h, u^0)$ and $\Psi(n_0 h, v^0)$ within the unit circle, and (5.23) follows from (5.22) since the Θ-method for (5.19) is the reverse of the $(1 - \Theta)$-method for $\dot{u} = -\lambda u$.

We notice that the scaled initial value v^0 depends on h, λ and u^0, but not on n in (5.23). This elementary result easily carries over to general linear systems

$$\dot{u} = Au,$$

where $A \in \mathbb{R}^{N,N}$ is hyperbolic and diagonalizable. We transform into a diagonal system, then adjust the initial values for unstable eigenvalues as in (5.24) and transform back to the original variables.

It was shown in Beyn[10] that estimates of the form (5.23) also hold in the neighbourhood of hyperbolic stationary points.

Theorem 5.3. *Let 0 be a hyperbolic stationary point of (5.1) and let it be a fixed point of the one-step mapping $\phi(h, \cdot)$ for all h. Moreover, let ϕ be smooth and consistent of order p (cf. (5.4)). Then there exist constants c, ϵ, h_0 such that for any $u^0 \in \mathbb{R}^N$, $||u^0|| \le \epsilon$ and $0 < h \le h_0$ there exists a $v^0 = v^0(u^0, h) \in \mathbb{R}^N$ with the property*

$$||\Phi(nh, u^0) - \Psi(nh, v^0)|| \le Ch^p \tag{5.25}$$

for all n such that $||\Phi(t, u^0)|| \le \epsilon$ for $t \in [0, nh]$. Here $\Psi(nh, v^0)$ denotes the solution of (5.2) with initial value v^0. Conversely, for any $v^0 \in \mathbb{R}^N$, $||v^0|| \le \epsilon$ and $h \le h_0$ there exists a $u^0 = u^0(v^0, h)$ such that (5.25) holds for all n with $||\Psi(jh, v^0)|| \le \epsilon$, $(j = 0, \ldots, n)$.

Loosely speaking this theorem shows that any discrete trajectory approximates *some* continuous trajectory (and vice versa), as long as it stays in some neighbourhood of the hyperbolic point. Thus in this case, the adjustment of initial values saves the uniform estimate. One should notice

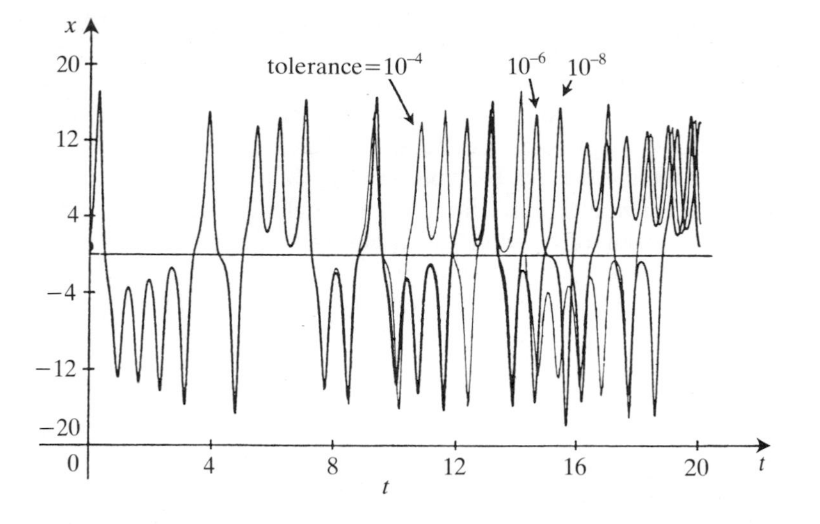

Fig. 5.20. Time diagram for the numerical approximation of trajectories in the Lorenz system ($\sigma = 10$, $\mu = 8/3$, $\lambda = 28$), initial value $u^0 = (1, 4, 9)$

that the set of n-values for which (5.25) holds can be arbitrarily large depending on u^0. In the extreme case when u^0 is on the stable manifold of the stationary point, then (5.25) holds for all $n \geq 0$. In fact, the fixed point of the one-step mapping $\phi(h, \cdot)$ has a stable manifold which approximates the continuous one within the order of the method (cf. Beyn[10] and also the various results on invariant manifolds in 5.5.2).

In the neighbourhood of periodic orbits, estimates of the form (5.25) are no longer possible, since the continuous and discrete trajectories inevitably run out of phase after some time (this is documented in Beyn[8]). If the periodic orbit is asymptotically stable, then all we can expect is convergence of the positive trajectories in the Hausdorff metric, (see Beyn[8]), i.e.,

$$H(\gamma_+(u^0),\ \gamma_{+h}(u^0)) \to 0 \ \text{ as } h \to 0,$$

where $\gamma_+(u^0) = \{\Phi(t, u^0) : t \geq 0\}$, $\gamma_{+h}(u^0) = \{\Psi(nh, u^0) : n \geq 0\}$.

For systems with strange attractors or chaotic behaviour such as the Lorenz system (Sparrow[90] or 5.3.6, Example 5) it seems no longer appropriate to try to approximate trajectories over large time intervals. This is clearly demonstrated in Figure 5.20 which shows three numerical time diagrams for the Lorenz equations ($\sigma = 10, \mu = \frac{8}{3}$, $\lambda = 28$) obtained with a standard ordinary differential equation-solver and three different tolerances. All of them go apart, and it can be estimated that even if the tolerance is taken to be the machine precision ($\sim 10^{-16}$) then accurate trajectories can be expected at most up to $t = 50$. As a consequence of this well-known effect one should rather compare certain quantities measured from the numerical trajectories with their continuous counterparts, such as Liapunov exponents. Although methods for calculating Liapunov exponents numerically have become quite popular (see e.g. Shimada and Nagashima[89], Kubiček and Marek[63], Seydel[87]), there seems to be no rigorous justification of these methods however simple the system may be. This does not come as a surprise since the proof of existence for these exponents is already a considerable task (see Oseledec[75]).

References

[1] E. L. Allgower and K. Georg, *Numerical continuation methods, an introduction*, Springer Series in Computational Mathematics, 13, Springer, New York, 1990.

[2] J. P. Abbott, *An efficient algorithm for the determination of certain bifurcation points*, J. Comput. Appl. Math. 4, 1978, 19-27.

[3] H. Amann *Gewöhnliche Differentialgleichungen*, De Gruyter, Berlin, 1983.

[4] A.A. Andronov, E. A. Leontovich, I.I. Gordon and A. G. Maier, *Theory of bifurcations of dynamical systems on a plane*, Wiley, New York, 1973.

[5] V. I. Arnold *Ordinary differential equations*, MIT Press, Cambridge, 1973.

[6] U. M. Ascher, R. M. M. Mattheij and R. D. Russell *Numerical solution of boundary value problems for ordinary differential equations*, Prentice Hall, 1988.

[7] W.-J. Beyn *Defining equations for singular solutions and numerical applications*, 42-56 in Küpper, Mittelmann, Weber (Eds. 1984).

[8] W.-J. Beyn *On invariant closed curves for one-step methods,* Numer. Math. 51, 1987, 103-122.

[9] W.-J. Beyn *The effect of discretization on homoclinic orbits,* 1-8 in Küpper, Seydel, Troger (Eds. 1987).

[10] W.-J. Beyn *On the numerical approximation of phase portraits near stationary points*, SIAM J. Numer. Anal. 24, 1987, 1095-1113.

[11] W.-J. Beyn *The numerical computation of connecting orbits in dynamical systems*, IMA J. Numer. Anal., 1990, 379-405.

[12] W.-J. Beyn *Global bifurcations and their numerical computation* , In B. De Dier, D. Roose and A. Spence (Eds. 1990).

[13] W.-J. Beyn *Global bifurcations near singular points*, In preparation, 1990.

[14] W.-J. Beyn, Lorenz, J. *Center manifolds of dynamical systems under discretization* , Numer. Funct. Anal. Optimiz. 9, 1987, 381-414.

[15] R.I. Bogdanov, *Versal deformations of a singular point on the plane in the case zero eigenvalues*, Functional Anal. and its Appl. 9, 1975, 144-145.

[16] M. Braun and J. Hershenov, *Periodic solutions of finite difference equations*, Quart. Appl. Math. 35, 1977, 139-147.

[17] F. Brezzi, H. Fujii and S. Ushiki, *Real and ghost bifurcation dynamics in difference schemes for ODEs*, 79-104 in Küpper, Mittelmann, Weber (Eds. 1984), 1984.

[18] J. Carr, *Applications of center manifold theory*, Springer, New York, 1981.

[19] T. F. Chan, *Deflation techniques and block-elimination algorithms for solving bordered singular systems*, SIAM J. Sci. Stat. Comput. 5, 1984, 121-134.

[20] S.-N. Chow and J. K. Hale, *Methods of bifurcation theory*, Grundlehren der math. Wiss. 251, Springer, New York , 1982.

[21] W. A. Coppel, *Dichotomies in stability theory*, Lecture Notes in Mathematics 629, Springer, 1978.

[22] F. R. De Hoog and R. Weiss, *An approximation theory for boundary value problems on infinite intervals*, Computing 24, 1980, 227-239.

[23] B. De Dier, D. Roose and P. Van Rompay, *Interaction between fold and Hopf curves lead to new bifurcation phenomena*, J. of Comp. Appl. Math. 26, 1989, 171-186.

[24] B. De Dier, D. Roose and A. Spence, (eds.), *Continuation and bifurcations: Numerical technique and applications*, to appear in NATO ASI Series Kluwer, Dordrecht, 1990.

[25] L. Dieci, J. Lorenz and R. D. Russell, *Numerical calculation of invariant tori*, to appear in SIAM J. Sci. Stat. Comput., 1990.

[26] H. T. Doan, *Invariant curves for numerical methods*, Quart. Appl. Math. 43, 1985, 385-393.

[27] E. J. Doedel, *AUTO: A program for the automatic bifurcation analysis of autonomous systems*, Congressus Numerantium 33, 1981, 115-146.

[28] E. J. Doedel and J. P. Kernévez, *AUTO: Software for continuation and bifurcation problems in ordinary differential equations*, Appl. Math. Technical Report, Caltech, 1986.

[29] E. J. Doedel and M. J. Friedman, *Numerical computation of heteroclinic orbits*, J. of Comp. Appl. Math. 26, 1989, 155-170.

[30] E. J. Doedel and M. J. Friedman, *Numerical computation and continuation of invariant manifolds connecting fixed points with application to computation of combustion fronts*, To appear in Proc. of 7th Int. Conf. on FEMIF, 1990.

[31] E. J. Doedel and M. J. Friedman, *Numerical computation and continuation of invariant manifolds connecting fixed points*, Preprint, University of Alabama, Huntsville, 1990.

[32] T. Eirola, *Invariant circles of one-step methods*, BIT 28, 1988, 113-122.

[33] T. Eirola, *Two concepts for numerical periodic solutions of ODE's*, Appl. Math. Comput. 31, 1989, 121-131.

[34] B. Fiedler, *Global Hopf bifurcation of two-parameter flows*, Arch. Rat. Mech. Anal. 94, 1986, 59-81.

[35] C. Foias, M. S. Jolly, I. G. Kevrekidis, G. R. Sell and E.S. Titi, *On the computation of inertial manifolds*, Physics Letters A 131, 1988' 433-436.

[36] E. Freire, E. Ponce and A. J. Rodriguez-Luis, *Un método de continuación de órbitas homoclinas en sistemas autonomos planos biparametricos* , Preprint, Universidad de Sevilla, 1989.

[37] T. Garrett, G. Moore and A. Spence *The detection of Hopf bifurcation points* , In B. De Dier, D. Roose and A. Spence, Ed. 1990.

[38] P. Glendinning, *Global bifurcations in flows*, In New Direction in *Dynamical Systems* (T. Bedford, J. Swift (Eds.)), London Math. Soc. Lecture Notes Series 127, Cambridge University Press, 1988, 120-149.

[39] W. Govaerts, *Stable solvers and block-elimination for bordered systems*, to appear in SIAM J. Matrix Analysis Appl.

[40] A. Griewank and G. W. Reddien, *The calculation of Hopf points by a direct method*, IMA J. Numer. Anal. 3, 1983, 295-303.

[41] A. Griewank and G. W. Reddien, *Characterization and computation of generalized turning points*, SIAM J. Numer. Anal. 21, 1984, 176-185.

[42] A. Griewank and G. W. Reddien, *Computation of cusp singularities for operator equations and their discretizations*, J. of Comp. Appl. Math. 26, 1989, 133-153.

[43] J. Guckenheimer and Ph. Holmes, *Nonlinear oscillations, dynamical systems, and bifurcations of vector fields*, Appl. Math. Sci. 42, Springer, New York, 1983.

[44] J. Hale, *Ordinary differential equations*, Wiley, New York , 1969.

[45] J. Hale, *Introduction to dynamic bifurcation*, in *Bifurcation Theory and Applications* (L. Salvadori, Ed.), Springer, Lecture Notes in Mathematics 1057, 1983.

[46] B. D. Hassard, N. D. Kazarinoff, and Y.-H. Wan, *Theory and applications of Hopf bifurcation* , London Math. Soc. Lecture Note Series 41, 1981.

[47] M. W. Hirsch, C. Pugh, and M. Shub, *Invariant manifolds*, Springer, Lecture Notes in Mathematics 583, 1977.

[48] M. W. Hirsch and S. Smale, *Differential equations, dynamical systems, and linear algebra*, Academic Press, New York, 1974.

[49] M. Holodniok, P. Knedlik, and M. Kubiček, M. *Continuation of periodic solutions in parabolic partial differential equations*, in Küpper, Seydel, Troger (Eds. 1987), 1987, 122-130.

[50] G. Iooss, *Bifurcation of maps and applications*, North-Holland, 1979.

[51] M. C. Irwin, *Smooth dynamical systems*, Academic Pres, New York, 1980.

[52] E. Isaacson and H. B. Keller, *Analysis of numerical methods*, Wiley, New York, 1966.

[53] A. Iserles, A. T. Peplow and A. M. Stuart, *A unified approach to spurious solutions introduced by time discretisation, Part I: Basic theory*, Numerical Analysis Report NA 4, DAMTP Cambridge, 1990.

[54] A. D. Jepson, *Numerical Hopf bifurcation*, Thesis, Caltech, Pasadena, 1981.

[55] A. D. Jepson and A. Spence, *Singular points and their computation,* in Küpper, Mittemann, Weber (Eds. 1984), 1984, 195-209.

[56] H. B. Keller, *Practical procedures in path following near limit points*, in *Computing Methods in Applied Sciences and Engineering* (R. Glowinski and J.L. Lions, Eds.), North Holland, 1982.

[57] H. B. Keller and A. D. Jepson, *Steady state and periodic solution paths: their bifurcations and computations*, in Küpper, Mittelmann and Weber, (eds. 1984).

[58] A. I. Khibnik, V. I. Bykov and G. S. Yablonskii, *Parametric portrait of the catalytic oscillator*, Preprint, Institute of Catalysis, USSR Academy of Sciences, Novosibirsk, (in Russian), 1986.

[59] A. I. Khibnik, L̂INBF: a program for continuation and bifurcation analysis of equilibria up to codimension three In B. De Dier, D. Roose and A. Spence (Eds. 1990).

[60] I. G. Kevrekidis, R. Aris, L. D. Schmidt and S. Pelikan, *Numerical computations of invariant circles of maps*, Physica 16 D, 1985, 243-251.

[61] P. E. Kloeden and J. Lorenz, *Stable attracting sets in dynamical systems and their one-step discretizations*, SIAM J. Numer. Anal. 23, 1986, 986-995.

[62] H.-O. Kreiss and J. Lorenz, *Initial-boundary value problems and the Navier-Stokes equations*, Academic Press, Boston, 1989.

[63] M. Kubiček and M. Marek, *Computational methods in bifurcation theory and dissipative structures*, Springer, New York, 1983.

[64] M. Kubiček and M. Holodniak, *Numerical determination of bifurcation points in steady and periodic solutions – numerical algorithms and examples,* in Küpper, Mittelmann, Weber (Eds. 1984), 1984, 247-270.

[65] T. Küpper, H. D. Mittelmann and H. Weber, (Eds.) *Numerical methods for bifurcation problems*, ISNM 70, Birkhäuser, Stuttgart, 1984.

[66] T. Küpper, R. Seydel and H. Troger, (Eds.) *Bifurcation: Analysis, Algorithms, Applications* , ISNM 79, Birkhäuser, Stuttgart, 1987.

[67] Y. Kuznetsov, *Computation of invariant manifold bifurcations*, In B. De Dier, D. Roose and A. Spence (Eds. 1990).

[68] M. Lentini and H. B. Keller, *Boundary value problems over semi-infinite intervals and their numerical solution*, SIAM J. Numer. Anal. 17, 1980, 577-604.

[69] X.-B. Lin, *Using Melnikov's method to solve Silnikov's problems* Carolina State University, 1989.

[70] J. Lorenz and Van de Velde *Concurrent computations of invariant manifolds*, Preprint, Caltech, Pasadena, 1989.

[71] V. K. Melnikov, *On the stability of the center for time periodic perturbations*, Trans. Moscow Math. Soc. 12, 1963, 1-57.

[72] G. Moore and A. Spence *The calculation of turning points of nonlinear equations*, SIAM J. Numer. Anal. 17, 1980, 567-576.

[73] S. Newhouse, D. Ruelle and F. Takens, *Occurence of strange axiom – A attractors near quasiperiodic flow on* T^n, $n \leq 3$, Comm. Math. Phys. 64, 1978, 35-40.

[74] J. M. Ortega and W. C. Rheinboldt, *Iterative solution of nonlinear equations in several variables*, Academic Press, New York, 1970.

[75] V. I. Oseledec, *A multiplicative ergodic theorem. Lyapunov characteristic numbers for dynamical systems*, Trans. Moscow Math. Soc. 19, 1968, 197-231.

[76] K. J. Palmer, *Exponential dichotomies and transversal homoclinic points*, J. Diff. Equ. 55, 1984, 225-256.

[77] G. Pönisch, *Computing hysteresis points of nonlinear equations depending on two parameters*, Computing 39, 1987, 1-17.

[78] G. Pönisch and H. Schwetlick, *Computing turning points of curves implicitly defined by nonlinear equations depneding on a parameter*, Computing 26, 1981, 107-121.

[79] W. C. Rheinboldt, *Numerical analysis of continuation methods for nonlinear structural problems*, Computing and Structures 13, 1981, 103-113.

[80] W. C. Rheinboldt, *Numerical analysis of parametrized nonlinear equations*, University of Arkansas, Lecture notes in the Math. Sci. 7, Wiley, 1986.

[81] A. J. Rodriguez-Luis, E. Freire and E. Ponce, *A method for homoclinic and heteroclinic continuation in two and three dimensions*, in B. De Dier, D. Roose and A. Spence (Eds. 1990).

[82] D. Roose, *Numerical computation of origins for Hopf bifurcation in a two-parameter problem*, in Küpper. Seydel, Troger (Eds. 1987), 1987, 268-273 .

[83] D. Roose and V. Hlavacek, *A direct method for the comnputation of Hopf bifurcation points*, SIAM J. Appl. Math. 45, 1985, 879-894.

[84] D. Roose and R. Piessens, *Numerical computation of turning points and cusps.*, Numer. Math. 46, 1985, 189-211.

[85] J. M. Sanz-Serna, *Studies in numerical nonlinear stability I. Why do leapfrog schemes go unstable?*, SIAM J. Sci. Stat. Comp. 6, 1985, 923-938.

[86] R. Seydel, *Numerical computation of branch points in nonlinear equations*, Numer. Math. 33, 1979, 339-352.

[87] R. Seydel, *From equilibrium to chaos: practical bifurcation and stability analysis*, Elsevier, New York, 1988.

[88] Shil'nikov *A case of the existence of a denunerable set of periodic motions*, Sov. Math. Dokl. 6, 1965, 163-16.

[89] I. Shimada and T. Nagashima, *A numerical approach to ergodic problems of dissipative type*, Progress of Theoretical Phys. 61, 1979, 1605-1616.

[90] C. Sparrow, *The Lorenz equations: Bifurcations, chaos and strange attractors*, Springer, New York, 1982.

[91] A. Spence and B. Werner, *Non-simple turning points and cusps* , IMA J. Num. Anal. 2, 1982, 413-427.

[92] A. Spence, K. A. Cliffe and A. D. Jepson, *A note on the calculation of paths of Hopf bifurcations*, J. Comp. Appl. Math. 26, 1989, 125-131.

[93] H. J. Stetter, *Analysis of discretization methods for ordinary differential equations*, Springer, Berlin, 1973.

[94] M. Stiefenhofer, *Zur mathematischen Analyse eines biochemischen Schleimpilzmodells*, Diplomarbeit (Universität Konstanz), 1988.

[95] J. Stoer and R. Bulirsch, *Introduction to numerical analysis*, Springer, New York, 1980.

[96] A. M. Stuart, *Nonlinear instability in dissipative finite difference schemes*, SIAM Review 31, 1989, 191-220.

[97] F. Takens, *Singularities of vector fields*, IHES 43, 1974, 47-100.

[98] R. Temam, *Infinite-dimensional dynamical systems in Mechanics and Phyics*, Appl. Math. Sci. 68, Springer, New York, 1988.

[99] M. van Veldhuizen, *A new algorithm for the numerical approximation of an invariant curve* , SIAM J. Sci. Stat. Comput. 8, 1987, 951-962.

[100] S. Wiggins, *Global bifurcations and chaos (analytical methods)* , Appl. Math. Sci. 73, Springer, New York, 1988.

[101] T. Yoshizawa, *Stability theory by Liapunov's second method*, The Mathematical Society of Japan, 1966.

Professor W.-J. Beyn
Falultät für Mathematik
Universität Bielefeld
Postfach 8640
4800 Bielefeld 1
Germany.

6

The Theory and Numerics of Differential-Algebraic Equations

Werner C. Rheinboldt

6.1 Preface

These notes introduce some typical applications and basic properties of differential-algebraic systems of equations (DAEs) and then present an overview of recent, new existence theories for such systems based on differential geometric considerations and on a numerical approach derived from these theories. In the presentation the stress is on general concepts, results and applications rather than on detailed proofs.

Differential-algebraic systems of equations (DAEs) arise in many applications in science and engineering. For some examples we refer, for instance, to the monographs [3,22] and the many references given there. Three typical applications are sketched in Section 6.2 below. Over the years, it has become well known that the solution behavior of DAEs may differ considerably from that of standard ordinary differential equations (ODEs). A valuable measure of the deviation of a DAE from an ODE is the concept of an index which was first introduced in [21] and has since been formalized in various ways. The index highlights also some of the differences between the existence behavior of DAEs and ODEs although it does not, by itself, provide for any existence results. In fact, up to now, existence theories for nonlinear DAEs are available only for a few selected classes of systems.

Since the solutions of any DAE are expected to be smooth paths in some space of dependent variables, we should expect the equations to define a dynamical system in a suitable domain of that space. While this connection with dynamical systems is immediately obvious for ODEs this is certainly not the case for DAEs and there appear to be only few studies that specifically address this connection (see e.g. [37] and [38]). Some aspects of the relationship between DAEs and dynamical systems will be discussed in Section 6.3.

In a series of papers [38],[41], and [36] a differential-geometric approach has been developed for the analysis of the dynamical system underlying a DAE and for the proof of general existence and uniqueness results for such systems. In Sections 6.4 and 6.5 the results in the two last-mentioned papers are summarized, moreover, Section 6.5 also addresses relations between some of these results and the index concept.

Comprehensive introductions to numerical methods for DAEs may be found in the cited monographs [3,22]. A brief survey of some of these methods is given in Section 6.6. Then in that section a new local parametrization approach for DAEs is presented which derives naturally from the differential-geometric existence theories and has been found to lead to very promising methods for the computational solution of higher-index DAEs. For the case of the Euler-Lagrange equations of constrained mechanical systems this approach includes the so-called method of generalized coordinate partitioning introduced first in [48].

6.2 Model Problems

This Section provides three illustrative examples of practical applications leading to differential-algebraic equations. As indicated before, there are numerous other areas were DAEs occur.

6.2.1 Constrained Dynamical Systems

A major source of DAEs is the kinematic and dynamic analysis of mechanical multi-body systems. This is a venerable field of mechanics and we give here only some very simple examples and refer for further details to the extensive literature (see e.g. [23,49]).

Suppose that, under the influence of a force Q, a particle with mass m slides on a two-dimensional surface in $\mathbb{R}^3$ specified by the real-valued equation $\Phi(x) = 0$. In order for the point to remain on the surface, a constraining force must act in the normal direction of the surface. If $D\Phi(x) \in L(\mathbb{R}^3, \mathbb{R}^1)$ denotes the derivative of Φ at any $x \in \mathbb{R}^3$, then this normal direction is given by the vector $D\Phi(x)^T \in \mathbb{R}^3$. Hence, by Newton's law we obtain here the DAE

$$\Phi(x) = 0, \quad mx'' + zD\Phi(x)^T = Q \tag{2.1}$$

where $z \in \mathbb{R}^1$ specifies the size of the constraining force. For example, suppose that the surface is a paraboloid and that gravity is the only force acting on the mass, then (2.1) becomes

$$\begin{aligned} x_1^2 + x_2^2 &= x_3 \\ mx_1'' + 2zx_1 &= 0 \end{aligned} \tag{2.2}$$

$$\begin{aligned} m x_2'' + 2 z x_2 &= 0 \\ m x_3'' - z &= mg. \end{aligned}$$

More generally, suppose that the vector $x \in \mathbb{R}^n$ characterizes the configuration of all bodies of a mechanical system and that the kinematic constraints acting on the system are modelled by the s-dimensional (holonomic) constraint equations

$$\Phi(x,t) = 0. \tag{2.3}$$

Here Φ is now a mapping from $\mathbb{R}^n \times \mathbb{R}^1$ into $\mathbb{R}^s$, $1 \le s < n$ and t represents time. Then the equations of motion are

$$M(x,t)x'' + D_x\Phi(x,t)^T z = Q(x,x',t) \tag{2.4}$$

where $M(x,t)$ is the mass matrix and $Q(x,x',t)$ the vector of applied forces.

As an example consider a simple, planar "slider crank" consisting of two bodies, namely, a bar of length 2 and a wheel of radius 1 centered at the origin of a (ξ,η) coordinate system in the plane. At one of its ends the bar pivots around a fixed point on the circumference of the wheel while its other end slides along the ξ-axis. Any configuration of the system is characterized by the vector $x = (\alpha_1,\ \alpha_2,\ \xi)^T$ consisting of the current coordinate ξ of the bar's sliding end and the two angles α_1 and α_2 formed by the ξ-axis and the direction from the origin to the pivot point and by the bar and the ξ-axis, respectively. Thus, if the wheel turns with a constant torque τ, then the equations (2.4), (2.3) have here the form

$$\begin{aligned} \cos\alpha_1 + 2\cos\alpha_2 &= \xi \\ \sin\alpha_1 - 2\sin\alpha_2 &= 0 \\ J_1\alpha_1'' - z_1\sin\alpha_1 + z_2\cos\alpha_1 &= \tau \\ m\xi'' - 2z_1\sin\alpha_2 - 2z_2\cos\alpha_2 &= 0 \\ J_2\alpha_2'' - z_1 &= 0 \end{aligned} \tag{2.5}$$

where J_1 and J_2 are the moments of inertia of the wheel and the bar, respectively, and m is the mass of the bar.

6.2.2 Electrical Circuits

A second extensive source of DAEs is the analysis of electrical circuits. Once again, we refer for details to the literature (see e.g. [10]) and discuss only the basic ideas and a simple example.

A circuit may be considered as an inter-connected collection of electrical devices, such as, resistors, inductors, capacitors, sources, etc. Its connection pattern is modelled by a finite, directed graph $\Omega = (N, \Lambda)$ with node set $N = \{1, 2, \ldots, p\}$ and branch set $\Lambda = \{\lambda_1, \ldots, \lambda_q\} \subset N \times N$ where single-node loops $(i, i) \in N \times N$ are excluded. Each branch corresponds to a specific component of the circuit. As an example consider a graph with four nodes and the following branches and components:

branch $(1, 2)$	linear resistor	resistance $= R_1$
branch $(2, 3)$	voltage source	voltage $= u_0$
branch $(1, 3)$	linear capacitor	capacitance $= C_1$
branch $(1, 4)$	linear inductor	inductance $= L_1$
branch $(4, 1)$	linear capacitor	capacitance $= C_2$
branch $(4, 3)$	linear capacitor	capacitance $= C_3$
branch $(3, 4)$	linear resistor	resistance $= R_2$.

Generally, the graph Ω can be characterized by its (node-arc) incidence matrix $A \in \mathbb{R}^{p \times q}$ with the elements

$$a_{ij} = \begin{cases} +1 & \text{if } (i, k) \in \Lambda \text{ for some } k \in N \\ -1 & \text{if } (k, i) \in \Lambda \text{ for some } k \in N \\ 0 & \text{otherwise.} \end{cases}$$

In our example the graph underlying the circuit then has the incidence matrix

$$A = \begin{pmatrix} 1 & 0 & 1 & 1 & -1 & 0 & 0 \\ -1 & 1 & 0 & 0 & 0 & 0 & 0 \\ 0 & -1 & -1 & 0 & 0 & -1 & 1 \\ 0 & 0 & 0 & -1 & 1 & 1 & -1 \end{pmatrix}.$$

With each branch $\lambda_j = (i, k)$ of Ω two electrical quantities are associated, namely a current y_j and a voltage-drop u_j. They are connected by a functional relation

$$\phi_j(y_j, u_j) = 0, \; j = 1, 2, \ldots, q, \tag{2.6}$$

the so-called *branch-characteristic* of λ_j. The specific form of (2.6) depends on the type of the device modelled by the branch, such as, for instance,

current source:	$y_j = \psi(t)$
voltage source:	$u_j = \psi(t)$
ideal diode:	$\max(u_j, -y_j) = 0$
linear resistor:	$u_j = Ry_j$
voltage driven resistor:	$y_j = \psi(u_j)$
current driven resistor:	$u_j = \psi(y_j)$
linear capacitor:	$C{u_j}' = y_j$
linear inductor:	$L{y_j}' = u_j$

In our example, the set of branch characteristics (2.6) is given by the equations

$$\begin{aligned} R_1 y_1 &= u_1 \\ u_2 &= u_0 \\ C_1 {u_3}' &= y_3 \\ L {y_4}' &= u_4 \\ C_2 {u_5}' &= y_5 \\ C_3 {u_6}' &= y_6 \\ R_2 y_7 &= u_7. \end{aligned}$$

Kirchhoff's first conservation-law requires that the (algebraic) sum of the currents on the branches starting at a node must equal the sum of the currents on the branches terminating at that node. In terms of the incidence matrix A this means that a permissible current flow is characterized by any vector $y = (y_1, \ldots, y_q)^T \in \mathbb{R}^q$ for which

$$Ay = 0. \tag{2.7}$$

Kirchoff's second law specifies that the (algebraic) sum of all the voltage drops on the branches of any closed path of Ω has to be zero. If we introduce the vector $u = (u_1, \ldots, u_q)^T \in \mathbb{R}^q$ of voltage drops, as well as the vector $w = (w_1, \ldots, w_p)^T \in \mathbb{R}^p$ of all nodal voltage levels, then the second law corresponds to the equation

$$u = A^T y \, , \; w_1 = 0 \tag{2.8}$$

where the last equation was introduced to fix the absolute values of the nodal voltages.

Thus altogether (2.6), (2.7), (2.8) form a DAE of $2q + p + 1$ equations in $2q + p$ unknown. The reason for this difference is that A does not have full rank. If Ω is a connected graph – which is certainly a reasonable assumption – then a standard theorem of graph theory (see e.g. [6]) ensures that $\text{rank} A = p - 1$. Thus one of the equations (2.7), is a linear combination of the others and hence may be dropped.

The equations (2.6), (2.7), (2.8) are called a descriptor form of the circuit. There are many ways of reducing the size of this system but we shall not enter into any details here.

6.2.3 Punch-Stretching of Sheet Metal

We end with a somewhat different example arising in connection with sheet metal stamping processes. Since in this case the formulation is somewhat more complex, we do not include all the details but refer instead to the literature (see e.g., [8,9]).

The processes to be considered involve the deformation of a sheet of metal in a forming press with a particular punch and die configuration. In order to ease the discussion we consider a simpler problem, namely the so-called hydrostatic bulge test used widely in metallurgy. An initially flat sheet of metal is clamped over one end of a cylindrical chamber into which hydraulic oil is then pumped. This creates a hydrostatic load on the sheet and causes it to bulge outward.

In line with the formulation presented in [47] the equation of virtual work for the hydrostatic bulge deformation is given by

$$h_0 \int_{A_0} \tau \delta \epsilon \, dA_0 = \int_{A_0} p \delta v \, dA_0. \tag{2.9}$$

Here h_0, A_0 denote the initial sheet thickness and surface area, respectively, ξ is the radial distance to a material point on the sheet at time zero, and v is a volume measure. Moreover, $\tau = (\tau_1, \tau_2)$ is the vector of the Kirchhoff stresses in the radial and circumferential directions, respectively, while $\varepsilon = (\varepsilon_1, \varepsilon_2)$ is the vector of the logarithmic strains in the corresponding directions, defined by

$$\varepsilon_1 = \ln[(1+u_\xi)^2 + w_\xi^2]^{1/2}, \quad \varepsilon_2 = \ln[1 + \frac{w}{\xi}],$$

where, u and w denote the radial and vertical displacements of the material point whose initial position is given by $(\xi, 0)$.

In addition, we have the material-dependent constitutive equations in rate form. These have the generic form

$$\tau' = L\varepsilon' + r(\tau, \bar{\varepsilon}), \quad \bar{\varepsilon}' = g(\tau, \bar{\varepsilon}) \tag{2.10}$$

where

$$L = \frac{E}{1-\nu^2} \begin{pmatrix} 1 & \nu \\ \nu & 1 \end{pmatrix}$$

and E and ν are Young's modulus and Poisson's ratio, respectively, $\bar{\varepsilon}$ is the effective strain, and the nonlinear functions r and g depend on the specific form of the hardening law.

The volume measure in (2.9) is defined by

$$v = w(1 + \frac{u}{\xi})(1 + u_\xi)$$

and

$$V(t) = \int_{A_0} v(\xi) \, dA_0$$

is the volume of the bulge at time t which may be assumed, for instance, to satisfy $V(t) = \gamma t$ with fixed γ.

For the computation, we introduce finite element approximations of the displacements u, w, the stress components τ_1, τ_2, and the effective stress $\bar{\varepsilon}$. Then the equation (2.9) of virtual work is approximated by a nonlinear equation of the form

$$F(x, y, q, t) = 0 \tag{2.11}$$

where the vectors x, y, q contain the approximations of (u, w), (τ_1, τ_2), and p, respectively. Correspondingly, the constitutive equations (2.10) are approximated by a differential equation of the form

$$\begin{aligned} y' - Lz' &= f_1(y, z) \\ z' &= f_2(y, z) \end{aligned} \tag{2.12}$$

where the vector z represents the approximation of the effective strain $\bar{\varepsilon}$. Thus, altogether the equations (2.11), (2.12) form a DAE.

6.3 DAEs and Dynamical Processes

6.3.1 DAEs and ODEs

The examples of Section 6.2 provide an indication of various possible forms of DAEs. In all cases the differential equations and algebraic equations turned out to be separated. Observe also that there may be variables for which no derivatives appear anywhere in the system. Of course, for the computation various specific properties of the equations are of particular interest. For instance, it is usually advantageous when the derivatives only occur linearly, etc.

In some applications the form of the equations may vary in different parts of the space, and, in particular, there may not exist a globally valid separation into algebraic equations and differential equations. Hence, in such cases, the DAEs have the generic form of an implicit differential equation

$$F(x, x', t) = 0 \tag{3.1}$$

which cannot be transformed into the form

$$x' = f(x, t) \tag{3.2}$$

of an explicit ordinary differential equation (ODE).

If, in (3.1), F is a sufficiently smooth map from $\mathbb{R}^{2n+1}$ into $\mathbb{R}^n$ and the derivative $D_pF(x, p, t) \in L(\mathbb{R}^n, \mathbb{R}^n)$ is an isomorphism at some solution (x, y, t) of $F(x, y, t) = 0$ then the implicit function theorem guarantees that (3.1) can be transformed locally into the form (3.2). Hence, in our setting we should assume that $D_pF(x,p,t)$ does not have full rank. More specifically, we shall call (3.1) an implicit DAE only if

$$\mathrm{rank} D_pF(x,p,t) = \text{constant} < n, \tag{3.3}$$

on the domain under consideration. This constant-rank assumption excludes various singular implicit equations (3.1) with a solution behavior that may differ radically from that of ODEs or DAEs (see e.g. [35]).

The existence and uniqueness theory for solutions of explicit ODEs (3.2) is a well-developed subject (see e.g. [11] and also the Appendix below). In particular, if $f : E \subset \mathbb{R}^{n+1} \to \mathbb{R}^n$ is of class C^1 on some open set E, then we know that for any point $(x_0, t_0) \in E$ there exists a C^2-solution $x : J \subset R^1 \to E$ of (3.2), defined on some open interval J containing t_0, which satisfies the initial condition $x(t_0) = x_0$. Moreover, any two such solutions satisfying the same initial condition are identical on the intersection of their domain.

This local solvability result for ODEs does not carry over directly to DAEs. In fact, consider the simple system

$$\begin{aligned} x_1 &= \cos x_2 \\ x_1' &= x_3 \\ x_2' &= 1, \end{aligned} \tag{3.4}$$

and suppose that $x : J \subset \mathbb{R}^1 \to \mathbb{R}^3$ is any C^1-solution of (3.4) on some open interval J. Then, by differentiating $x_1(t) - \cos x_2(t) = 0$ with respect to t and using the differential equations, we find that the solution must satisfy the algebraic condition

$$x_3 + \sin x_2 = 0, \tag{3.5}$$

whence necessarily

$$x(t) = \begin{pmatrix} \cos t \\ t \\ -\sin t \end{pmatrix}. \tag{3.6}$$

Conversely, (3.6) does define a C^∞ -solution $x : \mathbb{R}^1 \to \mathbb{R}^3$ of (3.4). In other words, (3.6) is the only solution of (3.4) and we certainly cannot prescribe any initial conditions other than, trivially, a point of (3.6).

This result indicates that for DAEs there may be some "hidden " constraints, such as (3.5), which all solutions have to satisfy. As a consequence, there may not exist any solution through every choice of initial point; that is, only certain initial conditions may be admissible.

Obviously, any "hidden" constraints may be expected to cause difficulties during the numerical solution of the DAE. This is indeed the case, and, in fact, it is by now well known that the degree of difficulty rises with the number of such additional constraints. This observation led W. Gear and L. Petzold ([21]) to introduce an index which measures the "deviation" of a DAE from an ODE. We shall discuss later some aspects of this concept; at this moment it will suffice to characterize the index of a DAE loosely as the total number of given algebraic and hidden constraints that are needed to specify the solution completely. In this sense, (3.4) is a DAE with index 2.

As another example consider the DAE (2.2) modelling the dynamics of a mass-point on a paraboloid. This system can be written in the first order form

$$\begin{aligned} \Phi(x) &\equiv x_1^2 + x_2^2 - x_3 = 0 \\ x' &= y \\ y' &= ge^3 - zD\Phi(x)^T, \end{aligned} \tag{3.7}$$

where $x, y \in \mathbb{R}^3$, e^3 is the third natural basis vector of $\mathbb{R}^3$ and we set $m = 1$. By differentiating the algebraic equation and using the differential equations we obtain as first "hidden" constraint

$$2x_1y_1 + 2x_2y_2 - y_3 = 0. \tag{3.8}$$

In turn, by differentiating (3.8) we are led to the further constraint

$$2(y_1^2 + y_2^2) - z(1 + 4(x_1^2 + x_2^2)) - g = 0. \tag{3.9}$$

It is not difficult to see that (3.7) together with these two constraints (3.8), (3.9) completely specifies the solutions. Thus in our terminology the DAE has index 3.

The two constraints have here a simple geometric meaning. In fact, (3.8) requires the velocity vector y to be tangential to the paraboloid while (3.9) means that the constraining force $zD\Phi(x)^T$ has to balance the other two forces.

This index-result is not restricted to the special example (2.2); in fact it turns out that all Euler-Lagrange systems (2.3), (2.4) have index 3.

6.3.2 Dynamical Processes

Ordinary differential equations are a fundamental tool in the study of dynamical processes that are finite-dimensional, differentiable and causal. By this we mean processes for which

(i) the states are characterized by finitely many degrees of freedom,

(ii) the changes of the states are described by differentiable functions, and

(iii) the future behavior is uniquely determined by the initial conditions.

Our examples suggest that differential-algebraic equations also represent models of such dynamical processes.

The theory of dynamical processes has been heavily influenced by mechanical considerations. Thus, we use a simple mechanical example to review some of the basic terminology. Consider the motion of k particles in $\mathbb{R}^3$ and let x_i and y_i, $i = 1, 2, \ldots, k$, denote the location and velocity of the particles at time t. Then with $q = (x_1, \ldots, x_k) \in \mathbb{R}^{3k}$ and $p = (y_1, \ldots, y_k) \in \mathbb{R}^{3k}$ the state of the system is given by (q, p). The states are usually restricted to some specified subset $S \subset \mathbb{R}^{3k} \times \mathbb{R}^{3k}$ – the state space of the system.

In many cases, the state space is an open subset of $\mathbb{R}^{3k} \times \mathbb{R}^{3k}$. For instance, if no two particles are ever allowed to be in the same place at the same time, then

$$S = \{(q,p) \in \mathbb{R}^{3k} \times \mathbb{R}^{3k}; x_i \neq x_j \text{ for } i \neq j,\ i,j = 1,2,\ldots,k\}$$

is certainly open.

On the other hand – for instance when there are angular variables or constraints – the state space need not be an open set. For example, consider a system of rigidly connected particles. Then the location vectors q have to belong to the set

$$C = \{q \in \mathbb{R}^{3k}; \| x_i - x_j \|_2 = c_{ij};\ \text{for } i \neq j,\ i,j = 1,2,\ldots,k\},$$

where c_{ij} are given constants. For $k \geq 4$ this is a six-dimensional submanifold [1] of $\mathbb{R}^{3k}$. This follows from the well-known fact that the position of a rigid body in $\mathbb{R}^3$ is uniquely characterized by the location of one point and the orientation of an orthonormal coordinate system fixed within the body. Then the state space may be identified with the tangent bundle $C \times \mathbb{R}^{3k}$ of C and hence is a 12-dimensional submanifold of $\mathbb{R}^{3k} \times \mathbb{R}^{3k}$.

In differential-geometric terms the two cases are not very different. In fact, any open subset of $\mathbb{R}^{3k} \times \mathbb{R}^{3k}$ is a $6k$-dimensional submanifold of that space and hence, in either case, the state space is a submanifold. This agrees with the fundamental assumption introduced by H. Poincarè ($\sim$ 1880) that the state space of a mechanical system should be a differentiable manifold. Correspondingly, the dynamical system is viewed as a field of vectors on this manifold such that a solution is a smooth curve tangent at each of its points to the vector attached to that point. We refer, e.g., to [1] and the historical references included there.

[1] Some differential-geometric concepts and results are collected in the Appendix.

However, from a computational viewpoint, there is indeed a substantial difference between the above two cases which, in fact, reflects again the earlier indicated differences between ODEs and DAEs. As noted before, in the classical theory of explicit ODEs (3.2) we assume f to be of class C^1 on some *open* subset E of $\mathbb{R}^{n+1}$ and, of course, on E the ODE induces the natural vector field

$$(x,t) \in E \quad \mapsto \quad ((x,t), f(x,t)) \in TE = E \times \mathbb{R}^{n+1}.$$

Since the points (x,t) represent the states of the system, this corresponds to the case when the state space is an open subset. On the other hand, as the above example of rigidly connected particles shows, for a DAE the state space is expected to be a lower dimensional manifold. For instance, in the trivial example (3.4) this is the one-dimensional submanifold

$$\{x \in \mathbb{R}^3;\ x_1 - \cos x_2 = 0,\ x_3 + \sin x_2 = 0\}$$

defined by the given and hidden constraints.

In the standard theory of numerical methods for solving ODEs of the form (3.2) it is critical that the domain E of the right-hand side is open and that there exists a locally unique solution of (3.2) through each point x of E. In fact, any such method generates a sequence of points $\{(x_k, t_k); k = 1, 2, \ldots\}$ that approximate points on the solution through the given initial point (x_0, t_0). At best, we know that these computed points belong to some open neighborhood of the exact solution contained in the open set E. Furthermore, for the step from (x_k, t_k) to (x_{k+1}, t_{k+1}) all methods are designed to approximate the *local* solution of the ODE through (x_k, t_k) in the sense that the error between (x_{k+1}, t_{k+1}) and this local solution converges to zero when the steplength $t_{k+1} - t_k$ tends to zero. It is by no means obvious how to extend this approach to the case when the domain E is no longer an open set but some lower-dimensional submanifold of $\mathbb{R}^n \times \mathbb{R}^1$.

6.4 Existence Theory for Implicit DAEs

As noted earlier, the literature on DAEs is growing rapidly but general existence theories have only begun to be developed relatively recently. The earliest such result appears to be the existence theory for gradient systems

$$\nabla_x g(x,y) = 0\ , \quad x' = f(x,y)$$

developed by F.Takens [45] who used the approximating, singularly perturbed system of differential equations

$$\varepsilon y' = \nabla_x g(x,y)\ , \quad x' = f(x,y)$$

with asymptotically small ε.

Best understood are probably the linear DAEs with constant coefficients

$$Ax' + Bx = g(t), \; x \in \mathbb{R}^n, \; A, B \in L(\mathbb{R}^n), \; \text{rank}A = r < n \tag{4.1}$$

for which existence results can be proved by means of the Kronecker canonical form for matrix pencils (see, e.g., [16]). For a presentation of this theory see e.g. [21] or [22]. For further references see also [26,46].

In [38] the indicated interpretation of DAEs as dynamical systems on manifolds was used to obtain existence results for semi-explicit systems

$$F_1(x) = 0, \quad A(x)x' = G(x).$$

These results were generalized in [41] to first and second order systems of the form

$$F_1(x) = 0, \quad F_2(x, x', z) = 0, \tag{4.2}$$

and

$$F_1(x) = 0, \quad F_2(x, x', x'', z) = 0, \tag{4.3}$$

respectively.

For general implicit equations (3.1) – of course, under the constant-rank assumption (3.3) – local existence results were first given in [22], however, under the restrictive condition that $\ker D_pF(x,p,t)$ is independent of x and p. Finally, without such a condition, a solution theory for implicit DAEs was presented in [36]. This theory will be outlined in the following subsection; for proofs and further details we refer to the original article.

6.4.1 Local Theory for Implicit DAEs

For ease of notation we shall consider (3.1) in the autonomous form

$$F(x, x') = 0, \tag{4.4}$$

under the three assumptions

(A1) $F : E \subset \mathbb{R}^n \times \mathbb{R}^n \to \mathbb{R}^n$ is C^2 on the open set E;

(A2) $\text{rank}DF(x,p) = n, \; \forall \; (x,p) \in E$;

(A3) $\text{rank}D_pF(x,p) = r < n, \; \forall \; (x,p) \in E$.

The condition (A2) requires that the equations (4.4) are independent and also implies that $F^{-1}(0)$ is an n-dimensional C^2-submanifold of $\mathbb{R}^n \times \mathbb{R}^n$ (see again the Appendix).

An instructive prototype for (4.4) is the semi-implicit DAE

$$F(x, x') = \begin{pmatrix} F_1(x) \\ F_2(x, x') \end{pmatrix}. \tag{4.5}$$

Here (A1) holds if $F_1 : E_x \to \mathbb{R}^{n-r}$ and $F_2 : E_x \times E_p \to \mathbb{R}^r$, are of class C^2 on open sets $E_x \subset \mathbb{R}^n$ and $E = E_x \times E_p \subset \mathbb{R}^n \times \mathbb{R}^n$, respectively. Moreover, (A2) and (A3) are satisfied if $\text{rank} DF_1(x) = n - r$, $\forall x \in E_x$ and $\text{rank} D_p F_2(x,p) = r$, $\forall (x,p) \in E$. Obviously, the first of these conditions implies that

$$M = \{x \in E_x; F_1(x) = 0\} \tag{4.6}$$

is an r-dimensional C^2-submanifold of $\mathbb{R}^n$.

A C^k-solution of the general equation (4.4) is any mapping

$$x : J \to R^n, \ (x(t), x'(t)) \in E, \ F(x(t), x'(t)) = 0, \ \forall t \in J,$$

which is of class C^k on some open interval J of $\mathbb{R}^1$. As noted in the previous Section, we cannot expect that there is a solution through each point $(x,p) \in E$ and the following lemma provides a necessary condition for this to hold:

Lemma 4.1. *For $(x,p) \in E$ the conditions*

$$F(x,p) = 0, \ D_x F(x,p)p \in \text{rge} D_p F(x,p), \tag{4.7}$$

are neccessary for the existence of a C^1-solution of (4.4) that passes through (x,p).

For any C^2-solution of (4.4) this follows directly from the fact that by differentiation of $F(x(t), x'(t)) = 0$ we obtain $D_x F(x(t), x'(t))x'(t) + D_p F(x(t), x'(t))x''(t) = 0$ for all t in J. Of course, for C^1-solutions this argument cannot be used and a more subtle proof is required (see [36]).

For a closer analysis of the set of points characterized by the necessary conditions (4.7) we introduce the orthogonal projections

$$\begin{aligned} P, Q : E \to L(\mathbb{R}^n, \mathbb{R}^n), \ P(x,p)\mathbb{R}^n &= \text{rge} D_p F(x,p), \\ Q(x,p) &= I_n - P(x,p), \ \forall (x,p) \in E. \end{aligned}$$

Because of the constant rank condition (A3) these projections are C^1-functions on E. Hence, also the reduced map

$$\hat{F} : E \to \mathbb{R}^n, \hat{F}(x,p) = P(x,p)F(x,p) + Q(x,p)D_x F(x,p)p, \ (x,p) \in E, \tag{4.8}$$

is of class C^1 on E. Then we can show that the set E_N of all points satisfying the necessary conditions (4.7) is given by

$$E_N = \{(x,p) \in E; \ F(x,p) = 0, \hat{F}(x,p) = 0\}. \tag{4.9}$$

In the special case of (4.5) the projections are independent of x and p. In fact we have

$$P = \begin{pmatrix} 0_{n-r} & 0 \\ 0 & I_r \end{pmatrix}$$

and therefore

$$\hat{F}(x,p) = \begin{pmatrix} DF_1(x)p \\ F_2(x,p) \end{pmatrix} \tag{4.10}$$

and

$$E_N = \{(x,p) \in E;\ F_1(x) = 0,\ DF_1(x)p = 0,\ F_2(x,p) = 0\}.$$

Generally, the mapping (4.8) defines the reduced equation

$$\hat{F}(x,x') = 0, \tag{4.11}$$

which, in essence, has the same solutions has the original DAE. More specifically the following result holds:

Lemma 4.2. *Any C^1-solution of the original equations (4.4) solves the reduced equation (4.11). Conversely, any C^2-solution of (4.11) that passes through some point of $F^{-1}(0)$ is a C^2-solution of (4.4).*

In the special case of (4.5) when F_2 is linear in p; that is, in the case of the DAE

$$F(x,x') \equiv \begin{pmatrix} F_1(x) \\ A(x)x' - G(x) \end{pmatrix} = 0, \tag{4.12}$$

$D_p\hat{F}(x,p)$ does not depend on p and the reduced equation (4.11) becomes the linear equation

$$B(x)x' = \begin{pmatrix} 0 \\ G(x) \end{pmatrix}, \quad B(x) = \begin{pmatrix} DF_1(x) \\ A(x) \end{pmatrix}. \tag{4.13}$$

Suppose that the subset

$$M_0 = \{x \in M; B(x) \in \mathrm{Isom}(R^n)\} \tag{4.14}$$

of the constraint manifold (4.6) is not empty. Then M_0 is an r-dimensional submanifold of $\mathbb{R}^n$ on which (4.13) induces the tangential C^1-vectorfield

$$x \in M_0 \quad \mapsto \quad (x,p) \in TM_0, \quad p = B(x)^{-1}\begin{pmatrix} 0 \\ G(x) \end{pmatrix}.$$

It is readily seen that the integral curves of this vectorfield are exactly the solutions of (4.13) and hence, by Lemma 4.2, also of the DAE (4.12). This corresponds to the approach used in [38] to develop an existence theory for DAEs of the form (4.12).

For the general DAE (4.4), we proceed analogously and assume that the set

$$E_A = \{(x,p) \in E_N;\ D_p\hat{F}(x,p) \in \mathrm{Isom}(\mathbb{R}^n)\} \tag{4.15}$$

is non empty. Clearly, by continuity E_A is (relatively) open in E_N. Moreover, by the implicit function theorem it follows that locally in some open neighborhood of any point $(x_0, p_0) \in E_A$ the reduced equation (4.11) can be transformed into an explicit ODE. Thus by applying the standard ODE-theory we can prove the following local existence result for the original DAE (4.4):

Theorem 4.3. *Given $t_0 \in R^n$, consider the initial value problem*

$$F(x,x') = 0,\ x(t_0) = x_0,\ x'(t_0) = p_0 \tag{4.16}$$

under the assumptions (A1,2,3). If a C^1-solution of (4.16) exists, then $(x_0, p_0) \in E_N$. Conversely, for any $(x_0, p_0) \in E_A$ there exists a C^1-solution of (4.16) which is unique on some open interval J containing t_0. Moreover, this solution is of class C^2 on J.

The set E_A of (4.15) has a manifold structure. This is self-evident in the case of (4.5) where, obviously, the derivative of the mapping

$$(x,p) \in E \quad \mapsto \quad \begin{pmatrix} F_1(x) \\ \hat{F}(x,p) \end{pmatrix} \in \mathbb{R}^{2n-r}$$

has full rank for any point of E_A and hence, E_A is either empty or an r-dimensional C^1-submanifold of $\mathbb{R}^n \times \mathbb{R}^n$. But the result also holds in general:

Lemma 4.4. *The set $E_A \subset E$ of admissible initial points of (4.4) is either empty or an r-dimensional C^1-submanifold of $\mathbb{R}^n \times \mathbb{R}^n$.*

From the implicit function theorem it follows that for any point (x_0, p_0) in E_A there exist an open neighborhood $U = S_x \times S_y \subset E$ and a unique C^1-mapping $\eta : S_x \to S_y$ with $\eta(x_0) = p_0$ such that $\hat{F}(x,p) = 0$ for $(x,p) \in U$ if and only if $p = \eta(x)$. Since E_A is open in E_N we may assume that $U_0 = U \cap E_A = U \cap E_N$. Let $\Pi : \mathbb{R}^n \times \mathbb{R}^n \to \mathbb{R}^n$ be the projection onto the first factor. Then, the result means that the restriction $\Pi|U_0$ is a C^1-diffeomorphism from U_0 onto ΠU_0. Hence

$$M = S_x \cap \Pi U_0$$

is an r-dimensional submanifold of $\mathbb{R}^n$ and

$$x \in M \quad \mapsto \quad (x, p) \in TM, \quad p = \eta(x),$$

is a tangential C^1-vectorfield on M for which it can be shown that the integral curves are exactly the solutions of (4.4) in E_0. In other words, the following result holds:

Theorem 4.5. *With the above terminology, $x : J \to \mathbb{R}^n$ is a C^1-solution of $F(x, x') = 0$ satisfying $(x(t), x'(t)) \in U$ for $t \in J$ if and only if $x(t) \in M$, $x'(t) = \eta(x(t))$, $\forall t \in J$.*

Thus, as expected, the DAE (4.4) is locally equivalent to an explicit ODE on an r-dimensional submanifold of $\mathbb{R}^n$.

The results in this section can be extended easily to the general nonautonomous case

$$F(x, x', t) = 0.$$

In fact, as usual we can transform this problem into autonomous form by introducing the mapping

$$G : \mathbb{R}^{n+1} \times \mathbb{R}^{n+1} \to \mathbb{R}^{n+1}, \quad G((x,t),(p,\tau)) = \begin{pmatrix} \tau - 1 \\ F(x,p,t) \end{pmatrix}.$$

Then under the required smoothness assumptions the necessary conditions (4.7) for G assume the form

$$F(x, p, t) = 0, \quad D_t F(x, p, t) + D_x F(x, p, t)p \in \mathrm{rge} D_p F(x, p, t).$$

Similary, we can derive the form of the reduced mapping and of the set E_A of admissible initial points (see [36]).

6.4.2 Globalizations

Consider again the implicit DAE (4.4) under the assumptions (A1,2,3). As we saw in Theorem 4.3, for any (x_0, p_0) in the set E_A of (4.15) the initial value problem (4.16) has a local C^1-solution on some open interval $J = (a, b) \in \mathbb{R}^n$ containing t_0. As in the standard ODE-theory, under appropriate conditions these local solutions can be continued.

In [36] the following basic continuation result was proved:

Theorem 4.6. *Suppose that $E_A = E_N$.*

(i) If for some $\epsilon \in (0, b-a)$ the set $\{p \in \mathbb{R}^n; p = x'(t), b - \epsilon < t < b\}$ is bounded then $\lim_{x \to b-} x(t) = x_b$ exists.

(ii) *If $\lim_{t\to b-} x(t) = x_b$ exists and for some sequence $\{t_k\} \in J$ with $lim_{k\to\infty} t_k = b$, the sequence $\{x'(t_k)\}$ has an accumulation point p^* for which $(x_b, p^*) \in E$ then $\lim_{t\to b-} x'(t) = p^*$ and hence, for $b < \infty$ the solution can be continued to the right.*

An analogous result holds for the left endpoint. Thus, in particular,any local solution of (4.16) can be extended to a maximal open interval $J^* = (a^*, b^*)$ where $-\infty \le a^* < b^* \le \infty$.

For an explicit initial value problem

$$x' = f(x), \quad x(0) = x_0, \quad x \in \mathbb{R}^n$$

with smooth f on all of $\mathbb{R}^n$, any maximally extended solution for which $\lim_{t\to b-} x(t)$ exists has bounded derivatives $x'(t)$ near b. This does not carry over to DAEs as the following result shows (see again [36]):

Theorem 4.7. *Suppose that $E = \mathbb{R}^n$ and $E_A = E_N$. If $b^* < \infty$, then $x'(t)$ is unbounded on the interval $(b^* - \epsilon, b^*) \in J^*$ for all sufficiently small $\epsilon > 0$. Hence, if $\lim_{t\to b^*-} x(t) = x_{b^*}$ exists, then $\lim_{t\to b^*-} ||x'(t)|| = \infty$.*

Again, an analogous result holds for the left endpoint.

Theorem 4.5 showed that the implicit DAE (4.4) is locally equivalent to a vector field on some r-dimensional submanifold of $\mathbb{R}^n$. These local vectorfields can be extended by applying the theory of covering spaces. This was first used in [41] in connection with the DAEs (4.2) and (4.3) and then generalized to the implicit case in [36].

We sketch only briefly the general approach. Clearly, the local result shows that the restriction $\Pi|E_A$ is a local homeomorphism between E_A and ΠE_A. Let E_A^* be some non-empty, arc-connected subset of E_A for which (E_A^*, Π_A^*), with $\Pi_A^* = \Pi|E_A^*$, is a covering space of ΠE_A^*. In other words, each point $x \in \Pi E_A^*$ is assumed to have an open, arc-connected neighborhood U such that each arc-component of $(\Pi_A^*)^{-1}U$ is not empty and is mapped topologically onto U by ΠE_A^*. Often $E_A^* = E_A$ can be used here. This is certainly the case when for fixed $x \in \Pi E_A$ there are only finitely many p with $(x, p) \in E_A$. For instance, this holds for the semi-linear DAE (4.12). In general, it is always possible to choose E_A^* as the closure of a non-empty, pre-compact, (relatively) open, and arc-connected submanifold of E_A.

For any given $(x_0, p_0) \in E_A^*$ let now M^* be a non-empty, (relatively) open, simply connected subset of ΠE_A^* that contains x_0. For any $x \in M^*$ choose a path $\xi : J \to M^*$ which connects x_0 with x. Then there exists a unique lifting $\xi^* : J \to E_A$ with initial point (x_0, p_0) for which $\Pi_A^* \xi^* = \xi$. This lifted path has a unique endpoint (x, p) in E_A because all paths in M^* between x_0 and x are homotopic. Since x was arbitrary in M^* our local result can now be used to prove that M^* indeed is an r-dimensional submanifold of $\mathbb{R}^n$ and that the DAE (4.4) induces a tangential vector field on M^* for which all integral curves in M^* are solutions of (4.4).

6.5 DAEs with Higher Index

6.5.1 Linear, Constant Coefficient DAEs

As mentioned before, the linear problems with constant coefficients (4.1) probably represent the most extensively studied DAEs in the literature. For sufficiently smooth g the necessary condition (4.7) turns out to have the form $Ap + Bx = g(t)$, $Bx - g'(t) \in \text{rge}A$.

As before, let $P \in L(\mathbb{R}^n)$ be the orthogonal projection onto rge$A$ and set $Q = I_n - P$. We differentiate again the DAE (4.1), multiply the resulting equation $Ax'' + Bx' = g'(t)$ by Q in order to remove the second derivative of x, and, finally add P times the original equation. This produces the reduced mapping

$$\begin{aligned} \hat{F}(x,p) &= A_1x' + B_1x - g_1(t), \\ A_1 &= PA + QB, \; B_1 = PB, \\ g_1(t) &= Pg(t) + Qg'(t). \end{aligned} \tag{5.1}$$

Thus we have $D_p\hat{F}(x,p) = A_1$; that is, $D_p\hat{F}$ is an isomorphism exactly if A_1 is invertible which is easily seen to be equivalent with

$$Au = 0 \quad \text{and} \quad Bu \in \text{rge}A \quad \text{imply} \quad u = 0. \tag{5.2}$$

Hence when this condition holds then the set E_A is non-empty and the linear DAE is equivalent to an explicit ODE.

The cited existence theory for (4.1) ensures solvability if and only if the matrix pencil (A, B) is regular; that is, if there is some $\lambda \in \mathbb{R}^1$ such that $B + \lambda A$ is invertible. A central concept in the solvability theory of (4.1) is its index which is defined to be the index of the – assumed to be – regular coefficient pencil. For any regular pencil (A, B) let λ be such that $B + \lambda A \in \text{Isom}(\mathbb{R}^n)$, then the index of the pencil is the smallest integer κ such that

$$\ker\left[(B+\lambda A)^{-1}A\right]^{\kappa+1} = \ker\left[(B+\lambda A)^{-1}A\right]^{\kappa}$$

It can be shown that κ is finite and independent of the choice of λ, (see e.g. [22]), and it is also readily seen that $\kappa = 0$ if and only of A is invertible; that is, if (4.1) is equivalent with an explicit ODE.

Suppose now that A_1 is singular. Then we may apply the same procedure repeatedly, as often as necessary, to obtain a sequence of DAEs of the form

$$A_jx' + B_jx = g_j'(t), \quad j = 0, 1, \ldots, \tag{5.3}$$

where $A_j,\ B_j,\ g_j$ are specified recursively by $A_0 = A,\ B_0 = B,\ g_0 = f$ and

$$\begin{aligned} A_{j+1} &= P_jA_j + Q_jB_j,\ B_{j+1} = P_jB_j, \\ g_{j+1}(t) &= P_jg_j(t) + Q_jg_j'(t),\quad j = 0, 1, \ldots . \end{aligned} \tag{5.4}$$

while P_j denote the orthogonal projections onto $\text{rge}A_j$ and $Q_j = I_n - P_j$. The process stops with the smallest integer k such that A_k is invertible.

The following result, proved in [36], shows that this integer k is exactly the index of the DAE:

Theorem 5.1. *If the matrix pencil (A, B) is regular and* $\text{rank}A < n$ *(so that $\kappa \geq 1$) then $k = \kappa < \infty$. Conversely, if $k < \infty$ then (A, B) is regular and $k = \kappa$.*

In [36] it was also shown that the theory of Section 6.4 provides all the solutions of (4.1) provided only that g is smooth enough.

6.5.2 Nonlinear Problems with Higher Index

The discussion of the previous section suggests that we may proceed analogously when the set E_A is empty for the general implicit initial value problem

$$F(x, x') = 0,\quad x(0) = x_0,\ x'(0) = p_0. \tag{5.5}$$

The first step in the construction of a sequence of problems corresponding to (5.3), (5.4) was already taken in Section 6.4. In fact, we differentiated the DAE and then applied the projections P and Q to obtain the reduced equation (4.11).

Our sufficient condition is that $D_p\hat{F}(x,p)$ is invertible at the given initial point $(x_0, p_0) \in E_N$. If this sufficient condition does not hold, then, as in the linear case, it is natural to construct recursively the sequence of mappings

$$\begin{aligned} F^0 &= F,\ F^1 = \hat{F}, \\ F^{j+1} &= P_j(x,p)F^j(x,p) + Q_j(x,p)D_xF^j(x,p),\ j = 0, 1, \ldots \end{aligned} \tag{5.6}$$

where again P_j is the orthogonal projection onto D_pF^j and $Q_j = I_n - P_j$. The process is repeated until D_pF^k is invertible at the point under consideration.

As before, one might consider calling this integer k the local index of the problem at the particular point. However, the situation differs here in a critical aspect from that of the linear case. The theory of Section 6.4 can be applied to the map F^k only if the conditions (A1,2,3) are valid for all the maps (5.6) in some neighborhood E_0 of the point (x_0, p_0). In particular, we require rank$D_pF^j(x,p)$ to be constant in such a neighborhood for the projections P_j, Q_j and hence F^{j+1} to be of class C^1. In addition, the three conditions are also needed to conclude from the non-singularity of D_pF^k that the system $F^k(x,x') = 0$ can be tranformed locally into an explicit ODE and that the original problem (5.5) has a unique solution. As mentioned earlier, the existence theory for (5.5) changes considerably when the constant-rank condition is violated (see [35]).

In line with this, the problem (5.5) will be defined to have local index k at (x_0, p_0) if there is some open neighborhood E_0 of that point such that for $j = 0, 1, \ldots, k-1$ the mappings (5.6) satisfy the conditions (A1,2,3) and $D_pF^k(x_0, p_0)$ is invertible. Note that this index does not merely depend on information at the given point and hence has a global nature. Obviously, the theory developed in Section 6.4 assumes that the implicit problem (4.4) has index one.

6.5.3 Semi-Implicit Problems with Higher Index

The recursive analysis outlined in the previous section is a powerful tool for the theoretical study of higher index problems. But for specific classes of equations it is often easier to derive existence and uniqueness results directly. As an example of this we consider in this section the special problems (4.2) and (4.3) both of which have, in general, index higher than one.

More specifically, for the system

$$F(x,x',z) = \begin{pmatrix} F_1(x) \\ F_2(x,x',z) \end{pmatrix} \tag{5.7}$$

suppose that

(B1) $F_1 : E_x \in \mathbb{R}^n \to \mathbb{R}^s$ is C^2,
(B2) $F_2 : E_2 = E_x \times E_p \times E_z \in \mathbb{R}^n \times \mathbb{R}^n \times \mathbb{R}^m \to \mathbb{R}^r$ is C^1,
(B3) $\begin{pmatrix} DF_1(x) & 0 \\ D_pF_2(x,p,z) & D_zF_2(x,p,z) \end{pmatrix} \in \mathrm{Isom}(\mathbb{R}^{n+m}), \quad \forall\ (x,p,z) \in E_2$

where $s < n \leq s + r = n + m$ and E_x, $E_p \subset \mathbb{R}^n$, $E_z \subset \mathbb{R}^m$ are non-empty, open sets.

Evidently, these conditions imply that (A1,2,3) are satisfied for (5.7). The assumption (B3) is often called the index-two condition since the index of (5.7) can be at most two. Of course, in the degenerate case $s = m = 0$, (B3) implies that (5.7) can be transformed into an explicit ODE and thus has index zero. Moreover, it is easily seen that for $s = 0$, $m \neq 0$ or $m = 0$, $s \neq 0$ the index is one.

From (B3) it follows that rank$DF_1(x) = s$ whence each member of the family of sets

$$M_b = \{x \in E_x;\ F_1(x) = b\},\ \ b \in F_1(E_x) \tag{5.8}$$

is an $(n-s)$-dimensional C^2-submanifold of $\mathbb{R}^n$. Evidently, any $x_0 \in E_x$ belongs to the unique constraint manifold (5.8) specified by $b = F_1(x_0)$. Moreover, for a solution of (5.7) to pass through this point we must have $x_0 \in M_0$.

Let $x : J \in E_x$ be any C^1-path defined on some open interval $J \in \mathbb{R}^1$. If x is a path on M_0; that is, if $F_1(x(t)) = 0,\ \forall t \in J$, then necessarily

$$DF_1(x(t))x'(t) = 0, \quad \forall t \in J; \tag{5.9}$$

that is, $t \in J \mapsto (x(t), x'(t))$ has to be a path on the tangent bundle of M_0. The explicit use of the differentiated constraint equation $DF_1(x)x' = 0$ is a basic step in the so-called index-reduction technique for rewriting the DAE as a lower index system (see [18]). But (5.9) can also be interpreted in another way. In fact, if (5.9) holds for some C^1-path $x : J \to E_x$, then it follows from the integral mean value theorem that $F_1(x(t)) = F_1(x(t_0))$ for any fixed $t_0 \in J$. Hence $t \in J \mapsto (x(t), x'(t))$ is a path on the tangent bundle of the manifold M_b specified by $b = F_1(x(t_0))$. This suggests that we imbed (5.7) into the family of DAEs

$$\begin{aligned} F_1(x) &= b,\ \ b \in F_1(E_x) \\ F_2(x, x', z) &= 0 \end{aligned} \tag{5.10}$$

indexed by the vectors of $F_1(E_x)$.

Let $x : J \to E_x$, $z : J \to E_z$ be a C^1-solution of a member of (5.10). Then for any point $(x_0, p_0, z_0) = (x(t_0), x'(t_0), z(t_0))$, $t_0 \in J$ the value $b = F_1(x_0)$ uniquely specifies the particular DAE. But, in addition, (x_0, p_0, z_0) must satisfy $DF_1(x_0)p_0 = 0$, as well as $F_2(x_0, p_0, z_0) = 0$. This suggests the definition of the C^1-map

$$H : E_2 \to \mathbb{R}^s \times \mathbb{R}^r, H(x,p,z) = \begin{pmatrix} DF_1(x)p \\ F_2(x,p,z) \end{pmatrix}, \ \forall (x,p,z) \in E_2 \tag{5.11}$$

as the initial data map of the family (5.10).

For any given $(x, p, z) \in E_2$ the derivative $D_{p,z}H$ of H with respect to (p, z) is exactly the linear operator in condition (B3). The nonsingularity of $D_{p,z}H$ implies the following result:

Lemma 5.2. *For any $(x_0, p_0, z_0) \in K$ we have an open neighborhood $U = S_x \times S_p \times S_z$ in E_2, and unique C^1-maps $\eta : S_x \to S_p$, $\zeta : S_x \to S_z$, with $\eta(x_0) = p_0$, $\zeta(x_0) = z_0$, such that for any $x \in S_x$ the only solution $(p, z) \in S_p \times S_z$ of $H(x, p, z) = 0$ is $p = \eta(x)$, $z = \zeta(x)$.*

Thus, on the open neighborhood $S_x \subset E_x$ of x_0,

$$\pi : S_x \to TS_x, \ \pi(x) = (x, \eta(x)), \ \ x \in S_x. \tag{5.12}$$

constitutes a C^1-vectorfield. Since $D_{p,z}H \in \mathrm{Isom}(\mathbb{R}^{n+m})$, the mapping H is a submersion and hence the solution set

$$K = \{(x, p, z) \in E_2; \ \ H(x, p, z) = 0\,\} \tag{5.13}$$

is an n-dimensional C^1-submanifold of $\mathbb{R}^{2n+m}$. This manifold turns out to be the state space of the family of DAEs (5.10). In fact, it can be shown that the vectorfield (5.12) is tangential to the constraint manifold through x; that is, that

$$\pi(x) \in T_x M_b, \ \ b = F_1(x) \ \ \forall x \in S_x.$$

Moreover, for any solution $x : J \to S_x$ of the explicit ODE

$$x' = \eta(x), \tag{5.14}$$

we obtain the C^1-solution $t \in J \mapsto (x(t), \zeta(x(t)))$ of the member of (5.10) specified by $b = F_1(x(t_0))$ for arbitrary fixed $t_0 \in J$.

Thus, the standard existence theory of initial value problems for ODEs provides the following result:

Theorem 5.3. *Suppose that the conditions (B1,2,3) hold and that K is non-empty. Then any point $(x_0, p_0, z_0) \in K$ has an open neighborhood $U \equiv S_x \times S_p \times S_z \subset E_2$ such that for any $x_c \in S_x$ there is exactly one point $(x_c, p_c, z_c) \in K \cap U$. Moreover, for any $x_c \in S_x$ there exists a unique, maximally extended C^1-solution $x : J \to S_x$, $z : J \to S_z$, on some open interval J with $0 \in J$, of the DAE (5.10) specified by $b = F_1(x_c)$ which satisfies $x'(J) \subset S_p$ and the initial conditions $x(0) = x_c$, $x'(0) = p_c$, $z(0) = z_c$.*

We refer to [41] for a globalization of this result based on the techniques from the theory of covering spaces mentioned at the end of Section 6.4.

The result extends to the second order DAE (4.3). In analogy with (B1,2,3) we suppose that the problem

$$F(x, x', x'', z) \equiv \begin{pmatrix} F_1(x) \\ F_2(x, x', x'', z) \end{pmatrix} = 0 \tag{5.15}$$

satisfies the conditions:

(C1) $F_1 : E_x \in \mathbb{R}^n \to \mathbb{R}^s$ is C^3,
(C2) $F_2 : E_2 = E_x \times E_y \times E_q \times E_z \in \mathbb{R}^{3n+m} \to \mathbb{R}^r$ is C^1,

(C3) $\begin{pmatrix} DF_1(x) & 0 \\ D_qF_2(x,y,q,z) & D_zF_2(x,y,q,z) \end{pmatrix}$ belongs to $\mathrm{Isom}(\mathbb{R}^{n+m})$ for each (x,p,q,z) in E_2,

where $s < n \leq s + r = n + m$ and E_x, E_y E_q $\subset$ $\mathbb{R}^n$, E_z $\subset$ $\mathbb{R}^m$ again are non-empty, open sets.

It is natural to reduce (5.15) to a first order system by introducing a new variable y and adding the equation $x' = y$. Then it turns out that – with (x, y) as new differential variable – the resulting system constitutes a DAE of the form (5.7) for which (B1) and (B2) are valid. However, in general, (B3) does not hold which is hardly surprising since we should expect (5.15) to induce local second order vectorfields instead of the local first order fields (5.12).

If $x : J \to E_x$ is a C^2-path on M_b for some $b \in F_1(E_x)$; that is, if $F_1(x(t)) = b$ for $t \in J$, then for all $t \in J$ then we must have

$$DF_1(x(t))x'(t) = 0,$$

as well as

$$DF_1(x(t))x''(t) + D^2F_1(x(t))(x'(t), x'(t)) = 0. \tag{5.16}$$

This shows that $t \in J \mapsto ((x(t), x'(t)), (x'(t), x''(t)))$ is a path on the second tangent bundle T^2M_b of M_b. Conversely, by the integral mean-value theorem we obtain the following result:

Lemma 5.4. *Let $x : J \to E_x$ be any C^2-path that satisfies (5.16). If there exists a t_0 in J such that $DF_1(x(t_0))x'(t_0) = 0$ and therefore $(x(t_0), x'(t_0))$ lies in TM_b for $b = F_1(x(t_0))$, then $((x(t), x'(t)), (x'(t), x''(t)))$belongs to T^2M_b, for all $t \in J$.*

As in the first order case this suggests that we imbed (5.15) into the family of DAEs

$$\begin{aligned} F_1(x) &= b, \quad b \in F_1(E_x) \\ F_2(x, x', x'', z) &= 0 \end{aligned} \tag{5.17}$$

and that we define the C^1-initial-data map:

$$\begin{aligned} H : E_2 &\to \mathbb{R}^s \times \mathbb{R}^r \\ H(x,y,q,z) &= \begin{pmatrix} DF_1(x)q + D^2F_1(x)(y,y) \\ F_2(x,y,q,z) \end{pmatrix}, \forall (x,y,q,z) \in E_2. \end{aligned} \tag{5.18}$$

By (C3) we have $D_{q,z}H(x,y,q,z) \in \mathrm{Isom}(\mathbb{R}^{s+r}, \mathbb{R}^{n+m})$ for $(x,y,q,z) \in E_2$, and hence H is a submersion and the solution set

$$K = \{(x,y,q,z) \in E_2; \;\; H(x,y,q,z) = 0\} \tag{5.19}$$

is a $2n$-dimensional C^1-submanifold of $\mathbb{R}^{3n+m}$ and the following result holds:

Lemma 5.5. *For any $(x_0, y_0, q_0, z_0) \in K$ there exists an open neighborhood $U = S_x \times S_y \times S_q \times S_z$ in E_2, and unique C^1-maps $\eta : S_0 \equiv S_x \times S_y \to S_p$, $\zeta : S_x \to S_z$, with $\eta(x_0, y_0) = q_0$, $\zeta(x_0, y_0) = z_0$, such that for any given $(x, y) \in S_0$ the only solution $(q, z) \in S_p \times S_z$ of $H(x, y, q, z) = 0$ is given by $q = \eta(x, y)$, $z = \zeta(x, y)$.*

Using this lemma we can now define on the open neighborhood S_0 of (x_0, y_0) the second-order C^1-vectorfield

$$\pi : S_0 \subset TS_x \to T^2 S_x, \quad \pi(x, y) = ((x, y), (y, \eta(x, y))), \ \forall (x, y) \in S_0. \tag{5.20}$$

Then, for any $(x, y) \in S_0$ such that $DF_1(x)y = 0$, it follows that $\pi(x, y) \in T^2 M_b$ for $b = F_1(x)$. Moreover, if $(x, y) : J \to S_0$ is any solution of the explicit ODE-system

$$x' = y, \quad y' = \eta(x, y), \tag{5.21}$$

satisfying $DF_1(x(t_0))y(t_0) = 0$ for some $t_0 \in J$, then $t \mapsto (x(t), \zeta(x(t)))$ is a C^1-solution of (5.17) for $b = F_1(x(t_0))$.

Thus the standard solution theory provides here the following local existence result:

Theorem 5.6. *Suppose that the conditions (C1,2,3) hold and that K is non-empty. Then any $(x_0, y_0, p_0, z_0) \in K$ has an open neighborhood $U = S_x \times S_y \times S_q \times S_z$ in E_2 such that for any $(x_c, y_c) \in S_0 = S_x \times S_y$ there is exactly one point $(x_c, y_c, p_c, z_c) \in K \cap U$. Moreover, for any $(x_c, y_c) \in S_0$ with $DF_1(x_c)y_c = 0$ there exists a unique, maximally extended C^1solution $x : J \in S_x$, $z : J \in S_z$ of (5.17) for $b = F_1(x_c)$ on some open interval $J \subset R^1$ containing the origin which satisfies $x'(J) \subset S_y$, $x''(J) \subset S_q$ and the initial condition $x(0) = x_c$, $x'(0) = y_c$, $x''(0) = q_c$, $z(0) = z_c$.*

Once again covering-space theory can be used to globalize this result (see [41]).

6.6 Numerical Methods for DAEs

6.6.1 Application of ODE Methods

The most frequently used approach to the computational solution of DAEs is the application of standard ODE methods. This idea appears to be due to W.Gear [17] who proposed the use of the backward-difference (BDF) methods developed for stiff ODEs.

Briefly, in an m-step BDF-method the derivative x' of the unknown function at the time t_k, $k \geq m$, is approximated by the derivative of the interpolation-polynomial through (x_k, t_k) and m earlier computed points

(x_{k-i}, t_{k-i}), $i = 1, 2, ..., m$. Hence, in the case of the implicit initial value problem

$$F(t, x, x') = 0, \quad x(t_0) = x_0, \; x'(t_0) = p_0 \tag{6.1}$$

the determination of x_k requires the solution of the nonlinear system of equations

$$F(t_k, x_k, \frac{1}{h_k} \sum_{i=0}^{m} \alpha_{k,i} x_{k-i}) = 0. \tag{6.2}$$

Here $\alpha_{k,i}$, $i = 0, 1, \ldots, m$, are the coefficients of the BDF formula at the k-th step which, of course, depend on k unless the stepsizes $h_i = t_i - t_{i-1}$ remain constant. For $m < 7$ the m-step BDF methods are known to be stable when applied to ODEs and hence we assume from now on that $1 \leq m \leq 6$.

When (6.1) represents a DAE; that is, when the constant-rank condition (3.3) holds, then the validity and performance of the process depends on several factors. In particular, the initial value problem (6.1) has to possess a solution and for $m > 1$ the required m starting points have to approximate this solution. Moreover, at each step the nonlinear system (6.2) has to have a feasible solution which is computable by a suitable iterative process such as some form of Newton's method. In general, the answers to these questions depend strongly on the index of the DAE.

For simplicity, we restrict ourselves here to the semi-implicit equation (4.5); that is, to the initial value problem

$$F(x, x') \equiv \begin{pmatrix} F_1(x) \\ F_2(x, x') \end{pmatrix}, \quad x(0) = x_0, \; x'(0) = p_0. \tag{6.3}$$

If the conditions (A1,2,3) hold, then, for (x_0, p_0) in the set E_A defined by (4.15), Theorem 4.3 ensures that (6.3) has a unique (local) solution. Moreover, on E_A the derivative $D_p\hat{F}$ of the reduced mapping (4.10) is non-singular.

The nonlinear system, to be solved at each step of the process, can be written in the form

$$G(x) = \begin{pmatrix} F_1(x) \\ \frac{h}{\alpha} F_2(x, \frac{\alpha}{h}(x + w)) \end{pmatrix} = 0,$$

where $h = h_k$, $\alpha = \alpha_{k,0}$ and w incorporates all information at the earlier computed points. All basic forms of Newton's method are locally convergent if the derivative DG is non-singular at the desired solution. Hence, with the abbreviation $p = x, \frac{\alpha}{h}(x + w)$ we are led to the matrix

$$\begin{pmatrix} DF_1(x) \\ D_pF_2(x, p) + \frac{h}{\alpha} D_xF_2(x, p) \end{pmatrix} = D_p\hat{F}(x, p) + \frac{h}{\alpha} \begin{pmatrix} 0 \\ D_xF_2(x, p) \end{pmatrix},$$

which, by definition of E_A, is clearly non-singular for (x, p) in some open neighborhood of any point on the exact solution in E_A and for all sufficiently small h.

Under these conditons it can be shown that when an m-step BDF method is used for the computational solution of (6.3), together with a fixed and sufficiently small stepsize h, then the convergence of the approximate points to the exact solution is of order $O(h^m)$ provided that all initial points are correct to order $O(h^m)$ and stopping criteria of order $O(h^{m+1})$ are applied in the Newton process at each step. A proof of this result for the general system (6.1) may be found in [3] where also its extension to the case of variable steps is discussed. These results about BDF methods for index-one systems form the theoretical basis for several highly successful numerical DAE-solvers notably the widely used codes DASSL [30] and LSODI [25].

Besides BDF-methods also other multistep have been considered in the DAE literature. In particular, an extensive analysis of general linear multistep methods for the index-one case is given in [22].

The situation changes considerably when the DAE has index higher than one and two basic difficulties arise. The first derives from the fact that for any particular multistep method[2] there exist DAEs with index exceeding one for which the method is unstable, [21]. The second difficulty is the appearance of a transient deterioration of the discretization errors following any change of the step-size in the method. This type of "boundary layer" was observed by several authors, see, e.g., [29,43].

More specifically, in [43] it was proved that when an m-step, fixed-stepsize BDF method is applied to a linear DAE (4.1) with index $\kappa \geq 1$, then the process converges with order $O(h^m)$ after $(\kappa - 1)m + 1$ steps. Moreover, in [19] it was shown that when variable stepsizes are used and the ratios of adjacent steps remain bounded then the global error has order $O(h^{\mu}_{\max})$ where $\mu = \min(m, m - \kappa + 2)$. Hence, for instance, for an index-three system the use of the implicit Euler method with variable stepsizes may lead to errors of order $O(1)$. However, note that for index-two problems we have $\mu = m$ and, in fact, it turns out, [2,4,27], that for semi-implicit systems (6.3) of index two after $m + 1$-steps the m-step BDF method with fixed steps is globally convergent of order $O(h^m)$ provided again that the initial points are correct to order $O(h^m)$ and the stopping-criteria of the iterative process at each step have order $O(h^{m+1})$. But, nevertheless, these iterative methods may well converge very poorly during the beginning steps. The variable-stepsize case of this result is discussed in [20].

Besides multistep methods various one-step methods have also been considered for the computational solution of DAEs. In particular, there

---

[2]In fact this holds also for Runge-Kutta methods.

exists a large literature on the use of implicit Runge-Kutta (IRK) methods. Any such method can be characterized by its Butcher-tableau

$$\begin{array}{c|cccc} c_1 & a_{11} & a_{12} & \cdots & a_{1m} \\ c_2 & a_{21} & a_{22} & \cdots & a_{2m} \\ \vdots & \vdots & \vdots & \ddots & \vdots \\ c_m & a_{m1} & a_{m2} & \cdots & a_{mm} \\ \hline & b_1 & b_2 & \cdots & b_m \end{array}$$

see e.g. [7]. When applied to the DAE (6.1) the basic algorithm assumes the form

$$\begin{aligned} &(i) \text{ solve} \quad && F(t_{k-1} + c_i h, x_{k-1} + h \sum_{j=1}^{m} a_{ij} Y_i, Y_j) = 0, \quad i = 1, 2, \ldots, m \\ & && \text{for } Y_1, \ldots, Y_m \in \mathbb{R}^n \\ &(ii) \text{ set} \quad && x_k = x_{k-1} + h \sum_{j=1}^{m} b_j Y_j. \end{aligned}$$

Implicit Runge-Kutta methods are useful for generating accurate initial data for higher order multistep methods; they are also advantageous for problems with multiple discontinuities. In general, the nonlinear system arising at each step has dimension mn and may be very costly to solve unless A has special properties. Thus the complexity of theses processes depends strongly on the form of the coefficient matrix $A = (a_{ij})$. Some important special cases include the DIRK-methods for which A is block-lower-triangular with equal diagonal blocks as well as the SIRK-methods where A has one real eigenvalue.

In the numerical integration of stiff ODEs it has become well-known that the computed solutions often exhibits a disappointingly low accuracy when compared with the order of consistency of the method. For Runge-Kutta methods applied to a class of stiff linear ODEs this was first observed in [34] where it was noted that for stiff problems the order of consistency should not be based on the classical Lipschitz condition. Instead in [15] and several subsequent papers (see also the monograph [12]) one-sided Lipschitz conditions were used to introduce the concepts of B-consistency and B-convergence which provide order results that correspond more closely to the observed behavior of stiff ODEs.

For an IRK method let the stage order be the largest integer $r \geq 1$ such that the conditions

$$\sum_{j=1}^{m} a_{ij} c_j^{k-1} = \frac{1}{k} c_i^k, \quad i = 1, 2, ..., m,$$

are valid for $k = 1, 2, \ldots, r$. Moreover, define the quadrature order as the largest integer $q \geq 1$ for which the conditions

$$\sum_{j=1}^{m} b_j c_j^{k-1} = \frac{1}{k},$$

hold for $k = 1, 2, \ldots, q$. Then $\tilde{p} = \min(r, q)$ is called the internal stage order and for $q > \tilde{p}$ the classical nonstiff-ODE- order p satisfies $q \geq p \geq \tilde{p} + 1$. There are examples of stiff ODEs and IRK-methods where $p > \tilde{p}$ and the observed order of convergence equals the internal stage order $\tilde{p}$ (see e.g. [12]).

This behavior is mirrored in the application of IRK methods to DAEs. In fact, DAEs have a close relationship with stiff ODEs, as is suggested, for instance, by the fact that singularly perturbed systems

$$x' = f_1(x, y), \quad \epsilon y' = f_2(x, y)$$

with small $\epsilon > 0$ become a DAE for $\epsilon = 0$.

Consider again (6.3) subject to the conditions (A1,2,3) and with (x_0, p_0) in E_A. For the approximation of the solution in E_A we consider an IRK method with a non-singular coefficient matrix A such that

$$|1 - b^T A^{-1} e| < 1, \quad e = (1, 1, \ldots, 1)^T \in \mathbb{R}^m,$$

which means that the method is A-stable (see e.g. [7]). If $q > \tilde{p} > 1$ then for a constant (sufficiently small) step-size $h > 0$ the global error is at least of order $O(h^{\tilde{p}+1})$ provided the error in the initial point and the termination criterion for Newton's method are of the same order.

A proof of this result is given in [3] where also examples of problems are found for which the achieved orders are higher than the stated lower bound. In essence, the results also carry over to semi-implicit index-two problems (see again [3] and also [5]).

Various other one-step methods have been used for the computational solution of DAEs. This includes, for example, the Runge-Kutta-Rosenbrock methods considered in [42] and the extrapolation methods studied by Deuflhard et al (see e.g. [13] or [14]). We shall not enter here into any further detail.

6.6.2 Local Parametrizations

In this section we turn to a local parametrization approach suggested by the earlier presented existence results. This approach was introduced in [41] and considered further in [32,31]. It is related to the generalized coordinate partitioning technique used in the numerical solution of Euler-Lagrange equations by E. Haug et al (see [28,48])

Consider first the system (5.7) subject to the conditions (B1,2,3), and, more specifically, suppose that we are in the setting of Theorem 5.3. Then, for given $(x_0, p_0, z_0) \in K$ we wish to compute the C^1-solution $x : J \to S_x$, $z : J \to S_z$, of the DAE (5.10) specified by $b = F_1(x_0)$ that satisfies $x'(J) \subset S_p$ as well as the initial conditions $x(0) = x_0$, $x'(0) = p_0$, $z(0) = z_0$. For any $x \in S_x$ the unique solution p, z of the equations

$$DF_1(x)p = 0, \quad F_2(x, p, z) = 0, \quad (x, p, z) \in K$$

is provided by the values $p = \eta(x)$ and $z = \zeta(x)$ of the mappings of Lemma 5.2. As we saw, once a procedure is available for computing $\eta(x)$ and $\zeta(x)$ for any needed $x \in S_x$, the problem of solving (5.10) in S_x reduces to that of solving the explicit ODE

$$x' = \eta(x), \quad x \in S_x. \tag{6.4}$$

Since the desired solution $x : J \to S_x$ of (6.4) through x_0 has to remain on the constraint manifold M_b through that point it is natural to work with a local coordinate system on M_b. For this we use a simple class of such coordinate systems applied earlier in other differential-geometric numerical methods (see e.g. [39,40]). Let $x_c \in M_b$ be any point on the $(n - s)$-dimensional constraint manifold M_b through x_0 and consider any linear subspace $T \subset \mathbb{R}^n$, with $\dim T = n - r$, such that

$$T^{\perp} \cap \ker DF(x_c) = \{0\}. \tag{6.5}$$

If $A \in L(\mathbb{R}^{n-s}, \mathbb{R}^n)$ is any matrix with orthonormal columns which span T, then (6.5) is equivalent with the assumption that

$$\begin{pmatrix} DF_1(x_c) \\ A^T \end{pmatrix} \in \mathrm{Isom}(\mathbb{R}^n). \tag{6.6}$$

Hence there exists an open neighborhood V_0 of the origin in $\mathbb{R}^{n-s}$ such that for all $u \in V_0$ the system

$$\begin{pmatrix} F_1(x) \\ A^T(x - x_c) \end{pmatrix} = \begin{pmatrix} 0 \\ u \end{pmatrix}$$

has a unique solution $x = \Psi(u) \in \mathbb{R}^n$. It is readily seen that the resulting mapping Ψ is a C^1 diffeomorphism from V_0 onto the relatively open neighborhood $\Psi(V_0) \subset M_b$ of x_c. This is the desired local coordinate mapping. Note that with $\omega(u) = \Psi(u) - x_c - Au$ we have $A^T \omega(u) = 0$ whence

$$\Psi : V_0 \to \mathbb{R}^n, \quad \Psi(u) = x_c + Au + \omega(u) \in M_b, \quad u \in V_0.$$

shows that the point $x = \Psi(u)$ on M_b is obtained by adding to x_c the vector $Au \in T$ and the orthogonal correction $\omega(u) \in T^{\perp}$. The local coordinate system at the point x_c of M_b is completely determined by the matrix A and hence we shall also speak of the local coordinate system induced by that matrix.

In practice, it is often useful to work with local coordinate mappings defined by the tangent space $T = \ker DF(x_c)$ at $x_c \in M_b$ (see e.g. [40]). Suppose that we compute the QR-factorization

$$DF_1(x_c)^T = (Q_1, Q_2)\begin{pmatrix} R \\ 0 \end{pmatrix}$$

where the matrices $Q_1 \in L(\mathbb{R}^s, \mathbb{R}^n)$ and $Q_2 \in L(\mathbb{R}^{n-s}, \mathbb{R}^n)$ have orthonormal columns and $R \in L(\mathbb{R}^s, \mathbb{R}^s)$ is upper-triangular and nonsingular. Then Q_2 can be used as the basis matrix A of T while the columns of Q_1 span $T^{\perp}$.

A second, practically useful choice of a local-coordinate space T and its basis matrix A consists in determining a permutation $e^{j_1}, e^{j_2}, \ldots, e^{j_n}$ of the natural basis of $\mathbb{R}^n$ such that the matrix

$$A = (e^{j_{s+1}}, e^{j_{s+2}}, \ldots, e^{j_n})$$

satisfies (6.6). This choice partitions the components of the vector $x = (x_1, x_2, \ldots, x_n)$ into a vector $(x_{j_{s+1}}, x_{j_{s+2}}, \ldots, x_{j_n})$ of independent coordinates and the complementary vector $(x_{j_1}, x_{j_2}, \ldots, x_{j_s})$ of dependent coordinates. This is the choice underlying the mentioned generalized coordinate partitioning approach of E.Haug et al (loc. cit.).

Once a local coordinate system has been chosen then it can be shown (see [41]) that the ODE (6.4) has the local representation

$$u' = A^T \eta(\Psi(u)), \quad u \in V_0 \subset \mathbb{R}^{n-s}. \tag{6.7}$$

This is an $(n-s)$-dimensional explicit ODE without constraints to which any standard ODE solver can be applied as long as the computed points remain in V_0.

This local coordinate approach can also be carried over to the second order DAEs (5.15). As before, suppose analogously that we are in the setting of Theorem 5.6. Let $(x_0, y_0, p_0, z_0) \in K$ be a given point for which $DF_1(x_0)y_0 = 0$ and suppose that we wish to compute the C^1-solution $x : J \to S_x$, $z : J \to S_z$ of (5.17) specified by $b = F_1(x_0)$ that satisfies $x'(J) \subset S_y$, $x''(J) \subset S_q$ and the initial conditions $x(0) = x_0$, $x'(0) = y_0$, $x''(0) = q_0$, $z(0) = z_0$. For any $(x, y) \in S_0 \equiv S_x \times S_y$ the unique solution q, z of

$$DF_1(x)q + D^2F_1(x)(y, y) = 0, \quad F_2(x, y, q, z) = 0, \quad (x, y, q, z) \in K$$

is given by the values $q = \eta(x,y)$ and $z = \zeta(x,y)$ of the mappings of Lemma 5.5. Thus, when a method for evaluating $\eta(x,y)$ and $\zeta(x,y)$ is available, then the problem of solving (5.17) in S_0 is reduced to that of solving the explicit first order system

$$x' = y, \quad y' = \eta(x,y). \tag{6.8}$$

As we know the desired solution $(x,y) : J \to S_0$ of (6.8) through (x_0, y_0) remains on the tangent bundle TM_b of the constraint manifold M_b through x_0. Thus we have to work here with a local coordinate system on TM_b. As before, let the matrix $A \in L(\mathbb{R}^{n-s}, \mathbb{R}^n)$ induce a local coordinate system at the point $x_c \in M_b$ of M_b. Then

$$\Theta : V_0 \times \mathbb{R}^{n-s} \to TM_b, \ \Theta(u,v) = (\Psi(u), D\Psi(u)v), \ u \in V_0, \ v \in \mathbb{R}^{n-s} \tag{6.9}$$

defines a local coordinate system on TM_b. By restricting V_0, if necessary, we can choose some neighborhood U_0 of the origin of $V_0 \times \mathbb{R}^{n-s}$ which Θ maps into E_0.

In this local coordinate system the differential equations (6.8) assume the local form

$$u' = v, \quad v' = A^T \eta(\Psi(u), D\Psi(u)v), \quad (u,v) \in U_0 \tag{6.10}$$

and hence, once again a standard ODE solver can be applied as long as the computed points remain in U_0

6.6.3 Euler-Lagrange Equations

In this section we sketch briefly how a multistep method might be applied when the local parametrization approach is used in the numerical solution of the Euler-Lagrange equations (2.3), (2.4); that is the constrained equations of motion

$$\begin{aligned} \Phi(x,t) &= 0 \\ M(x,t)x'' + D_x\Phi(x,t)^T z &= Q(x,x',t). \end{aligned} \tag{6.11}$$

For further detail we refer to [24,33].

In the case of (6.11) it is easily seen that (C3) is equivalent with the assumption

$$\text{rank} D_x\Phi(x,t) = s, \quad y^T M(x,t) y \neq 0, \ \forall y \in \ker D_x\Phi(x,t) \tag{6.12}$$

which is equivalent with the non-singularity of the matrix of the linear system

$$\begin{pmatrix} M(x,t) & D_x\Phi(x,t)^T \\ D_x\Phi(x,t) & 0 \end{pmatrix} \begin{pmatrix} q \\ z \end{pmatrix} = \begin{pmatrix} Q(x,y,t) \\ g(x,y,t) \end{pmatrix}. \tag{6.13}$$

Thus under the condition (6.12) the general existence theory applies to (6.11), (see e.g. [41]).

For given (x, y, t), set

$$g(x,y,t) = -(D^2_{xx}\Phi(x,t)(y,y) + 2D^2_{xt}\Phi(x,t)y + D^2_{tt}\Phi(x,t)).$$

in (6.13) and let $q = \eta(x,y,t)$, $z = \zeta(x,y,t)$ be the unique solution of that system. Then the problem of solving (6.11) is reduced to solving the explicit first order system

$$x' = y, \quad y' = \eta(x,y,t). \tag{6.14}$$

As in the previous section let $A \in L(\mathbb{R}^{n-s}, \mathbb{R}^n)$ induce a local coordinate system at the current point of M_b and introduce the corresponding local coordinate system (6.9) on the tangent bundle TM_b. Then it can be shown that the local represention of (6.14) is given by

$$u' = v, \quad v' = A^T\eta(\Psi(u,t), D_u\Psi(u,t)v + D_t\Psi(u,t), t). \tag{6.15}$$

Suppose that for its solution we use a (consistent) explicit multistep method of the form

$$u_k = \sum_{j=1}^{m} \alpha_j u_{k-j} + h\sum_{j=1}^{m} \beta_j v_{k-j}, \quad v_k = \sum_{j=1}^{m} \alpha_j v_{k-j} + h\sum_{j=1}^{m} \beta_j v'_{k-j}$$

with constant step $h > 0$. For the computation it is advantageous not to work with the local variables u, v but to transform all formulas immediately back to the original variables x, y.

If the approximations $x_{k-j},, y_{k-j}, y'_{k-j}, z_{k-j}$, $j = 1, 2, \ldots, m$ of the solution are already available, then the algorithm for computing x_k, y_k, y'_k, z_k has the form:

(i) Set $t_k = t_{k-1} + h$;

(ii) Evaluate

$$\begin{aligned} a_k &= A^T\{\sum_{j=1}^{m} \alpha_j x_{k-j} + h\sum_{j=1}^{m} \beta_j y_{k-j}\} \\ a'_k &= A^T\{\sum_{j=1}^{m} \alpha_j y_{k-j} + h\sum_{j=1}^{m} \beta_j y'_{k-j}\}; \end{aligned}$$

(iii) Solve the nonlinear system

$$\begin{pmatrix} \Phi(x, t_k) \\ A^T x \end{pmatrix} = \begin{pmatrix} 0 \\ a_k \end{pmatrix}$$

and set $x_k = x$;

(iv) Solve the linear system

$$\begin{pmatrix} D_x\Phi(x_k, t_k) \\ A^T \end{pmatrix} y = \begin{pmatrix} -D_t\Phi(x_k, t_k) \\ a'_k \end{pmatrix}$$

and set $y_k = y$;

(v) Solve the linear system

$$\begin{pmatrix} M(x_k, t_k) & D_x\Phi(x_k, t)^T \\ D_x\Phi(x, t) & 0 \end{pmatrix} \begin{pmatrix} w \\ z \end{pmatrix} = \begin{pmatrix} Q(x_k, y_k, t_k) \\ g(x_k, y_k, t_k) \end{pmatrix}$$

and set $y'_k = w$ and $z_k = z$.

In stage (ii) the multistep formula is evaluated in terms of the original variables x, y. Then the stages (iii) and (iv) determine the local coordinate mapping (6.9) and finally in stage (v) the linear system (6.13) is solved to obtain the accelerations y'_k and the algebraic variable z_k defining the constraint force.

In stage (iii) a chord-Newton process can be used involving the matrix obtained in stage (iv) of the previous solution-step. When the computed points leave the domain of validity of the current local coordinate system, then the matrix A has to be updated. The need for this can be detected by monitoring the number of iteration-steps of the nonlinear solver in stage (ii) or the condition of the linear system in stage (iii).

When an implicit multistep method is to be used then all three steps (iii)-(v) have to be combined into one. Now special attention has to be given to the inherent structure of the resulting large nonlinear system in order to keep the computational complexity at an acceptable level. For some detail we refer to [31] where also a numerical example is given.

6.7 Appendix

In this Appendix we collect some background material used throughout the presentation. For further details, especially on the differential geometric aspects, we refer to standard text such as [44] or [1].

As usual, a mapping $F : U \to \mathbb{R}^m$ on the open set $U \subset \mathbb{R}^n$ is of class C^r, $r \geq 0$, on U if F is continuous and for $r > 0$ all its partial derivatives up to and including order r exist and are continuous on U. More generally, a map $F : S \to \mathbb{R}^m$ on an arbitrary set $S \subset \mathbb{R}^n$ is of class C^r if for each $x \in S$ there exists an open set $U \subset \mathbb{R}^n$ containing x and a C^r-mapping $\hat{F} : U \to \mathbb{R}^m$ that coincides with F throughout $U \cap S$. A map $F : S \subset \mathbb{R}^n \to T \subset \mathbb{R}^m$ is a homeomorphism between the sets S and T if F is a one-to-one mapping from S onto T and both F and its inverse $F^{-1} : T \to S$ are continuous. Finally, a map $F : S \subset \mathbb{R}^n \to T \subset \mathbb{R}^m$ is a

C^r-diffeomorphism if F is a homeomorphism between S and T and if both F and F^{-1} are of class C^r.

A subset $M \subset \mathbb{R}^n$ is a d-dimensional C^r-sub-manifold of $\mathbb{R}^n$ if for each point $x \in M$ there exists an open set $U \subset \mathbb{R}^n$ containing x such that the neighborhood $U \cap M$ of x on M is C^r-diffeomorphic to an open subset V of $\mathbb{R}^d$. Any particular such diffeomorphism $\phi : U \cap M \to V$ is called a chart and its inverse a local coordinate system on $U \cap M$.

By this definition any open subset $U \subset \mathbb{R}^n$ is an n-dimensional C^∞-sub-manifold of $\mathbb{R}^n$. The tangent space T_xU of this manifold U at any point $x \in U$ is defined as the n-dimensional linear space $\{x\} \times \mathbb{R}^n$, and its tangent bundle TU is the $2n$-dimensional submanifold $U \times \mathbb{R}^n$ of $\mathbb{R}^{2n}$.

Let $F : U \mapsto \mathbb{R}^m$, $n > m$, be some C^r-mapping, $r \geq 1$, on the open set $U \subset \mathbb{R}^n$. A point $x \in U$ is a regular point of F if $DF(x)\mathbb{R}^n = \mathbb{R}^m$; that is, if the derivative $DF(x)$ has full rank m. If all points of a set $S \subset \mathbb{R}^n$ are regular points then F is a submersion on S. A point $b \in \mathbb{R}^m$ is a regular value of F if all points of the inverse image $F^{-1}(b) = \{x \in U, F(x) = b\}$ are regular; that is, if F is a submersion on $F^{-1}(b)$. A fundamental result then states that for any regular value $b \in \mathbb{R}^m$ the inverse image $M_b = F^{-1}(b)$ is either empty or a $d = (n-m)$-dimensional C^r-sub-manifold of $\mathbb{R}^n$. The tangent space T_xM_b at any point x of this manifold M_b may be identified with the set

$$T_xM_b = \{(x,p) \in T_x\mathbb{R}^n; DF(x)p = 0\}.$$

Clearly, T_xM_b is a d-dimensional linear subspace of the n-dimensional linear space $T_x\mathbb{R}^n$. The tangent bundle TM_b of M_b is the disjoint union of all tangent spaces T_xM_b for $x \in M_b$; that is,

$$TM_b = \{(x,p) \in TU; F(x) = b, DF(x)p = 0\},$$

and TM_b is a $2d$-dimensional C^{r-1}-sub-manifold of $T\mathbb{R}^n$. Evidently, the tangent bundle of TM_b is then the $4d$-dimensional C^{r-2}-sub-manifold

$$T^2M_b = \left\{ \begin{array}{ll} ((x,y)(p,q)) \in T^2U; & F(x) = b, \\ DF(x)p = 0, & DF(x)q + D_2F(x)(y,p) = 0 \end{array} \right\}.$$

of $\mathbb{R}^{4n}$.

A C^s-vectorfield on some open subset $U \subset \mathbb{R}^n$ is a C^s-mapping on U such that

$$\pi : U \to TU; \ \pi(x) = (x, \eta(x)), \ \forall x \in U. \tag{7.1}$$

An integral curve of π through a point $x_0 \in U$ is any C^s-path $x : J \to U$, defined on an open interval $J \subset \mathbb{R}^1$ containing the origin, for which $x(0) = x_0$ and $(x(t), x'(t)) = \pi(x(t))$ for $t \in J$; that is, which solves the initial value problem

$$x' = \eta(x), \ x \in U, \ x(0) = x_0.$$

For a vectorfield (7.1) of class C^s, $s \geq 1$, on the (non-empty) open subset $U \subset \mathbb{R}^n$, the following results hold:

(i) There exists a C^s-integral curve $x : J \to U$ of π through each $x \in U$ defined on an open interval J. Moreover, any two such curves are equal on the intersection of their domains.

(ii) The union of the domains of all integral curves of π through a point $x \in U$ is an open, possibly unbounded interval J_x^* and there exists a C^s-integral curve $x^* : J_x^* \to U$ of π through x. Moreover, J_x^* is the largest interval on which such an integral curve exists.

(iii) The set $D(\pi) = \{(t, x) \in \mathbb{R}^1 \times U;\ t \in J_x^*\}$ is open in $\mathbb{R}^1 \times U$ and contains $\{0\} \times U$. Moreover, the global flow $\xi : D(\pi) \to U$, $\xi(t, x) = x^*$, $t \in J_x^*$ of π is of class C^s on $D(\pi)$.

Consider now a second order ODE $x'' = \eta(x, x')$, $x \in U$ where η is of class C^s on some open set $E \subset \mathbb{R}^{2n}$. When this problem is written in the first order form $x' = y$, $y' = \eta(x, y)$, $(x, y) \in E$, then we encounter a vector-field of the form

$$\pi : E \subset TU \to T^2U;\ \pi(x, y) = ((x, y), (y, \eta(x, y))),\ \forall (x, y) \in E. \tag{7.2}$$

Note that the second and third component of the image vector are identical; in other words, (7.2) represents a sub-class of all tangential vector fields on TU, namely, the vector fields that are consistent with second order ODE's.

An integral curve of (7.2) through a point $(x_0, y_0) \in E$ is now a C^s-path $x : J \to \mathbb{R}^n$, defined on some open interval $J \subset \mathbb{R}^1$ containing the origin, for which $(x(t), x'(t)) \in E$ and $((x(t), x'(t)), (x'(t), x''(t))) = \pi((x(t), x'(t))$ for all $t \in J$; that is, which is a solution of the original second order ODE.

Acknowledgements

This work was in part supported by ONR-grant N-00014-90-J-1025 and NSF grant CCR-8907654.

References

[1] R. Abraham, J. E. Marsden, and T. Ratiu, *Manifolds, Tensor Analysis, and Applications*, Springer Verlag, New York, NY, 1988.

[2] K. E. Brenan, *Stability and Convergence of Difference Approximations for Higher-Index Differential-Algebraic Systems with Applications in Trajectory Control*, PhD thesis, University of California at Los Angeles, 1983.

[3] K. E. Brenan, S. L. Campbell, and L. R. Petzold, *Numerical Solution of Initial Value Problems in Differential-Algebraic Equations*, North-Holland Publ. Co., New York, NY, 1989.

[4] K. E. Brenan and B. E. Engquist, *Backward differentiation approximations of nonlinear differential/algebraic equations*, Math. of Comp., 51 (1988), pp. 659–676 and S7–S16.

[5] K. Burrage and L. R. Petzold, *On order reduction for Runge-Kutta methods applied to differential-algebraic systems and to stiff systems of ODEs*, Tech. Report UCRL-98046, Lawrence Livermore National Laboratory, January 1988.

[6] R. G. Busacker and T. L. Saaty, *Finite Graphs and Networks, An Introduction with Applications*, McGraw Hill Co, New York, NY, 1965.

[7] J. C. Butcher, *The Numerical Analysis of Ordinary Differential Equations*, John Wiley and Sons, New York, NY, 1987.

[8] J. C. Cavendish, M. J. Wenner, J. Burkardt, C. A. Hall, and W. C. Rheinboldt, *Dem: A new computational approach to sheet metal forming problems*, Int. J. for Num. Meth. in Eng., 23 (1986), pp. 847–862.

[9] ——, *Punch stretching of sheet metal and differential equations on manifolds*, Int. J. for Num. Meth. in Eng., 25 (1988), pp. 269–282.

[10] L. O. Chua and P. M. Lin, *Computer-Aided Analysis of Electronic Circuits*, Prentice Hall, Englewood Cliffs, NJ, 1975.

[11] E. A. Coddington and N. Levinson, *Theory of Ordinary Differential Equations*, McGraw Hill Co, New York, NY, 1955.

[12] K. Dekker and J. G. Verwer, *Stability of Runge-Kutta Methods for Stiff Differential Equations*, North Holland, Amsterdam, The Netherlands, 1984.

[13] P. Deuflhard, E. Hairer, and J. Zugck, *One-step and extrapolation methods for differential-algebraic systems*, Numer. Math., 51 (1987), pp. 501–516.

[14] P. Deuflhard and U. Nowak, *Extrapolation integrators for quasilinear implicit ODE*, in Large-Scale Scientific Computing, P. Deuflhard and B. Engquist, eds., Birkhauser Verlag, Basel, Switzerland, 1987, pp. 37–50.

[15] R. Frank, A. Schneid, and C. W. Ueberhuber, *Stability properties of implicit Runge-Kutta methods*, SIAM J. Numer. Anal., 22 (1985), pp. 515–534.

[16] F. R. Gantmacher, *The Theory of Matrices, 2 Volumes*, Chelsea Publ. Co., New York, NY, 1959.

[17] C. W. Gear, *Simultaneous numerical solution of differential-algebraic equations*, IEEE Trans. Circuit Theory, CT-18 (1971), pp. 89–95.

[18] ——, *Differential-algebraic equation index transformations*, SIAM J. Sci. and Stat. Comp., 9 (1988), pp. 39–47.

[19] C. W. Gear, H. H. Hsu, and L. E. Petzold, *Differential/algebraic systems revisited*, in Proc. of ODE Meeting, Mathem. Forschungsinst. Oberwolfach, Germany, 1982,

[20] C. W. Gear, B. Leimkuhler, and G. K. Gupta, *Automatic integration of Euler-Lagrange equations with constraints*, J. Comp. Appl. Math., 12/13 (1985), pp. 77–90.

[21] C. W. Gear and L. R. Petzold, *Differential-algebraic systems and matrix pencils*, in Matrix Pencils, B. Kagstrom and A. Ruhe, eds., Lect. Notes in Math. Vol. 973, Springer Verlag, Heidelberg, Germany, 1983.

[22] E. Griepentrog and R. Maerz, *Differential-Algebraic Equations and Their Numerical Treatment*, B.G.Teubner Verlag, Leipzig, Germany, 1986.

[23] E. J. Haug, *Computer Aided Kinematics and Dynamics of Mechanical Systems, Vol. I: Basic Methods*, Allyn and Bacon, Boston, MA, 1989.

[24] E. J. Haug and J. Yen, *Implicit numerical integration of constrained equations of motion via generalized coordinate partitioning*, Tech. Report R-39, Univ. of Iowa, Ctr. for Simul. and Design Optim. of Mech. Systems, February 1989.

[25] A. C. Hindmarsh, *LSODE and LSODI, two new initial value ordinary differential equation solvers*, ACM-SIGNUM Newsletter, 15 (1983), pp. 10–11.

[26] J. W. Hooker and C. E. Langenhop, *On regular systems of linear differential equations with constant coefficients*, Rocky Mountain J. of Math., 12 (1982), pp. 591–614

[27] P. Lötstedt and L. R. Petzold, *Numerical solution of nonlinear differential equations with algebraic constraints. Part 1: Convergence results for backward differentiation formulas*, Math. of Comp., 46 (1986), pp. 491–516.

[28] N. K. Mani, E. J. Haug, and K. E. Atkinson, *Application of singular value decomposition for analysis of mechanical systems dynamics*, Trans. ASME, 107 (1985), pp. 82–87.

[29] L. E. Petzold, *Differential/algebraic equations are not ODEs*, SIAM J. Sci. Stat. Comput., 3 (1982), pp. 367–384.

[30] L. R. Petzold, *A description of DASSL: A differential/algebraic system solver*, in Scientific Computing, R. S. Stepleman, ed., North Holland, Amsterdam, The Netherlands, 1983, pp. 65–68.

[31] F. A. Potra and W. C. Rheinboldt, *On the numerical solution of Euler-Lagrange equations*, Mech. of Struct. and Machines, (1990) to appear.

[32] ——, *Differential-geometric techniques for solving differential algebraic equations*, in Real-Time Integration Methods for Mechanical Systems Simulation, NATO Advanced Workshop, 1989, E. Haug and R. Deyo, eds., Springer Verlag, Heidelberg, Germany, 1990.

[33] F. A. Potra and J. Yen, *Implicit numerical integration for Euler-Lagrange equations via tangent space parametrization*, Tech. Report R-56, Univ. of Iowa, Ctr. for Simul. and Design Optim. of Mech. Systems, June 1989.

[34] A. Prothero and A. Robinson, *On the stability and accuracy of one-step methods for solving stiff systems of ordinary differential equations*, Math. of Comp., 28 (1974), pp. 145–162.

[35] P. J. Rabier, *Implicit differential equations near a singular point*, J. Math. Anal. and Appl., 144 (1989), pp. 425–449.

[36] P. J. Rabier and W. C. Rheinboldt, *A general existence and uniqueness theory for implicit differential-algebraic equations*, Tech. Report ICMA-90-145, Univ. of Pittsburgh, Inst. for Comp. Math. and Appl., January 1990, Diff. and Int.Eq., submitted.

[37] S. Reich, *On a geometrical interpretation of differential-algebraic equations*, Tech. Report, Techn. Univ. Dresden, Sekt. Inf. Technik, 1989.

[38] W. C. Rheinboldt, *Differential-algebraic systems as differential equations on manifolds*, Math. of Comp., 43 (1984), pp. 473–482.

[39] ——, *Numerical Analysis of Parametrized Nonlinear Equations*, John Wiley and Sons, New York, NY, 1986.

[40] ——, *On the computation of multi-dimensional solution manifolds of parametrized equations*, Numer. Math., 53 (1988), pp. 165–181.

[41] ——, *On the existence and uniqueness of solutions of nonlinear semi-implicit differential-algebraic-equations*, Tech. Report ICMA-90-139, Univ. of Pittsburgh, Inst. for Comp. Math. and Appl., August 1989, Nonlin. Anal. Theory and Appl., (1990) to appear.

[42] M. Roche, *Rosenbrock methods for differential-algebraic equations*, Numer. Math., 52 (1988), pp. 45–63.

[43] R. F. Sincovec, A. M. Erisman, E. L. Yip, and M. A. Epton, *Analysis of descriptor systems using numerical algorithms*, IEEE Trans.Aut.Control, AC-26 (1990), pp. 139–147.

[44] M. Spivak, *A Comprehensive Introduction to Differential Geometry, 5 Volumes*, Publish or Perish, Inc., Berkeley, CA, 1979.

[45] F. Takens, *Constrained Equations; A Study of Implicit Differential Equations and their Discontinuous Solutions*, Lect. Notes in Math. Vol. 525, Springer Verlag, Heidelberg, Germany, 1976, pp. 143–234.

[46] G. Verghese and B. Levy and T. Kailath, *A generalized state space for singular systems*, IEEE Trans. on Autom. Control, AC-26 (1981)

[47] N. M. Wang and B. Budiansky, *Analysis of sheet metal stamping by a finite-element method*, J. Appl. Mech., Trans. AMSE, 45 (1978).

[48] R. A. Wehage and E. J. Haug, *Generalized coordinate partitioning for dimension reduction in analysis of constrained dynamic systems*, Trans. ASME, 104 (1982), pp. 247–255.

[49] J. Wittenburg, *Dynamics of Systems of Rigid Bodies*, B. G. Teubner, Stuttgart, Germany, 1977.

Professor W. C. Rheinboldt
Department of Mathematics
University of Pittsburgh
Pittsburg
Pennsylvania, PA 15260
USA